# OPTIMIZATION IN QUALITY CONTROL

# OPTIMIZATION IN QUALITY CONTROL


EDITED BY

**Khaled S. Al-Sultan, P.E.**
*Associate Professor of Systems Engineering and Dean, College of Computer Sciences and Engineering King Fahd University of Petroleum and Minerals*

*Dhahran, Saudi Arabia*

■

**M. A. Rahim**
*Professor of Quantitative Methods University of New Brunswick*

*Fredericton, Canada*



KLUWER ACADEMIC PUBLISHERS
Boston/London/Dordrecht

**Distributors for North America:**
Kluwer Academic Publishers
101 Philip Drive
Assinippi Park
Norwell, Massachusetts 02061 USA

**Distributors for all other countries:**
Kluwer Academic Publishers Group
Distribution Centre
Post Office Box 322
3300 AH Dordrecht, THE NETHERLANDS

---

**Library of Congress Cataloging-in-Publication Data**

Optimization in quality control / Khaled S. Al-Sultan, M.A. Rahim.
p. cm.
Includes bibliographical references and index.
ISBN 0-7923-9889-0
1. Quality control--Statistical methods. 2. Process control.
I. Al-Sultan, Khaled S. II. Rahim, M. A.
TS156.067 1997
658.5'62--dc21 97-150
CIP

---



*Printed on acid-free paper.*

Printed in the United States of America

# Dedication

To the memory of my father and to my mother, wife, and son
Khaled

To my wife, Bilkis Rahim, and two sons, Iftekhar Rahim and Abid Rahim
M. A. Rahim

# CONTENTS

# PREFACE

Quality plays a very important role in today's highly competitive industrial environment. Indeed, quality has become a unifying theme of business strategies of many firms in recent years. Consequently, problems of optimization in quality control have received considerable attention in the last few decades and have generated a large volume of publication. Quality leads to an improvement in productivity. By improving quality, the methods of optimization reduce (minimize) process operational costs and variation in the product. Quality, productivity and cost of operations are the most important determinants of national competitiveness. The major goals of optimization in quality control and variation reduction are to reduce manufacturing variation, reduce usage variation, and reduce variation due to deterioration. In order to improve product quality, statistical process control techniques have been used widely to monitor the production process. If the process randomly produces unacceptable product quality due to random process failure, the mechanism of improving product quality is then to provide preventive maintenance inspection. In the case where tools gradually wear out with age or use, calibration or replacement of the tools is necessary to maintain the desired performance standard of the machine and concurrently maintain the quality of the product at the desirable level.

There are a wide variety of optimization and variation reduction problems. One of these is characterized by the existence of target value. Another type of optimization and variation reduction problem is uniformity around the target value. W. Edwards Deming says: "Optimization means cooperation." The birth of this book is the result of cooperation between several people, particularly the contributors and the referees. This book presents a broad survey of the state of the art in optimization in quality cotrol. It can serve as an important source of reference in the field of optimization in quality control for academicians, professionals and researchers. Each chapter submitted for publication has been refereed anonymously by at least two or three experts in the area of optimization in quality control. The papers are published in 5 sections in 14 chapters.

In the first chapter, K. S. Al-Sultan covers the fundamentals of optimization techniques. It will give the readers of the book first-hand introductory knowledge of optimization techniques before getting into the depth of application of optimization methods in quality control. Algorithms for various types of optimization problems are presented. Successful applications of optimization methods to quality control problems are highlighted.

Chapter 2 by M. A. Rahim and K. S. Al-Sultan focuses on some contemporary approaches to optimization models in process control. An overview of some of the recent works on optimization models in quality control, joint control of production quantity and quality, economic selection of target means and optimal determination of production runs is presented. Some new directions for further research are highlighted.

In Chapter 3, Elart von Collani presents a simplified approach for the determination of the economic design of control charts. The chapter outlines the fact that a large number of different input parameters make optimization very difficult. Therefore, a simplified approach is taken to mitigate this cumbersome situation. This simplified approach teaches us where the fundamentally difficult problems can be solved fairly easily.

Chapter 4 by G. Tagaras presents an economic design of time-varying and adaptive control charts. This paper offers a provocative look at the issue of dynamic economic design of control charts and yields exciting new insights from recent years. The paper concludes by summarizing the findings so far and proposing fruitful areas for further research.

M. A. Rahim, in Chapter 5, addresses an economic design of $\overline{x}$-control charts assuming gamma distributed in-control times. Various inspection schemes are compared, and the results are indeed very interesting to practitioners.

In Chapter 6, T. P. McWilliams presents research on the design of constrained economic control charts. A variety of numerical examples are presented which illustrate applications of constrained designs and yields a number of new insights for further reserach.

Chapter 7 is a joint contribution of J. Liu, K. Tang and Y. H. Chun. The paper considers the problem of economic selection of the mean and upper limit for a container filling process under capacity constraints. They illustrate the effectiveness of their model with an example problem and compare the result with other models applied to the capacitated case.

Chapter 8 by Olle Carlsson determines optimal target values in multiple criteria economic selection models. Examples from the pulp and paper industry are used.

In Chapter 9, the author, F. J. Arcelus addresses the issue of uniformity of production versus conformity to specifications in the canning problem. The primary objective of the chapter is to access the viability of combining the twin quality objectives of minimizing rejection rates and maximizing the uniformity of production of the resulting items.

Chapter 10 by B. J. Melloy, M. A. Coffin and P. C. Kiessler provides a stepwise-optimal setup procedure for setting machines and adjusting processes. The objective of this chapter is to develop a supplementary methodology that will optimize the intermediate settings, while maintaining the desirable characteristic of the final setting.

Chapter 11 is a joint effort of E.A. Elsayed, M. Gultekin and J. H. Byun. The chapter addresses shift detection in process mean using regression and cross-correlation analysis. The authors compare their model with other models in the literature, and show that the new model is effective in detecting the shift in a process mean under some conditions.

In Chapter 12, J. Yang and V. Makis, present optimal control and monitoring of deteriorating production processes. A study of effective monitoring of a controlled production process subject to variation from both deterministic tool-wear drift and random shocks are accessed in this chapter.

I. Djamaludin, R. J. Wilson, and D. N. P. Murthy in Chapter 13 present their work paper on lot sizing and life testing for quality improvement of items sold with a warranty. In this chapter, the authors develop a model that examines quality improvement schemes by trading off between the expected warranty servicing costs and increasing manufacturing costs.

In the concluding Chapter 14, T. C. E. Cheng, M. S. D. Lau and S. O. Duffuaa provide a concise review of lot by lot acceptance sampling by attributes and provide some of the optimization models and methods in this area to design optimal inspection plans. The chapter also provides a comprehensive list of references in this area.

K. S. Al-Sultan
**Dhahran.**

M. A. Rahim
**Fredericton.**

# ACKNOWLEDGMENTS

Many individuals have contributed to this book. We would like to thank the contributing authors for their valuable contributions, their timely revision of papers and their whole-hearted cooperation through all stages of this project. We would also like to acknowledge the great and sincere efforts of the referees, whose names are mentioned at the end of the book. Their comments have ensured the high quality of all the chapters.

We would also like to thank the Rector of King Fahd University of Petroleum and Minerals, H. E. Dr. Abdulaziz A. Al-Dukhayyil, for his continuous support and encouragement.

The tireless efforts of Dr. Sadiq M. Sait, M. Moizuddin, S. Anas Vaqar, S. K. Mukarram, Shahid Parvez, N. Quraishi, M. Alam, and S. M. Adil of KFUPM in the presentation of the book are highly appreciated. Appreciations are also due to Mr. Robert Barr, the English editor and to Mr. Beng, who constantly helped the second editor throughout the project.

We are grateful to Gary Folven, Editor of Operations Research and Management Sciences at Kluwer Academic Publishers for his enthusiasm in publishing this book and for all his cooperation. His editorial assistant, C. Wilson, has also been of great help.

Lastly, we would like to mention our gratitude to our wives (Amal and Bilkis) for their patience, understanding and many sacrifices.

This project has been funded by King Fahd University of Petroleum and Minerals under project number SE/Quality/178 granted to the first-editor. This support is highly appreciated. The financial assistance of the National Council of Canada to the second-editor is also warmly acknowledged.

# CONTRIBUTORS

**Khaled S. Al-Sultan**
Department of Systems Engineering,
King Fahd University of Petroleum
and Minerals, Dhahran - 31261,
Saudi Arabia

**F. J. Arcelus**
Faculty of Administration,
University of New Brunswick,
Fredericton, New Brunswick,
Canada, E3B 5A3

**Jai-Hyun Byun**
Department of Industrial Engineering
Gyeongsang National University
Chinju, Gyeongnam 660-701
Korea

**Olle Carlsson**
ESA, Department of Statistics,
University of Örebro,
S-70182, Örebro,
Sweden

**T. C. E. Cheng**
Office of the Vice President,
(Research & Postgraduate Studies),
The Hong Kong Polytechnic University,
Kowloon,
Hong Kong

**Young H. Chun**
Department of Information Systems
and Decision Sciences,
College of Business Administration,
Louisiana State University,
Baton Rouge, LA 70803-6316,
U. S. A.

**M. A. Coffin**
College of Engineering and Science,
Clemson University,
South Carolina,
U. S. A.

**Elart von Collani**
Institut für Angewandte Mathematik
und Statistik,
Universität Würzburg,
Sanderring 2, D-97070 Würzburg,
Germany

**I. Djamaludin**
Technology Management Centre,
The University of Queensland,
Brisbane, Qld, 4072,
Australia

**S. O. Duffuaa**
Department of Systems Engineering,
King Fahd University of Petroleum
and Minerals, Dhahran 31261,
Saudi Arabia

**E. A. Elsayed**
Department of Industrial Engineering,
Rutgers University,
P.O. Box 909,
Piscataway, NJ 08855-0909,
U.S.A.

**Muge Gultekin**
Department of Industrial Engineering,
Gyeongsang National University,
Chinju, Gyeongnam 660-701,
Korea.

**P. C. Kiessler**
College of Engineering and Science,
Clemson University,
South Carolina,
U. S. A.

**M. S. D. Lau**
Department of Actuarial and
Management Sciences,
University of Manitoba,
Winnipeg,
Manitoba,
Canada R3T 2N2

**Jaiwen Liu**
Department of Information Systems
and Decision Sciences,
College of Business Administration,
Louisiana State University,
Baton Rouge, LA 70803-6316,
U. S. A.

**Viliam Makis**
Dept. of Mechanical and Industrial
Engineering,
University of Toronto,
Toronto, Ontario,
Canada M5S 1A4

**Thomas P. McWilliams**
School of Management,
Arizona State University West,
U. S. A.

**B. J. Melloy**
College of Engineering and Science,
Clemson University,
South Carolina,
U. S. A.

**D. N. P. Murthy**
Department of Mechanical Engineering,
The University of Queensland,
Brisbane, Qld, 4072,
Australia

**M. A. Rahim**
University of New Brunswick,
Fredericton, New Brunswick,
Canada E3B 5A3

**George Tagaras**
Department of Mechanical Engineering,
Aristoteles University of Thessaloniki,
54006 Thessaloniki,
Greece

**Kwei Tang**
Department of Information Systems
and Decision Sciences,
College of Business Administration,
Louisiana State University,
Baton Rouge, LA 70803-6316,
U. S. A.

**R.J. Wilson**
Department of Mathematics,
The University of Queensland,
Brisbane, Qld, 4072,
Australia

**Jiangbin Yang**
Dept. of Mech. & Industrial Engineering,
University of Toronto,
Toronto, Ontario,
Canada M5S 1A4

# PART I

## INTRODUCTION

# 1

# INTRODUCTION TO OPTIMIZATION

**K. S. Al-Sultan**

*Systems Engineering Department,*
*King Fahd University of Petroleum and Minerals,*
*Dhahran-31261,*
*Saudi Arabia.*

## ABSTRACT

In this chapter, we discuss *optimization* as an important tool for aiding decision making and managing complex systems. Elements of the optimization study are highlighted, followed by necessary mathematical background. Algorithms for various types of optimization problems are then presented. Available computer codes for solving optimization problems are discussed. Finally, successful applications of optimization methods to quality control problems are highlighted.

**Key Words:** optimization, algorithms, mathematical models, unconstrained optimization problems, constrained optimization problems

## 1 THE OPTIMIZATION STUDY

In this section, we discuss the optimization study in general and its elements. We give brief classification of optimization problems. Then we discuss modelling issues. Finally, we present some mathematical preliminaries.

### 1.1 Introduction

Optimization is the field of study aimed at finding the best allocation of scarce resources among competing activities. Thus, it is considered a valuable tool for decision making which aids managers to select the "*best*" alternative out of many possible courses of actions.

These days, managers and engineers have to take decisions that are related to managing, operating, and maintaining complex systems. These systems could either be manufacturing, process or service industries. Due to the complexity of these systems, it is no longer possible that a manager, or an engineer can take the best course of action by common sense, nor is it possible for him or her to pick the *best* alternative among the possible ones by trial and error. It is in these situations that optimization serves as an effective tool for finding the best decision to take without having to enumerate all possible actions.

Optimization is a branch of mathematics that has found successful applications in business, economics, engineering, medicine, to name a few. In particular, optimization has been effectively applied in the areas of logistics, maintenance management, quality control, production systems, inventory control, economic planning, transportation, manufacturing systems, scheduling, power systems, finance and management, to name a few applications. For its successful applications, knowledge in computer science and the specific field under study are usually necessary.

Optimization is also considered as one of the branches of *Operations Research* (known as operational research in Europe) which is a more general field concerned with aiding managers in the decision making process for managing complex systems. *Management Science* is a synonym for operations research.

The field of optimization is deeply rooted in the early stages of civilization, but it was not until the second world war that it became a respected field of study. Over the past fifty years, it has attracted many researchers, and its methodologies have been developed and successfully applied to many real life situations.

In this chapter, we discuss optimization as a tool for decision making. We start by discussing the elements of the optimization study. Then, we introduce mathematical concepts needed to conduct the study. We, then, summarize various algorithms developed for solving optimization problems. The remainder of this chapter is organized as follows: in Section 1.2, we discuss elements of the optimization study, followed by classification of optimization problems in Section 1.3. The modelling process is discussed in Section 1.4, and a guide to successful implementation of optimization is presented in Section 1.5. Mathematical preliminaries are discussed in Section 1.6, while optimality conditions are highlighted in Section 2. Line search techniques for single valued functions are presented in Section 3, followed by methods for functions of several variables in Section 4. In Section 5, approaches for constrained optimization problems are discussed. In Section 6, we provide some computer packages for solving optimization problems, while in Section 7, we discuss some successful applications of optimization methods to quality control problems. Conclusions are summarized in Section 8, followed by a comprehensive list of references.

## 1.2 Elements of the Optimization Study

The key elements in any optimization study are the following:

1. System boundary
2. Criteria
3. Decision variables
4. Interrelationships among variables

Next, we discuss the above elements in details.

- *System Boundary*
  Before embarking into any optimization study, one has to clearly define the boundary of the system under study. This is important because relationships with the outside world are considered frozen, and hence the interest of the study is limited to the system within the boundary. Clearly, a solution to an optimization study with reference to a defined boundary may be different if the boundary is enlarged or shrinked.

  Consider a production system in which one is interested in optimizing tool replacement strategies. One can assume that only the replacement, and quality costs are involved while the feed rate, and speed of the machine are fixed (i.e., they are out of the system boundary). If one optimizes this system, and comes up with a solution, then this solution may change if he or she decides to include the feed rate, speed of the machine into the model (i.e., expand the boundary of the system).

- *Criteria*
  Once the boundary of the system under study is defined, the *best* alternative among those within the boundary has to be selected. However, the *best* has to be defined for the specific purpose of the study. In a production system, if one considers the finance department, then the best alternative is to have zero inventories, while the marketing department would like high levels of inventories to satisfy customers' requirements. The quality control

department would like tools to be replaced too often to reduce the number of defective items produced, while the maintenance department would like to replace tools infrequently to reduce the investment in tools and their replacement cost.

In summary, the criterion has to be defined with respect to the specific purpose of the study. Mostly, the criterion is economic, like return on investment, profits, annual sales, annual expenditures, etc. If it is profit, then one maximizes it, otherwise one has to minimize the criterion (objective) function. In many situations, the criteria involve conflicting objectives in which case one seeks a compromise or can prioritize these objectives. An example is inventory systems where one finds the economic order quantity by striking a balance between the ordering cost (which decreases as the ordered quantity increases) and the holding cost (which increases as the ordered quantity increases).

- *Decision Variables*
  These are the variables that are under the decision maker's control. They represent alternatives for decisions. Examples are cycle time for production systems, order quantities in inventory systems, and production quantities in a blending problem.

- *Interrelationship Among Variables (Constraints)*
  This element reflects the interactions among various activities in the system. Various business, scientific, engineering, social, and other principles are used to come up with these relationships. These relationships, which are usually called constraints, are important since they control the interactions among the decision variables to make sure that the system boundary is observed.

## 1.3 Classification of Optimization Problems

Optimization problems can be classified according to the mathematical characteristics of the objective function and constraints. These could either be deterministic or stochastic. In this chapter we will be concerned with deterministic optimization. The following are the classes of deterministic optimization problems:

- *Linear Programming Problems*
  In this class, the objective function and the constraints are all linear. This is the simplest type of optimization problem. The simplex method for

linear programming is well developed and it can solve very large scale problems. For more details on this class see Dantzig (1963) Murty (1976), Murty (1983) and Bazaraa et al.(1990).

- *Integer Programming Problems*
  In this class, problems resemble linear programming problems except that decision variables are restricted to be integer. Of course, this makes this class of problems a lot harder than their linear programming counterparts. A subclass of integer programming problems is the class of mixed integer programming problems in which only some of the variables are restricted to be integers. For more details on this class, see Salkin (1975), Murty (1976), Nemhauser and Wolsey (1988).

- *Nonlinear Programming Problems*
  In this class, the objective function and/or at least one of the constraints are nonlinear. This is the most general form of optimization problem. Solution procedures for this class are the least developed compared to other classes. However, the exceptions to this are the following subclasses of nonlinear programming problems (NLP) which are relatively easier to solve:

  - *Quadratic Programming Problems*
    In this class, the objective function is quadratic and the constraints are linear. For more details on this class, see Murty (1988), and Bazaraa et al.(1993).
  - *Convex Programming Problems*
    In this class, the objective function and constraints are all convex, where convex functions are defined in Section 1.6. For more details on this subclass, see Murty (1988), Mangasarian (1991).

  For more details on this class see Leunberger (1984), Fletcher (1987), Murty (1988), Bazaraa et al.(1993).

All the above problems are, in general, constrained optimization problems. However, in some situations, the decision variable(s) is (are) allowed to take any values and hence denoted as ***unconstrained optimization problems.*** Classification of optimization problems is shown in Figure 1.

In this chapter, we will only cover algorithms for nonlinear programming problem due to their wide applications in quality control and related fields. For other problems, interested readers can refer to references cited above.

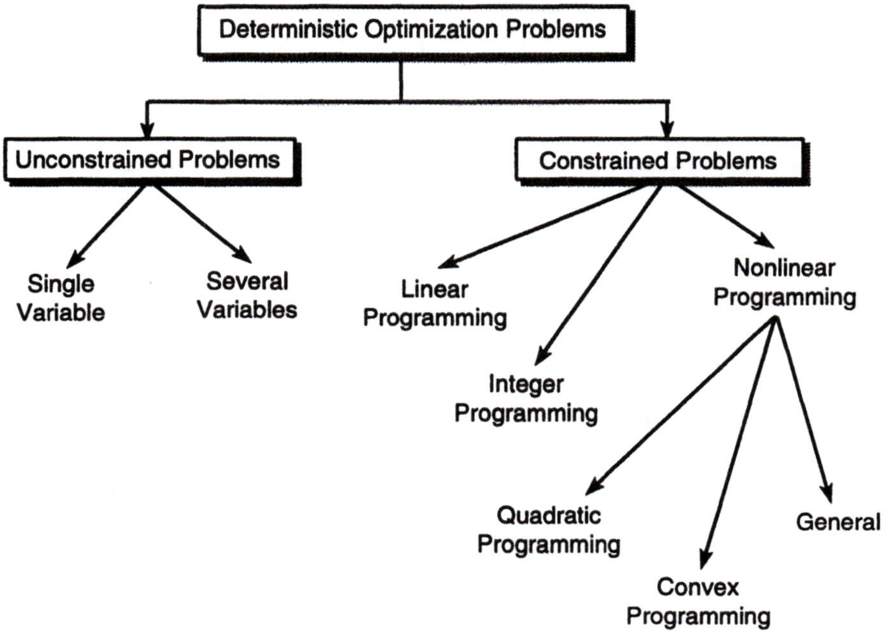


**Figure 1** Classification of deterministic optimization problems

## 1.4 The Modelling Process

In order to use optimization to solve real life problems, one has to develop a mathematical model that represents the real life situation. Clearly, one can not usually represent all real life complexities by a mathematical model. Therefore, one has to resort to some approximations by making some assumptions. The process of abstracting the essentials of the real life case and translating it into a mathematical model is called *problem formulation*.

In this stage, mathematical functions that represent the criterion and the interactions among decision variables are developed. The function that represents the criterion is called the *objective function*, while the functions that represent the interactions among variables and the boundary are called the *constraints*. The region encompassed within the boundary is called the *feasible region*. Problem formulation represents the real life problem by a mathematical model of the following form

$$\text{minimize (maximize)} \quad f(x) \tag{1.1}$$

subject to

$$\begin{aligned} h_i(x) &= 0 \quad i = 1, 2, ..., p \\ g_j(x) &\geq 0 \quad j = 1, 2, ..., m \end{aligned}$$

After formulation, a solution for the developed model can be obtained by optimization techniques (see Sections 2-6 ). One should remember, that if a detailed model is developed for the problem, then it may be difficult to solve it, and one has to resort to approximations later at the solution stage, and vice versa, i.e., if a simple model is developed then an exact solution may be possible. Therefore, one has to choose ***either an exact model and an approximate solution, or an approximate model and an exact solution.*** After solving the model, ***sensitivity analysis*** is applied to see how sensitive are the obtained solutions to parameter values, and to be careful about those parameters to which the solution is sensitive. The final stage is to implement the solution obtained, and finally provide feedback for updating the model (see Figure 2)

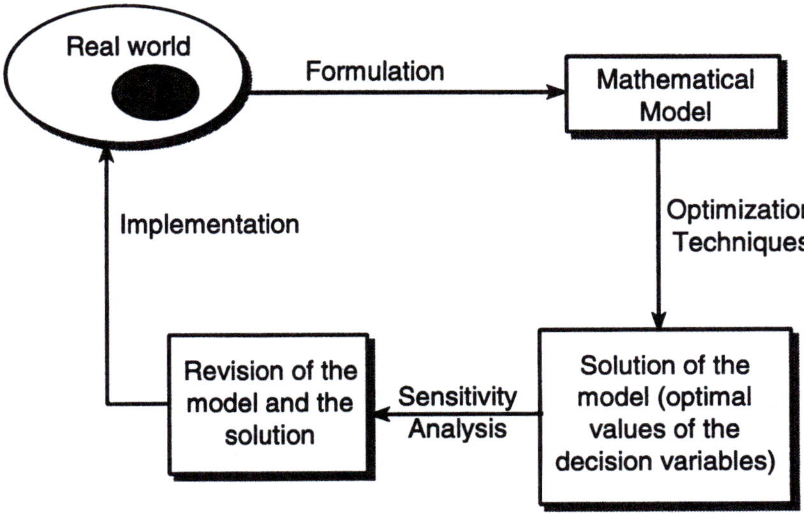


**Figure 2** The modelling process

## 1.5 Guide to a Successful Optimization Study

Ravindran et al.(1987) suggest the following general principles which are useful in guiding the modelling process:

1. Do not build a complex model when you find a simpler one.
2. Do not mould the problem to fit an available technique.
3. Model validation should be executed before its implementation.
4. A model should only be taken as an approximation of reality.
5. A model should only be used for the purpose it was intended.
6. Do not oversell a model.
7. The process of developing a model carries by itself some benefits.
8. A model can not be better than the information that goes into it.
9. Models can aid but never replace decision makers.

## 1.6 Mathematical Preliminaries

In this section, we consider some Mathematical Preliminaries about sets and functions.

**Definition 1.1**
A set of points $D$ is called convex if for any $x_1, x_2 \in D$, then $x \in D$ where $x = \lambda x_1 + (1 - \lambda)x_2$ , and $0 \leq \lambda \leq 1$.

The above definition states that a set is convex if for any two points in the set, the line joining them lies entirely in the set. See Figure 3 for illustration of convex sets. Convex sets are important in deriving optimality conditions for optimization problems, and in designing optimization algorithms.

Let $f : R^n \rightarrow R$, or $f(x)$ be a function that takes vectors in $n$-dimensional space and maps them into the real line (i.e., single valued function), then consider the following definitions.

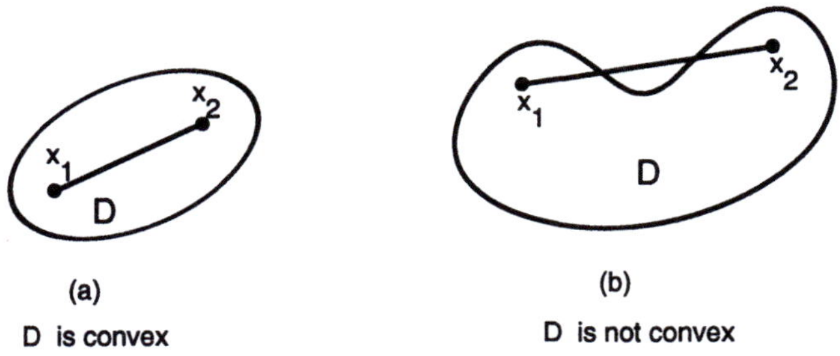


**Figure 3** Convex sets

**Definition 1.2**
Let $f : D \to R$, where $D$ is a convex set in $R^n$. Then $f(x)$ is called convex if the following inequality (called *Jensen's inequality*) holds

$$f(\lambda x_1 + (1-\lambda)x_2) \leq \lambda f(x_1) + (1-\lambda)f(x_2) \qquad (1.2)$$
$$\text{for } x_1, \text{ and } x_2 \in D, \text{ and } 0 \leq \lambda \leq 1$$

Another way of stating the above definition is that given any two points in the domain of the function, the line segment joining the function values at the two points will be on or above the function itself. See Figure 4 for illustration.

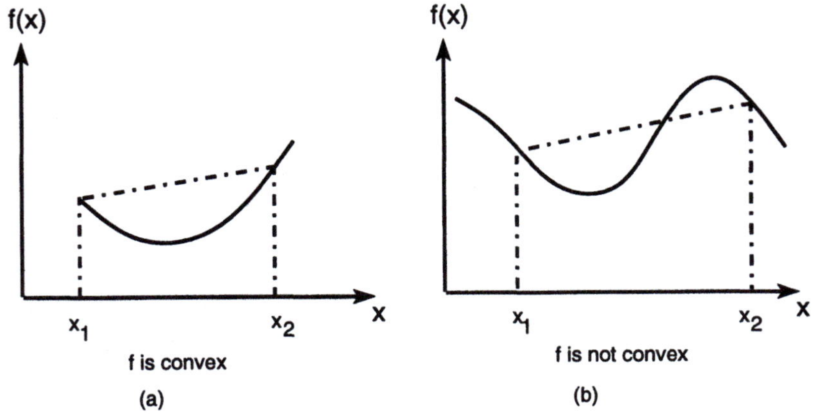


**Figure 4** Convex functions

**Definition 1.3**
Let $f : D \to R$, where $D$ is a convex set in $R^n$. Then, $f(x)$ is called concave if the reverse of inequality (1.2) holds, or $f(x)$ is concave if and only if $-f(x)$ is convex.

Another way of stating the above definition is that given any two points in the domain of the function, the line segment joining the function values at the two points will be on or below the function itself. See Figure 5 for illustration. The

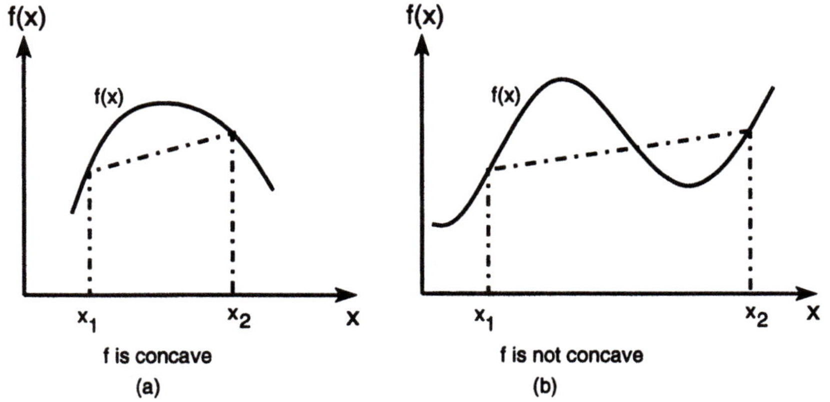


**Figure 5** Concave functions

notion of convexity is extremely important for optimization for reasons that will be apparent in the next section.

**Definition 1.4**
A nonlinear programming problem is defined as a convex programming problem if $f(x)$ is convex, $h_i(x)$, $i = 1, 2, \cdots, p$ are linear, and $g_j(x)$, $i = 1, 2, \cdots, m$ are concave.

Another way of stating the above definition is that if the objective function of an NLP is convex and the feasible region is convex, then the NLP is a convex program. Importance of convex programs will be clear in the next section.

**Definition 1.5**
$\nabla f(x) : R^n \rightarrow R^n$ is defined to be *the gradient of f* where

$$\nabla f(x) = \begin{bmatrix} \frac{\partial f(x)}{\partial x_1} \\ \frac{\partial f(x)}{\partial x_2} \\ \vdots \\ \frac{\partial f(x)}{\partial x_n} \end{bmatrix}$$

or $\nabla f(x)$ is a column vector of the partial derivatives of $f$ with respect to $x_i$'s.

The gradient vector is important in designing optimization algorithms because, when evaluated at a certain point, it represents the direction of the maximum increase (steepest ascent) of the function at the current point (its negative is the direction of the maximum decrease (steepest descent) of the function at the current point).

**Definition 1.6**
$H(f(x)) : R^n \rightarrow R^{n^2}$ is defined to be *the Hessian of f* where

$$H(f(x)) = \begin{bmatrix} \frac{\partial^2 f(x)}{\partial x_1^2} & \cdots & \frac{\partial^2 f(x)}{\partial x_1 \partial x_n} \\ \vdots & & \vdots \\ \frac{\partial^2 f(x)}{\partial x_n \partial x_1} & \cdots & \frac{\partial^2 f(x)}{\partial x_n^2} \end{bmatrix}$$

or $H(f(x))$ is an $n \times n$ matrix of the second partial derivatives of $f$ with respect to $x_i$'s.

The Hessian matrix is important as will be seen later in characterizing the curvature of function and hence in checking convexity. It is also used in characterizing optimality as will be seen in Section 2.

Next, we state some results that help in checking convexity of functions

*Results 1.1*
Let $f : D \rightarrow R$, $f$ is a *concave function* if and only if $H(f(x))$ is a positive semidefinite matrix for all $x \in D$. For *Proof*: see Bazaraa et al.(1993)

The above results can be used to show convexity of functions. Positive semidefiniteness can be shown by variety of ways. One of them is to show that all eigenvalues of $H(f(x))$ are nonnegative. Another way is to show that

$x^T H(f(x))x \geq 0$ for all $x \in R^n$. $H(f(x))$ is also said to be positive definite matrix if all eigenvalues of $H(f(x))$ are positive. Another way is to show that $x^T H(f(x))x > 0$ for all $x \in R^n$, (see Murty (1988) for other tests of definiteness of matrices).

*Result 1.2*
Let $f : D \to R$, then $f$ is a *concave function* if and only if $H(f(x))$ is a negative semi-definite matrix, for all $x \in D$. *Proof*: Obvious

Strict convexity and concavity of functions are useful ideas (strict convex functions and strict concave functions are defined exactly in the same ways as in Definitions 1.2, and 1.3 except that inequality (1.2) holds as strict inequality).

**Definition 1.7**
Let $f : D \to R$, then $f$ is *a unimodal function* if there exists $x^* \in D$ such that for any two points $x_1$ and $x_2 \in D$, $x^* \leq x_1 \leq x_2$ implies $f(x^*) \leq f(x_1) \leq f(x_2)$ and $x^* \geq x_1 \geq x_2$ implies $f(x^*) \leq f(x_1) \leq f(x_2)$. Clearly if the function is unimodal, then it has only one minimum or one maximum. See Figure 6 for illustration of unimodal functions.

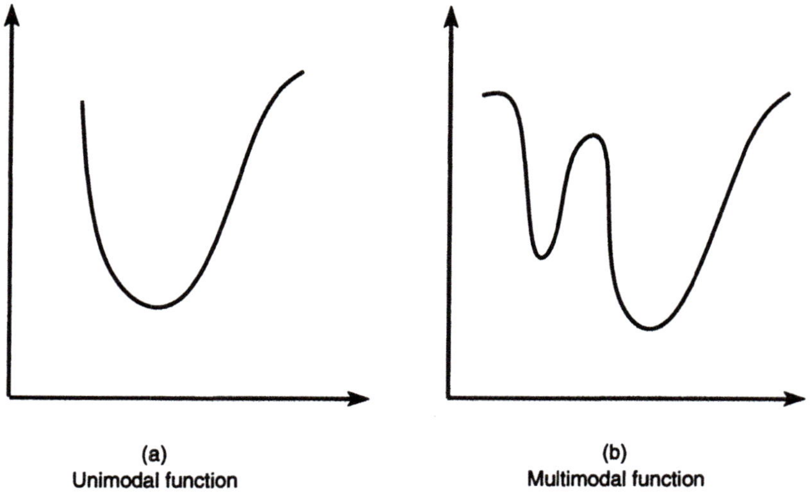


**Figure 6** Unimodal and multimodal functions

Clearly the notion of unimodality is useful since one knows that he/she expects only one minimum or maximum.

# 2 OPTIMALITY CONDITIONS

Consider the following optimization problem

$$\min_{x \in D} f(x) \tag{1.3}$$

where $D \subset R^n$.

If $D = R^n$ then problem (1.3) is *unconstrained*, otherwise it is called *constrained optimization problem.*

Consider the following definitions

**Definition 3.1**
$x^* \in D$ is said to be a local minimum of problem (1.3) above if there exists an $\epsilon > 0$ such that $f(x^*) \leq f(x)$ for all $x \in D$ and $||x - x^*|| < \epsilon$, where $||.||$ denotes *Euclidean*-norm, i.e., $x^*$ is the minimum value of the function in a ball centered at $x^*$ and has a radius $\epsilon$.

The above definition states that a local minimum of an optimization problem is a point that is the minimum in a small neighborhood of the feasible region around it.

**Definition 3.2**
$x^* \in D$ is said to be a unique local minimum of problem (1.3) above if there exists an $\epsilon > 0$ such that $f(x^*) < f(x)$ for all $x \in D$, and $||x - x^*|| < \epsilon$.

**Definition 3.3**
$x^*$ is said to be a global minimum of problem (1.3) above if $f(x^*) \leq f(x)$ for all $x \in D$. $x^*$ is said to be the unique global minimum of problem (1.3) if $f(x^*) < f(x)$ for all $x \in D$

Notice that a global minimum of an optimization problem is a point where the function attains its minimum over the whole set $D$. The following points are now in order

1. Every global minimum is a also local minimum.
2. At a local minimum, there is no local information that tells whether this is global or not. This is what makes global optimization very difficult.

3. Local, unique local, global, and global maxima are defined analogous to the definitions above and hence will not be considered in the remainder of this chapter. Any conclusions about maxima can be directly obtained from their minima counterpart.

4. If $f$ is convex and $D = R^n$ (i.e., the problem is unconstrained), then a local minimum is global. If $f$ is strictly convex and $D = R^n$ then a local minimum is the unique global minimum. This conclusion is also true for unimodal and strict unimodal (to be defined in Section 3) functions respectively.

Local, unique local, global, unique global minima and maxima are shown in Figure 7.

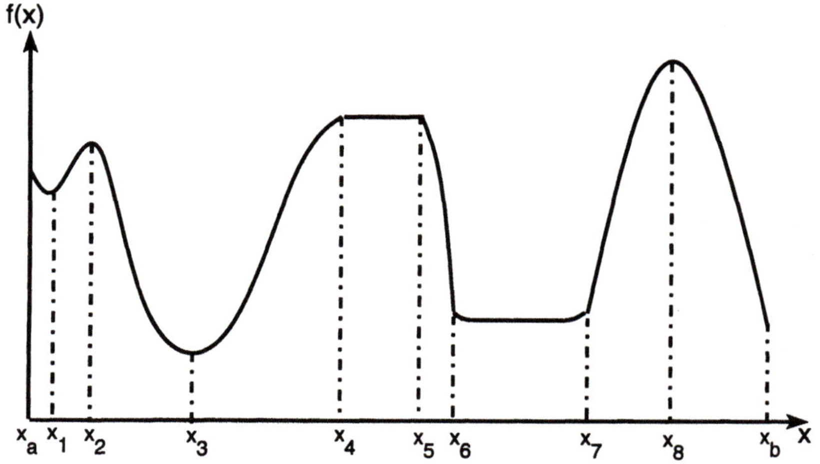


**Figure 7** Local and global minima and maxima

$x_1, x_3, x_6, x_7, x_b$ are local minima
$x_a, x_2, x_4, x_5, x_8$ are local maxima
$x_1$ is the unique local minimum
$x_2$ is the unique local maximum
$x_3$ is the unique global minimum
$x_8$ is the unique global maximum

Next, we consider optimality conditions for unconstrained optimization problems or

$$\min_{x \in R^n} f(x) \tag{1.4}$$

*Result 3.1* (*First order necessary condition.*)
Let $f$ be a differentiable function. If $\overline{x} \in R^n$ is a local minimum for (1.4) then $\nabla f(\overline{x}) = 0$ where $\nabla f(\overline{x})$ is the gradient of $f$.

*Result 3.2* (*Second order necessary conditions.*)
Let $f$ be twice differentiable function. If $\overline{x} \in R^n$ is a local minimum for (1.4) such that $\nabla f(\overline{x}) = 0$, then $H(f(\overline{x}))$ must be positive semidefinite where $H(f(\overline{x}))$ is the Hessian matrix of $f$.

Clearly any local minimum must satisfy the above first and second order necessary conditions (i.e., a point that does not satisfy the above conditions is not a local minimum). However, a point that satisfies the above conditions is not guaranteed to be a local minimum. Therefore, we need the sufficiency conditions presented below.

*Result 3.3* (*Second order sufficiency conditions*)
Let $f$ be twice differentiable function. If $\overline{x}$ is a point such that $\nabla f(\overline{x}) = 0$, and $H(f(\overline{x}))$ is positive definite then $\overline{x}$ is a local minimum for (1.4).

For proofs of the above results see Murty (1988) or Bazaraa et al.(1993).

Optimality conditions for constrained optimization problems are more involved than their counterparts for the unconstrained case. They can be summarized as follows: a point that satisfies some constraint qualifications and at which there exists no direction which is both feasible and improving, is a local minimum. For details on these conditions see Murty (1988), and Bazaraa et al.(1993).

In the next three sections, we present some nonlinear programming algorithms. Classification of these algorithms is depicted in Figure 8.

# 3 LINE MINIMIZATION ALGORITHMS

Consider the following problem

$$\min_{x} f(x) \tag{1.5}$$

where $L \leq x \leq U$, and $x \in R$

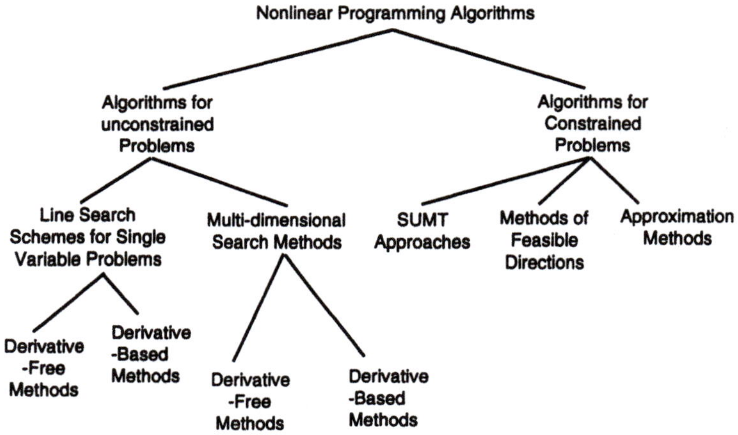


**Figure 8** Classification of nonlinear programming algorithms

Clearly (1.5) is a problem in one variable. The easiest way to approach (1.5) is to solve $\frac{df(\overline{x})}{dx} = 0$ for $\overline{x}$, if $f$ is differentiable. However, the following may constitute some difficulties

1. The function $f$ may not be differentiable.
2. Solving $\frac{df(\overline{x})}{dx} = 0$ may be as difficult as the original problem (1.5) itself.
3. The value $\overline{x}$ may not lie in the range $L \leq x \leq U$.

Therefore, one has to resort to other approaches for solving (1.5). These approaches are called *line search techniques.*

Most line search schemes assume that the function is strictly unimodal (i.e., if $x^*$ is the minimum of $f$, then $x_1 \geq x_2 \geq x^*$ implies $f(x_1) > f(x_2) > f(x^*)$ and $x_1 \leq x_2 \leq x^*$ implies $f(x_1) > f(x_2) > f(x^*)$), and are divided into two steps:

- **Bracketing Phase (Bounding)**
  In this phase, a bracket (or an interval) is found which contains the minimum.

- **Interval Refinement Phase**
  In this phase, the interval found in the first phase is refined (i.e., reduced in width) till it becomes very small (i.e., of acceptable accuracy).

Next, we present schemes for executing the above phases.

- **Bracketing Phase**

  Consider an example of this phase, Swann's method (1964) in which we select the starting point, $x_0$ arbitrarily, then given the $k$th point the next point, $(k+1)st$, is generated using the following equation

  $x_{k+1} = x_k + 2^k\Delta$ for $k = 0, 1, 2, \cdots$

  where $\Delta$ is called the *step* size parameter and is selected arbitrarily but of suitable magnitude. Its sign is made positive if the minimum lies to the right of the current point, otherwise, it is made negative. The scheme goes like this:

  1. Evaluate $f(x_0)$, $f(x_0 + |\Delta|)$,$f(x_0 - |\Delta|)$, let $k = 1$
  2. If $f(x_0 - |\Delta|) \geq f(x_0) \geq f(x_0 + |\Delta|)$ then $\Delta$ must be positive. Let $x_1 = x_0 + \Delta$, and go to step 3. Else if $f(x_0 - |\Delta|) \leq f(x_0) \leq f(x_0 + |\Delta|)$ then $\Delta$ must be negative, let $x_1 = x_0 - \Delta$ and go to step 3. Else $f(x_0 - |\Delta|) \geq f(x_0) \leq f(x_0 + \ \Delta|)$ then the minimum is bracketed, or $x^* \in [x_0 - |\Delta|, x_0 + |\Delta|]$ stop.
  3. Let $x_{k+1} = x_k + 2^k\Delta$, evaluate $f(x_{k+1})$
  4. Compare $f(x_{k-1})$, $f(x_k)$, $f(x_{k+1})$. If $f(x_{k-1}) \geq f(x_k) \leq f(x_{k+1})$ stop, if $\Delta$ is positive, $x^* \in [x_{k-1}, x_{k+1}]$. If $\Delta$ is negative $x^* \in [x_{k+1}, x_{k-1}]$; otherwise, let $k = k + 1$ and repeat steps 3 and 4.

  **Example 1**

  Consider the problem of minimizing $f(x) = (50 - x)^2$ given the starting point $x_0 = 20$, and a step size $|\Delta| = 5$. Use Swann's method to find a bracket that contains the minimum.

  **Solution:**

  1. The sign of $\Delta$ is determined by comparing

  $$f(x_0) = f(20) = 900$$

  $$f(x_0 + |\Delta|) = f(25) = 625$$

  $$f(x_0 - |\Delta|) = f(35) = 1225$$

  $$k = 1$$

  2. Since $f(x_0 - |\Delta| \geq f(x_0) \geq f(x_0 + |\Delta|)$, $\Delta$ must be positive and the minimum point $x^*$ must be greater than 20. Let $x_1 = x_0 + \Delta = 25$.

  3. $x_2 = 25 + 2(5) = 35$; $f(35) = 225$

4. $f(25) = 625 > f(35) = 225$, hence $k = 2$.
3. $x_3 = 35 + 2^2(5) = 55$; $f(55) = 25$
4. $f(35) = 225 > f(55) = 25$, hence $k = 3$.
3. $x_4 = 55 + 2^3(5) = 95$; $f(95) = 2025$
4. $f(95) \geq f(55)$, stop.
$x^* \in [x_{k-1}, x_{k+1}]$ or $x^* \in [35, 95]$

- **Interval Refinement Phase**
Starting from an interval obtained from the bounding phase which contains the minimum, the interval is refined in this phase using one of several schemes. Schemes for interval refinement phase are divided into two groups: derivative-free and derivative-based methods as follows:

**Derivative-based Line Search Schemes**
These methods assume that the function is differentiable, and they use the information of the derivative of the function (and the second derivatives for some methods ) to guide the search. Among these schemes are the following:

- The Bisection Search Method
- Newton's Method
- Modified Newton's Method
- Secant Method

The first method starts by an interval that contains the minimum, while the last three methods start by an arbitrary point.

Next, we summarize the steps of the bisection method as an example of the above methods.

**Statement of Bisection Search Method**

**Initialization Step**
Let $[L_1, U_1]$ be in the initial interval of uncertainty obtained from the bounding phase, let $\epsilon$ be the final interval of uncertainty, let $m$ be the smallest positive integer such that $(\frac{1}{2})^m \leq \frac{\epsilon}{U_1 - L_1}$, let $k = 1$, and go to the main step.

**Main Step**

1. Let $x_k = \frac{1}{2}(L_1 + U_1)$. Compute $f'(x_k)$. If $f'(x_k) = 0$ stop, $x_k$ is the optimal solution; otherwise, go to step 2 if $f'(x_k).f'(U_k) > 0$, and go to step 3, if $f'(x_k).f'(U_k) < 0$.

2. Let $L_{k+1} = L_k$, and $U_k = x_k$. Go to step 4.

3. Let $L_{k+1} = x_k$, and $U_{k+1} = U_k$. Go to step 4.

4. If $k = m$ stop, the minimum lies in the interval $[L_{k+1}, U_{k+1}]$ (one can take the middle point of this interval which makes the maximum error $\frac{\epsilon}{2}$), otherwise, let $k = k + 1$ and go to step 1.

For more details of this method and other derivative-based methods, see Bazaraa et al.(1993), and Murty (1988).

**Example 2**

Use the bisection search method to solve the following problem:

$$\min f(x) = (50 - x)^2 \text{ use } \epsilon = 0.5$$

**Solution:**

**Initialization Step**

The bracketing phase of Example 1 yielded $x^* \in [35, 95]$, hence $L_1 = 35$, $U_1 = 95$. Let $k = 1$, $(\frac{1}{2})^m \leq \frac{0.5}{95-35} = \frac{0.5}{60}$, or $m = 7$.

**Main Step**

1. $x_1 = \frac{1}{2}(35 + 95) = 65$
   $f'(x) = 2(50 - x)$
   $f'(65) = 2(15) = 30 \neq 0$
   $f'(65).f'(95) > 0$
2. $L_2 = 35$, $U_2 = 65$
4. $k = 1 \neq m, k = 2$
1. $x_2 = \frac{1}{2}(35 + 65)$
   $f'(50) = 0$ stop, $x^* = 50$ is the optimal solution (note that if this did not happen, one should continue to execute the steps till $k = 7$).

**Derivative-Free Line Search Scheme**

When the function is only implicitly available, or explicitly available but not differentiable, or differentiable but difficult to evaluate the derivative, then one has to resort to derivative-free line search schemes. Among these schemes are the following:

- Dichotomous Search Method
- the Golden Section Search Method
- the Fibonacci Search Method

Next, we present the statement of the Golden Section method as an example of the above methods.

**Statement of the Golden Section method**

**Initialization Step**

Let $[L_1, U_1]$ be the initial interval of uncertainty, and $\epsilon > 0$ be the final length of uncertainty. Let $x_1 = L_1 + (1-\alpha)(U_1 - L_1)$ and $y_1 = L_1 + \alpha(U_1 - L_1)$ where $\alpha = 0.618$. Compute $f(x_1)$, and $f(y_1)$, let $k = 1$, and go to the main step.

**Main Step**

1. If $U_k - L_k < \epsilon$ stop, the optimal solution lies in the interval $[L_k, U_k]$ (with maximum error $\epsilon/2$), otherwise, if $f(x_k) > f(y_k)$ go to step 2. If $f(x_k) \leq f(y_k)$ go to step 3.

2. Let $L_{k+1} = x_k$, and $U_{k+1} = U_k$. Let $x_{k+1} = y_k$, and let $y_{k+1} = L_{k+1} + \alpha(U_{k+1} - L_{k+1})$. Compute $f(y_{k+1})$. Let $k = k+1$, and go to step 1.

3. Let $L_{k+1} = L_k$, and $U_{k+1} = y_k$. Let $y_{k+1} = x_k$, and $x_{k+1} = L_{k+1} + (1-\alpha)(U_{k+1} - L_{k+1})$. Compute $f(x_{k+1})$. Let $k = k+1$ and go to step 1.

For more details on the derivative-based methods see Bazaraa et al.(1993), and Murty (1988).

**Example 3**

Use the Golden Section technique to solve the following problem:

$$\min f(x) = (50 - x)^2, \epsilon = 1/2$$

**Solution:**

**Initialization Step**

From example 1, $L_1 = 35, U_1 = 95$.
Hence $x_1 = 35 + (1 - 0.618)(95 - 35) = 57.92$
$y_1 = 35 + 0.618(95 - 35) = 72.08$
$k = 1$

**Main Step**

1. $U_1 - L_1 = 60 > \epsilon$
   $f(x_1) = f(57.92) = 62.73 < f(y_1) = f(72.08) = 487.5$
3. $L_2 = 35, U_2 = 72.08$
   $y_2 = x_1 = 57.92$
   $x_2 = 35 + (1 - 0.618)(72.08 - 35) = 49.16$
   $f(x_2) = 0.71, k = 2$
1. $U_2 - L_2 = 37.08 > \epsilon \quad f(x_2) = f(49.16) = 0.71 < f(57.92) = 62.73$
3. $L_3 = 35, U_3 = 57.92$
   $y_3 = x_2 = 49.16$
   $x_3 = 35 + (1 - 0.618)(57.92 - 35) = 43.76$
   $f(x_3) = 38.94, k = 3$
1. $U_3 - L_3 = 22.92 > \epsilon \quad f(x_3) = f(43.76) = 38.94 > f(49.16) = 0.71$
2. $L_4 = 43.76, U_4 = 57.92 \quad x_4 = 49.16$
   $y_4 = 43.76 + 0.618(57.92 - 43.76) = 52.51$
   $f(y_4) = f(52.51) = 6.30$
   $k = 4$
1. $U_4 - L_4 = 14.16 > \epsilon$.

The above iterations are repeated till $U_k - L_k < \epsilon$.

# 4 MULTIDIMENSIONAL SEARCH TECHNIQUES

When the function involved has several variables, then one has to employ multidimensional search techniques to find a local minimum. These methods transform the multivariable search into a sequence of single dimensional search problems, in which the line search techniques of the last section can be used. These techniques are either derivative-based or derivative-free methods.

**Derivative-Based Multidimensional Search Techniques**

These methods use gradient and Hessian information to find a descent direction, (i.e., a direction along which the function decreases) and then use line search techniques to find the minimum along this direction. This process is repeated until a prespecified criterion is satisfied. Examples of these methods are:

1. Steepest Descent Method
2. Newton's Method
3. Quasi Newton's Method
4. Conjugate Gradient Methods

These methods are only different in the way they use gradient and Hessian information to generate search directions. Next, we state Newton's method as an example of the above algorithms.

**Statement of Newton's method**

**Initialization Step**

Let $\epsilon > 0$ be a prespecified accuracy. Let $k = 0$, choose a starting point $x_k$ and go to the main step.

**Main Step**

If $||\nabla f(x_k)|| < \epsilon$, stop, otherwise, let $d_k = -H(f(x_k))^{-1}\nabla f(x_k)$.
Let $x_{k+1} = x_k + d_k$, $k = k + 1$, and repeat this step.

Notice that the steepest descent method is similar to Newton's method except that the direction $d_k = -\nabla f(x_k)$ and $x_k = x_k + \theta_k d_k$ where $\theta_k$ is obtained by line search. In the modified Newton's method, the direction used is the same as that of Newton's method but with a line search. For more details on the above methods see Murty (1988), or Bazaraa et al.(1993).

**Example 4**

Use Newton's method to solve the following problem

$$\min\ (x_1 - 2)^2 + (x_2 - 3)^2$$

**Solution:**

**Initialization Step**

$\epsilon = 0.1$ let $k = 0$,$x_0 = (1, 2)^T$

**Main Step**

$$\nabla f(x) = \begin{bmatrix} 2(x_1 - 2) \\ 2(x_2 - 3) \end{bmatrix}$$

$$H(f(x)) = \begin{bmatrix} 2 & 0 \\ 0 & 2 \end{bmatrix}$$

$$||\nabla f(x_1)|| = \sqrt{(2^2 + 2^2)} = \sqrt{8} \geq 0.1$$

$$d_0 = -\begin{bmatrix} 2 & 0 \\ 0 & 2 \end{bmatrix}^{-1} \begin{bmatrix} -2 \\ -2 \end{bmatrix}$$

$$= -\frac{1}{4}\begin{bmatrix} 2 & 0 \\ 0 & 2 \end{bmatrix}\begin{bmatrix} -2 \\ -2 \end{bmatrix} = \begin{bmatrix} 1 \\ 1 \end{bmatrix}$$

$$x_1 = x_0 + d_0 = \begin{bmatrix} 1 \\ 2 \end{bmatrix} + \begin{bmatrix} 1 \\ 1 \end{bmatrix} = \begin{bmatrix} 2 \\ 3 \end{bmatrix}$$

$$||\nabla f(x_1)|| = 0$$

*stop*

**Derivative-Free Multidimensional Search Methods**

These methods use search directions that are either decided *apriori* (e.g., in the coordinate cyclic method and exploratory search of Hooke and Jeeve's method coordinate directions are used), or directions that are dependent on the outcomes of the search (e.g., Rosenbrock's and pattern search of the Hooke and Jeeve's methods). In either case, the derivative and Hessian information are not used in finding the search directions. Examples of these methods include

1. The Coordinate Cyclic Method
2. The Hooke and Jeeve's Method
3. The Rosenbrock's Method

We next present a summary of the Hooke and Jeeve's method as an example of the above algorithms.

### Statement of the Hooke and Jeeve's method

**Initialization Step**

Let $\epsilon > 0$, be a prespecified accuracy, let $x_1$ be the starting point (selected arbitrarily), let $y_1 = x_1$. Let $k = j = 1$, let $d_1, \cdots, d_n$ be the coordinate directions, and go to the main step.

**Main Step**

1. Let $\theta_j$ be the optimal solution to the line search problem $\min f(y_j + \theta d_j)$ let $y_j \;\; = y_j + \theta_j d_j$. If $j < n$, let $j = j + 1$ and repeat this step. If $j = n$, let $x_{k+1} = y_{j+1}$. If $||x_{k+1} - x_k|| < \epsilon$, stop, otherwise, go to step 2.

2. Let $d = x_{k+1} - x_k$, and let $\hat{\theta}$ be an optimal solution to the problem $\min f(x_{k+1} + \theta d)$, let $y_1 = x_{k+1} + \hat{\theta} d$, let $k = k + 1$, and $j = 1$, and go to step 1.

Notice that the method of Hooke and Jeeve's goes through several cycles, where each cycle consists of single dimensional searches along all coordinate directions (called *exploratory searches*), and one dimensional search along a direction that points from the first point to the last point obtained from the exploratory searches (called *pattern search*). For more detail on the above methods, see Murty (1988) and Bazaraa et al.(1993).

**Example 5**

Use Hooke and Jeeve's method to solve the following problem

$$\min \; (x_1 - 2)^2 + (x_2 - 3)^2$$

**Solution:**

**Initialization Step**

$$\epsilon = 0.1, \, x_1 = (1,2)^T, \, y_1 = (1,2)^T, \, d_1 = \begin{bmatrix} 1 \\ 0 \end{bmatrix}, d_2 = \begin{bmatrix} 0 \\ 1 \end{bmatrix}, \, k = 1, \, j = 1$$

**Main Step**

1. Solve the following line search problem

$$\min \; f(1 + \theta, 2), \theta \geq 0$$

The solution is attained at $\theta_1 = 1$ using the bisection method.
$y_2 = y_1 + \theta_1 d_1$

$y_2 = \begin{bmatrix} 1 \\ 2 \end{bmatrix} + 1 \begin{bmatrix} 1 \\ 0 \end{bmatrix} = \begin{bmatrix} 2 \\ 2 \end{bmatrix}$
$j = 1 < n = 2$ , let $j = 1$ and repeat step 1
Solve the following line search problem
$\min f(2, 2 + \theta), \theta \geq 0$
The solution is attained at $\theta_2 = 1$ using the bisection method
$y_3 = y_2 + \theta_2 d_2$
$y_3 = \begin{bmatrix} 2 \\ 2 \end{bmatrix} + 1 \begin{bmatrix} 0 \\ 1 \end{bmatrix} = \begin{bmatrix} 2 \\ 3 \end{bmatrix}$

$j = 2 = n$
$x_2 = y_3$
$||x_2 - x_1|| = \sqrt{(2-1)^2 + (3-2)^2} = \sqrt{2} \geq \epsilon$

2. $d^T = (2,3)^T - (1,2)^2 = (1,1)^T$
Let $\hat{\theta}$ be an optimal solution to the problem $\min f(2 + \theta, 3 + \theta)$
$\hat{\theta} = 0$, $y_1 = (2,3)^T$, $k = 2$, $j = 1$
1. Clearly, $\theta_1$ and $\theta_2$ will be zero which means that $x_3 = x_2$, and hence the algorithm stops with $x^* = (2,3)^T$.

# 5 METHODS FOR CONSTRAINED OPTIMIZATION

In the last section, methods for multivariable unconstrained optimization problems were discussed. For constrained problems, there are several approaches for them which include:

1. *Sequential Unconstrained Minimization Techniques (SUMT)*
   These methods include penalty and barrier function approaches. That transform the constrained problem into a sequence of unconstrained optimization problems which eventually converge to the solution of the original problem. (For more detail, see Fiacco and McCormick (1968)).

2. *Approximation Methods*
   These methods approximate the nonlinear optimization problem by a sequence of linear or quadratic programming problems which are supposedly easier to solve. (For more detail, see Fletcher (1987), and Bazaraa et al.(1993)).

3. *Methods of Feasible Directions*
   These methods construct directions that are both feasible and improving, and implement the methods developed in the last section to solve the resulting sequence of subproblems. (For more details, see Lueneberger (1984) and Bazaraa et al.(1993)).

Next, we state the penalty function algorithm as an example of SUMT approaches to nonlinear programming problems(see Problem (1.1) in Section 1.4). In this algorithm, the constrained optimization problem is transformed into an unconstrained problem, by appending a penalty function to the original objective function. This penalty function heavily penalizes any infeasible point (i.e., it is a penalty parameter multiplied by a measure of the infeasibility in the constraints), and hence the minimum of the modified objective function will be a feasible point.

**Statement of the Penalty Function Method**

**Initialization Step**
Let $\epsilon > 0$ be a prespecified accuracy level. Choose an initial point $x_1$ and a penalty parameter $\mu_1 > 0$, choose $\rho > 1$, let $k = 1$ and go to the main step.

**Main Step**
1. Starting from $x_k$, use one of the multidimensional search techniques (discussed in the last section) to solve the problem

$$f(x) + \mu_k \left( \sum_{j=1}^{m} \max\{0, -g_i(x)\}^2 + \sum_{i=1}^{p} (h_i(x))^2 \right)$$

Let $x_{k+1}$ be the optimal solution of the above problem. Go to step 2.

2. If $\mu_k \left( \sum_{j=1}^{m} \max\{0, -g_i(x)\}^2 + \sum_{i=1}^{p} (h_i(x))^2 \right) < \epsilon$ stop, otherwise let $\mu_{k+1} = \rho\mu_k$, let $k = k + 1$ and go to step 1.

For details on the above method see Bazaraa et al.(1993).

**Example 6**
Solve the following problem by the penalty function method

$$\min \left( (x_1 - 2)^2 + (x_2 - 3)^2 \right)$$

subject to

$$x_1 + x_2 = 0$$

**Solution:**

**Initialization Step**

Let $\epsilon = 0.1, x_1 = (0,0)^T, \mu_1 = 1, \rho = 5, k = 1$

**Main Step**

1. $\min\ (x_1 - 2)^2 + (x_2 - 3)^2 + 1(x_1 + x_2)^2$
We can use the Hooke and Jeeve's method to solve the above problem, which yields $x_1 = (.333, 1.333)^T$.
2. $\mu_1 h_1^2 = 1(1.67)^2 = 2.79 \geq \epsilon$

$$\mu_2 = 5\mu_1 = 5, k = 2$$

and the process is repeated till the stopping criterion is satisfied.

# 6 SOFTWARE FOR OPTIMIZATION ALGORITHMS

Due to the need for optimization algorithms to tackle problems in various fields, many software packages have been developed for this purpose. There are some packages that can be used for the modelling purposes, among them are the following:

- GAMS (Brooke et al.(1985))
- LINGO (Cunningham and Scharge (1989))
- AMPL (Fourer et al.(1990))

The following are some packages that can be used to solve certain classes of optimization problems:

- LINDO for linear, quadratic and integer programming problems (Scharge (1991)).
- GINO for nonlinear programming problems (Liebman et al.(1986))

- MINOS for linear and quadratic programming problems (Murtagh and Saunders (1982))
- IMSL is general mathematical and statistical package but has subroutines to solve linear and nonlinear quadratic programming (Math Library Manual 1987).
- CONCOPT for nonlinear programming problems (Drud (1985))
- LSGRG for large sparse nonlinear programming problems (Smith and Lasdon (1992))

# 7 APPLICATIONS OF OPTIMIZATION METHODS IN QUALITY CONTROL

Recently, there has been a lot of interest among researchers in the economics of quality control which can be explained partly by the tough global competition. Optimization methods have been developed to aid at selecting the most economical levels of performing a quality control function. Optimization models have been successfully applied to the following classes of quality control problems.

1. *Targeting Problems*
   In this class of quality control problems, the most economical level(s) of operating a machine are selected. Targeting problem may be stated as follows: Given a process which produces products with a certain variation for one of its quality characteristics, and specification limits for that quality characteristic (either lower specification limit (LSL), upper specification limit (USL) or both), and given the costs of not conforming to the given limit(s), unit cost of material used, and scrap and rework costs and policy, *find the best process setting(s) that will minimize the total expected cost.* One can see that setting the target too high will minimize the number of undersized items but at the expense of more material cost, and more oversized items are produced (if applicable) and vice versa.

   A classical example of the targeting problems is the canning problem where the manufacturer is interested in filling out a can with a certain fluid (e.g., beverages), and interested in finding the optimal setting which minimizes the total expected costs. If the process does not deteriorate with time, then we call the resulting model a *static model.* If the process deteriorates with time (i.e., tool wear), then we call the resulting model a *dynamic model.* Springer (1951) was the first to build a model for the static case by Hunter and Kartha (1977). Gibra (1967) was the first to develop a model for the dynamic case. See the survey by Al-Sultan and Rahim (1994) for static models and Al-Fawzan and Al-Sultan (1996) for dynamic models and references at the end of this chapter.

2. *Economic Design of Control Charts*
   Control charts are the most effective tools that can be used to monitor manufacturing and service processes to ensure that they are in statistical control (i.e., stable with time). To design a control chart, one has to decide on its parameters which include sample size, sampling frequency, and control limits for the chart. Control charts have been traditionally designed with statistical criteria in mind. These criteria include minimizing the probability of failing to detect a shift in the process when it happens (type I error), and concluding that a shift in the process has happened when it has not (type II error). However, since many costs are involved in using control charts (i.e., cost of detecting an assignable cause and correcting, cost of sampling, cost of undiscovered defective items), it is only logical to use models that minimize these costs and this is what is called *economic design of control charts.* Hence, the parameters of control charts are selected such that a prespecified cost function is minimized. Many models have been developed for this problem as it has been extensively studied by researchers. Simple search techniques have been used to optimize these models. See Montgomery (1991) for an excellent discussion of this problem, and the surveys by Montgomery (1980), Ho and Case (1994), and Collani (1997), and other references at the end of this chapter.

3. *Sampling Plans*
   In this class of quality control problems, the optimal parameters of the plan need to be determined with respect to certain criteria. These criteria are either statistical or economic. In the design of a single sampling plan Bennett et al.(1974) developed necessary conditions for the parameters to be optimal and then used a simple incremental procedure to determine the optimal sample size and the critical value. The models developed in the literature minimize the expected cost using optimization methods such as multidimensional search. In the double sampling plan, this has been applied to find optimal parameters of such plans (Stewart et al.(1978)). Other models have also been developed for using sequential sampling plan.

   In repeat inspection plans, Duffuaa and Raouf (1989) developed three optimization models to find the optimal number of repeat inspections. They also developed an optimal rule for sequencing characteristics for inspection

(1990). Later, Duffuaa and Nadeem (1994) extended these models for the case of statistical dependency. Also Duffuaa and Al-Najjar (1995) developed alternative models for repeat inspection plans. Several optimization models were developed to investigate the effect of inspection error (e.g., Tang and Schneider (1988, 1993), Bennett et al.(1974), Duffuaa (1996)). For more details, see other references at the end of this chapter.

4. *Taguchi's Quality Control Models*
   Customers are usually more satisfied with products that are more robust and tolerant to variations in the environments and conditions. Only customers in the field can determine the actual degree of product robustness which is highly affected by product-process design (Kolarik (1995)). Taguchi has developed the concept of loss functions and signal-to-noise ratios, which are a combination of classical statistics and economy. These tools are applied in the off-line stage, generally at the parameter and tolerance design stages. Taguchi's approach to quality is to use a quadratic loss function which quantifies the total loss to society and optimizes the design based on that criteria. For more details on Taguchi's approach, see references at the end of this chapter.

5. *Economic Models combining quality, production and maintenance*
   In industrial processes, production, quality and maintenance policies are usually interdependent. Many authors have attempted to develop models that integrate two of these policies. Recently, some researchers have developed models that integrate all the three policies. The following is a summary of the models that integrate quality control policies to other policies.

   - *Models which integrate production and quality:* Makhdoum (1996) stated that researchers have used two different approaches to integrate quality cost with the determination of economic production quantity. The first approach is by computing the percentage of defective items produced in a production system and the second by using control chart for performance monitoring of the process ( e.g., Rosenblatt and Lee (1986), Porteus (1986, 1990), Liou et al.(1994), Cheng (1989, 1991a, 1991b), Peters et al.(1988), Khouja and Mehrez (1994), Goyal and Gunasekaran (1990), Yum and McDowell (1987, 1981)). For an extensive survey see Goyal et al.(1993).
   - *Models which integrate quality and maintenance:* Many models have been developed which consider the effects of failure mechanisms, preventive maintenance policies, on the economic design of control chart (e.g., Lorenzen and Vance (1986), Baker (1971), Montgomery and

Heikes (1976), Benerjee and Rahim (1988), Ben Daya and Rahim (1996), Chiu and Huang (1996), and Ben Daya and Duffuaa (1995)).

- *Models which integrate production, maintenance and quality:* This is an area where less research has been done when compared to the above situations. However, there have been some attempts to develop models that integrate production planning, inspection, and preventive maintenance (Huang and Chiu (1995)), and to study the effects of maintenance policies on the determination of the economic production quantity and the economic design of control charts (e.g., Rahim(1994) and Ben Daya (1996)).
For an elaborate survey of the above models see Makhdoum (1996).

# 8 CONCLUSION

In this chapter, importance of optimization as a tool for solving real life problems has been highlighted. The modelling process has been discussed. Optimality conditions for optimization problems have been presented. Various methods for optimization including single dimensional line search, multidimensional search techniques, and methods for constrained optimization have been highlighted. Computer packages for solving optimization problems have been presented. Finally, successful applications of optimization algorithms in quality control have been discussed.

# Acknowledgment

The author is thankful to the three referees for comments that have improved the presentation of this chapter.

# REFERENCES

## References on Optimization


1. Avriel, M., *Nonlinear Programming: Analysis and Methods*, Prentice Hall, Englewood Cliffs, N.J. 1976.
2. Bazaraa, M.S., J.J. Jarvis, and H.D. Sherali, *Linear Programming and Network Flows*, 2nd Edition, John Wiley and Sons, New York, 1990.
3. Bazaraa, M.S., H. Sherali, and C.M. Shetty, *Nonlinear Programming: Theory and Algorithms*, John Wiley, New York, 1993.
4. Brooke, A., D. Kendrick, and A. Mieerans, "GAMS - A user Guide", Scientific Press, Redwood City, CA, 1988.
5. Cunningham, K., and L. Schrage, "The LINGO Modelling Language", LINDO Systems, Inc., Chicago, IL, 1989.
6. Dantzig, G.B., *Linear Programming and Extensions*, Princeton University Press, 1963.
7. Dennis, J.E., Jr., and R.B. Schnabel, Numerical Methods for Unconstrained Optimization and Nonlinear Equations, Prentice Hall, Englewood Cliffs, N.J., 1983.
8. Dixon, L.C. W., *Nonlinear Optimization*, The English Universities Press, London, 1972.
9. Dixon, L.C.W. (Ed.), *Optimization in Action*, Academic Press, New York, 1976.
10. Drud, D., "CONOPT: A GRG code for large sparse dynamic nonlinear optimization problems", Technical Note No. 21, Development Research Center, The World Bank, 1818 H Street, N.W., Wahsington D.C., 1984.
11. Fiacco, A.V. and G.P. McCormick, *Nonlinear Programming: Sequential Unconstrained Minimization Techniques*, John Wiley and Sons, New York, 1968.
12. Fletcher, R., *Practical Methods of Optimization* , (2nd Edition), John Wiley and Sons, N.Y., 1987.
13. Fletcher R., *Practical Methods of Optimization 1: Unconstrained Optimization*, John Wiley and Sons, Chicester, 1980.

14. Fletcher, R., *Practical Methods of Optimization 2: Constrained Optimization*, John Wiley and Sons, Chichester, 1981.

15. Fourer, R., D. Gay, and B. Kernighan, "Modelling Language for Mathematical Programming", *Management Science*, 36(5), pp 519-554, 1990.

16. Fox, R.L., *Optimization Methods for Engineering Design*, Addison Wesley Reading, Mass., 1971.

17. Gill, P.E., W. Murray, and M.H., Wright, *Practical Optimization*, Academic Press, London and New York, 1981.

18. Himmelblau, D.M., *Applied Nonlinear Programming*, McGraw Hill, New York, 1972.

19. IMSL Math Library Manual, 1987.

20. Lasdon, L.S., *Optimization Theory for Large Systems*, Macmillan, NY, 1970.

21. Lasdon, L.S., A. D. Waren, A. Jain, and M. Ratner, "Design and testing of a GRG code for Nonlinear Optimization", *ACM Transactions on Mathematical Software*, 4, pp 34-50, 1978.

22. Leibman, J., L. Lasdon, L. Schrage, and A. Waren, *Modelling and Optimization with GINO*, Scientific Press, Palo Alto, California, 1986.

23. Lootsma, F.A., (Ed.), *Numerical Methods for Nonlinear Optimization*, Academic Press, NY, 1972.

24. Luenberger, D.G., *Linear and Nonlinear Programming*, Addison Wesley, Reading, Mass., Second Edition, 1984.

25. Mahidhara, D., and L.S. Lasdon, "An SQP Algorithm for Large Sparse Nonlinear Programs", Working Paper, MSIS Dept., School of Business, The University of Texas, Austin, TX, 1990.

26. Mangasarian, O.L., *Nonlinear Programming*, McGraw Hill, New York, 1991.

27. Managasarian, O.L., R.R. Meyer, and S.M. Johnson (Eds.), *Nonlinear Programming*, Academic Press, NY, 1975.

28. Murtagh, B.A. and M.A. Saunders, "MINOS 5.0 Users Guide", Technical Report SOL 83.20, Systems Optimization Laboratory, Stanford University, Stanford, California, 1983.

29. Murty, K.G., *Linear and Combinatorial Programming*, John Wiley, 1976.

30. Murty, K.G., *Linear Programming*, John Wiley, 1983.

31. Murty, K.G., *Linear Complementarity, Linear and Nonlinear Programming*, Elsevier Publishing Company, 1988.

32. Martos, B., *Nonlinear Programming: Theory and Methods*, American Elsevier, NY, 1975.

33. McMillan, C., Jr., *Mathematical Programming*, John Wiley and Sons, NY, 1970.

34. Nemhauser, G.L., and L.A. Wolsey, *Integer and Combinatorial Optimization*, John Wiley and Sons, NY, 1988.

35. Nemhauser, G.L., and A.H.G. Rinnoy, Y. Kan, and M.J. Todd (editors), *Handbooks in Operations Research and Management Science*, Vol. 1, Optimization, North Holland, 1989.

36. *Optimization Subroutines Library*, International Business Machine, Release 2, 1991.

37. Ravindran, A., D.T. Philips, and J.J. Solberg, *Operations Research: Principles and Practice*, Second Edition, John Wiley, NY, 1987.

38. Reklaitis, G.V., A. Ravindran, and K.M. ragsdell, *Engineering Optimization: Methods and Applications*, John Wiley, NY, 1983.

39. Rockafeller, R.T., *Convex Analysis*, Princeton University Press, Princeton, NJ, 1970.

40. Salkin, *Integer Programming*, John Wiley, New York, 1975.

41. Schittkowski, K., *Nonlinear Programming Codes-Information, Tests, Performance*, in Lecture Notes in Economics and Mathematical Systems, Vol. 183, Springer-Verlag, NY, 1980.

42. Scharge, L., *LINDO: An Optimization Modelling System*, The Scientific Press, Fourth Edition, 1991.

43. Smith, S., and L. Lasdon, "Solving Large Sparse Nonlinear Programs Using GRG", *ORSA Journal of Computing*, 4(1), pp 2-15, 1992.

44. Swann, W. H., "Report on the development of a direct search method of optimization" *ICI Ltd., Central Instr. Res. Lab.*, Note 64/3, London, 1964.

45. Zangwill, W.I., *Nonlinear Programming: A Unified Approach*, Prentice Hall, Englewood Cliffs, NJ, 1969.

**References on Static Targetting in Quality Control**

1. Al-Sultan, K.S., "An Algorithm for the Determination of the Optimum Target Values for Two Machines in Series With Quality Sampling Plans", *International Journal of Production Research*, 32(1), pp 37-45.

2. Al-Sultan, K.S., and M.A. Rahim, "Economic Selection of Process Parameters: A Literature Survey", Submitted (1994).

3. Arcelus, F.J. and M.A. Rahim, "Joint Determination of Optimum Variable and Attribute Target Means", *Naval Research Logistics*, 138, pp 851-864, 1991.

4. Arcelus, F.J. and P.K. Banerjee, "Optimal Production Plan in a Tool-Wear Process with Rewards for Acceptable, Undersized, and Oversized Parts", *Engineering Costs and Production Economics*, 11, pp 13-19, 1987.

5. Arcelus, F.J. and Banerjee, P.K., "Selection of the Most Economical Production Plan in a Tool-Wear Process", *Technometrics*, 27(4), pp 433-437, 1985.

6. Bettes, D.C., "Finding an Optimum Target Value in Relation to a Fixed Lower Limit and an Arbitrary Upper Limit", *Applied Statistics*, pp 202-210.

7. Bisgaard, S., W.G. Hunter, and L. Pallesen, "Economic Selection of Quality of Manufactured Product", *Technometrics*, 26(1), pp 9-18, 1984.

8. Boucher, T.O., and M.A. Jafari, "The Optimum Target Value for Single Filling Operations with Quality Sampling Plans", *Journal of Quality Technology*, 23(1), pp 447-47, 1991.

9. Burr, I.W., "A New Method for Approving a Machine or Process Setting - Part I", *Industrial Quality Control*, pp 12-18, 1949.

10. Burr, I.W., "A New Method for Approving a Machine or Process Setting - Part II", *Industrial Quality Control*, pp 15-19, September, 1949.

11. Burr, I.W., "A New Method for Approving a Machine or Process Setting - Part III", *Industrial Quality Control*, pp 13-16, November, 1949.

12. Carlsson, O., "Determining the Most Profitable Level for a Production Process Under Different Sales Condition", *Journal of Quality Technology*, 16, pp 69-78, 1978.

13. Carlsson, O., "Economic Selection of a Process Level Under Acceptance Sampling by Variables", *Engineering Costs and Production Economics*, 16, pp 69-78, 1989.

14. Carlsson, O., "Quality Selection of a Two-Dimensional Process Level Under Single Acceptance Sampling by Variables", *International Journal of Production Economics*, 27, pp 43-56, 1992.

15. Chen, R., D. Strong, and O. Hawaleshka, "An Economic Model for Raw Material Selection", *International Journal of Production Research*, 31(10), pp 2275-2285, 1993.

16. Dodson, B.L., "Determining the Optimal Target Value for a Process with Upper and Lower Specifications", *Quality Engineering*, 5(3), pp 393-402, 1993.

17. Elsayed, E.A. and A. Chen, "Optimal Levels of Process Parameters with Multiple Characteristics", *International Journal of Production Research*, 31(5), 1117-1132, 1993.

18. Fathi, Y., "Producer-Consumer Tolerances", *Journal of Quality Technology*, 22(2), pp 138-145, April 1990.

19. Golhar, D.Y., "Determination of the Best Mean Contents for a Canning Problem", *Journal of Quality Technology*, 19(2), pp 82-84, 1987.

20. Golhar, D.Y., "Computation of the Optimal Process Mean and the Upper Limit for a Canning Problem", *Journal of Quality Technology*, 20(3), pp 193-195, 1988.

21. Golhar, D.Y. and S.M. Pollock, "Determination of the Optimal Process Mean and the Upper Limit for a Canning Problem", *Journal of Quality Technology*, 20(3), pp 188-192, 1988.

22. Golhar, D.Y., and S.M. Pollock, "Cost Savings Due to Variance Reduction in a Canning Process", *IIE Transactions*, 24(1), pp 89-92, 1992.

23. Goyal, S.K., and G. Rajamannar, "Determination of Confidence Intervals for the Economic Tool Life", *Engineering Costs and Production Economics*, 11, pp 49-52, 1987.

24. Grubbs, F.E., "An Optimal Procedure for Setting Machines or Adjusting Processes", *Journal of Quality Technology*, 15(4), pp 186-189, 1983.

25. Harrington, J.H., *Poor Quality Cost*, Marcel Dekker, New York, N.Y., 1987.

26. Ho, C., and K.E. Case, "Economic Design of Control Charts: A Literature Review", *Journal of Quality Technology*, 26(1), 1994.

27. Hunter, W.G. and, C.P., Kartha, "Determining the Most Profitable Target Value for a Production Process", *Journal of Quality Technology*, 9(4), pp 176-181, 1977.

28. Ladany, S.P. and Y. Alperovitch, "An Optimal Set-up Policy for Control Charts", *OMEGA*, 3(1), pp 113-118, 1975.

29. Melloy, B.J., "Determining the Optimal Process Mean and Screening Limits for Packages Subject to Compliance Testing", *Journal of Quality Technology*, 23(4), pp 318-323, October, 1991 (2 copies).

30. Montgomery, D.C., "The economic design of control charts: a review and literature survey", *Journal of Quality Technology*, 12, 75-87, 1980.

31. Murthy, D.N.P. and I. Dajamaludin, "Quality Control in a Single State Production System: Open and Closed Loop Policies", *International Journal of Production Research*, 28(12), pp 2219-2242, 1990.

32. Natarajan, R. Nat, "Determining Input Parameters Under Process Uncertainties", *International Journal of Production Economics*, 29 pp 203-210, 1993.

33. Nelson, L.S., "Best Target Value for a Production Process", *Journal of Quality Technology*, 10(2), pp 88-89, 1978.

34. Nelson, L.S., "Nomograph for Setting Process to Minimize Scrap Costs". *Journal of Quality Technology*, 11(1), pp 48-49, 1979.

35. Pugh, G.A., "An Algorithm for Economically Setting A Uniformally-Shifting Process", *Computers and Industrial Engineering*, 14(3), pp 237-240, 1988.

36. Rose, J.S., "The Newsboy With Known Demand and Uncertain Replenishment: Applications to Quality Control and Container Fill", *Operations Research Letters*, 11, pp 111-117, 1992.

37. Schmidt, R.L., and P.E. Pfeifer, "An Economic Evaluation of Improvements in Process Capability for a Single-Level Canning Problem", *Journal of Quality Technology*, 21(1), pp 16-19, 1989.

38. Schmidt, R.L. and P.E. Pfeifer, "Economic Selection of the Mean and Upper Limit for a Canning Problem with Limited Capacity", *Journal of Quality Technology*, 23(4), pp 312-317, October, 1991.

39. Springer, C.H., "A Method for Determining the Most Economic Position of a Process Mean", *Industrial Quality Control*, pp 36-39, July 1951.

40. Tseng, S. T. and T. Y. Wu, "Selecting the Best Manufacturing Process", *Journal of Quality Technology*, 23(1), pp 53-62, January, 1991.

41. Vidal, R.V., "A Graphical Method to Select the Optimum Target Value of a Process", *Engineering Optimization*, 13, pp 285-291, 1988.

**References on Dynamic Targetting in Quality Control**

1. Al-Fawzan, M. A., and K. S. Al-Sultan, "The optimal control of a production process subject to drift and shift in the process mean: A survey" *Proceedings of the 20th International Conference on Computers and Industrial Engineering*, Kyongjo, Korea, October 1996.

2. Albright, S.C. and R. S. Collins, "A Bayesian Approach to the Optimal Control of Continuous Industrial Processes," *International Journal of Production Research*, 15(1), pp 37-45, 1977.

3. Arcelus, F.J. and P.K. Banerjee, "Selection of the Most Economical Production Plan in a Tool-Wear Process", *Technometrics*, 27(4), pp 433-437, 1985.

4. Arcelus, F.J. and P.K. Banerjee, "Optimal Production Plan in a Tool-Wear Process with Rewards for Acceptable, Undersized, and Oversized Parts", *Engineering Costs and Production Economics*, 11, pp 13-19, 1987.

5. Arcelus, F.J., P.K. Banerjee, and R. Chandra, "The optimal production run for a normally distributed quality characteristic exhibiting nonnegative shifts in the process mean and variance," *IIE Transactions*, 14, pp 90-98, 1982

6. Arcelus, F.J., P.K. Banerjee, and R. Chandra, "The optimal schedule to produce a given number of acceptable parts with a specified confidence level", *International Journal of Production Research*, 23(1), pp 185-196.

7. Bisgaard, S., W.G. Hunter, and L. Pallesen, "Economic Selection of Quality of Manufactured Product", *Technometrics*, 26(1), pp 9-18, 1984.

8. Drezener, Z. and G.O. Wosolowsky, "Optimal Control of a Linear Trend Process with Quadratic Loss", *IIE Transactions*, 21(1), pp 66-72, 1989.

9. Gibra, I.N., "Optimal Control of Processes Subject to Linear Trends", *Journal of Industrial Engineering*, 18(1), pp 35-41, 1967.

10. Gibra, I.N., "Optimal Production Runs of Processes Subject to Systematic Trends", *International Journal of Production Research*, 12(4), pp 511-517.

11. Hall, R.I. and S. Eilon, "Controlling Production Processes Which are Subject to Linear Trends", *Operational Research Quarterly*, 14(3), pp 179-189, 1963.

12. Jeang, A. and K. Yang, "Optimal Tool Replacement with Nondecreasing Tool-Wear", *International Journal of Production Research*, 30(2), pp 299-314.

13. Kamat, S.J., "A Smoothed Bayes Control Procedure for the Control of a Variable quality characteristic with linear shift", *Journa of Quality Technology*, 8(2), pp 98-104, 1976.

14. Kubat, P. and C.Y.T. Lam, "Optimal Monitoring Strategies for Slowly Deteriorating Repairable Systems", *IEEE Transactions on Communications*, 40(4), pp 661-665, 1992.

15. Lee, H.L. and M.J. Rosenblatt, "Economic Design and Control of Monitoring Mechanism in Automated Production Systems", *IIE Transactions*, 20(2), pp 201-208, 1988.

16. Lee, H.L., K. Moinzadeh, and G. Tagaras, "A Model for Continuous Production Control with Warning Signals to Fault Occurrances", *Journal of Operational Research Society*, 37(5), pp 515-523, 1986.

17. Pate-Cornel, M.E., H.L. Lee, and G. Tagaras, "Warnings of Malfunction: The Decision to Inspect and Maintain Production Processes on Schedule or on Demand", *Management Science*, 33(10), pp 1277-1290, 1987.

18. Pugh, G.A., "An Algorithm for Economically Setting A Uniformally-Shifting Process", *Computers and Industrial Engineering*, 14(3), pp 237-240, 1988.

19. Quesenberry, C.P., "An SPC Approach to Compensating a Tool-Wear Process", *Journal of Quality Technology*, 20(4), pp 220-229, 1988.

20. Rahim, M.A. and P.K. Banerjee, "Optimal Production Run for a Process with Random Linear Dirft", *Omega*, 16(4), pp 347-351, 1988.

21. Rahim, M.A. and R. S. Lashkari, "Optimal Decision Rules for Determining the Length of Production Run", *Computers and Industrial Engineering*, 9(2), pp 195-202, 1985.

22. Rahim, M.A. and A. Raouf, "Optimal Production Run for a Process Having Multilevel Tool Wear", *International Journal of Systems Science*, 19(1), pp 139-149, 1988.

23. Schneider, H., K. Tang, and C. O'Cinneide, "Optimal Control of a Production Process Subject to Random Deterioration", *Operations Research*, 38(6), pp 1116-1122, 1990

24. Smith, B.E. and R.R. Vemuganti, "A Learning Model for Process With Tool-Wear", *Technometrics*, 10(2), pp 379-387, 1968.

25. Taha, H.A., "A policy for determining the optimal cycle length for a cutting tool", *Journal of Industrial Engineering*, 17(3), pp 157-162.

**References on Economic Design of Control Charts**

1. Arnold, B. F., "Optimal Control Charts and Discrimination Between 'Acceptable' and 'Unacceptable' States", *Sankhya* 51, Series B, pp 375-389, 1989.

2. Arnold, B. F., "An Economic $\bar{x}$-Chart Approach to the Joint Control of the Means of Independent Quality Characteristics", *ZOR - Methods and Models of Operations Research* 34, pp 59-74, 1990.

3. Arnold, B. F. and E.V. Collani, "Economic Process Control," *Statistica Neerlandica* 41, pp 89-97, 1987.

4. Arnold, B. F. and E.V. Collani, "On the Robustness of $\bar{x}$-Charts", *Statistics* 20, pp 149-159, 1989.

5. Banerjee, P. K. and M. A. Rahim, "The Economic Design of Control Charts: A Renewal Theory Approach", *Engineering Optimization* 12, pp 63-73, 1987.

6. Banerjee, P. K. and M. A.Rahim, "Economic Design of $\bar{x}$ Control Charts Under Weibull Shock Models", *Technometrics* 30, pp 407-414, 1988.

7. Chung, K. J., "A Simplified Procedure for the Economic Design of $\bar{x}$ Charts", *International Journal of Production Research* 28, pp 1239-1246, 1990.

8. Chung, K. J., "An Efficient Procedure for the Economic Design of *np*-Charts", *International Journal of Quality and Reliability Management* 9, pp 58-68, 1992.

9. Chung, K. J., "An Economic Study of $\bar{x}$ Charts with Warning Limits", *Computers in Industrial Engineering* 24, pp 1-7, 1993.

10. Chung, K. J., "An Algorithm for Computing the Economically Optimal $\bar{x}$-Control Chart for a Process with Multiple Assignable Causes", *European Journal of Operational Research* 72, pp 350-363, 1994.

11. Chung, K. J. and C.-N. Lin, "The Economic Design of Dynamix $\bar{x}$ - Control Charts Under Weibull Shock-Model", *International Journal of Quality an Reliability Management* 10, pp 41-56, 1993.

12. v. Collani, E., "Determaination of the Economic Design of Control Charts Simplified", in K. S. Al-Sultan, and M. A. Rahim, editors *Optimaization in Quality Controls, Kluwer Academic Publishers, USA,*, 1997

13. v. Collani, E., "A Simple Procedure to Determine the Economic Design of an $\bar{x}$ Control Chart", *Journal of Quality Technology* 18, pp 145-151, 1986.

14. v. Collani, E., "Economic Control of Continuously Monitored Production Processes", *ep. Stat. Appl. Res., JUSE,* 34, pp 1-18, 1987.

15. v. Collani, E., "A Unified Approach to Optimal Process Control", *Metrika* 35, pp 145-159, 1988.

16. v. Collani, E., "The Economic Design of Control Charts. Stuttgart: Teubner", 1989.

17. v. Collani, E., "Economically Optimal *c*- and *np*-Control Charts", *Metrika* 36, pp 215-232, 1989.

18. v. Collani, E. and V. Dräger, "Determination of the Economic Design of CUSUM Charts for Controlling a Normally Distributed Process Mean", *Technical Reports of the Würzburg Research Group on Quality Control,* 69, 1966.

19. v. Collani, E. and K. Dräger, "A Simplified Economic Design of Control Charts for Monitoring the Nonconforming Probability", *Economic Quality Control* 10, pp 231-292, 1995.

20. v. Collani, E., E. M. Saniga, and Ch. Weigand, "Economic Adjustment Designs for $\bar{x}$ Control Charts", *IIE Transactions* 26, pp 37-43, 1994.

21. v. Collani, E. and J. Sheil, "An Approach to Controlling Process Variability", *Journal of Quality Technology* 21, pp 87-96, 1989.

22. v. Collani, E. and J. Treml, "Control of a Two-Dimensional Process-Quality-Indicator by Means of a Screening Procedure", *Economic Quality Control* 8, pp 167-194, 1993.

23. v. Collani, E. and Ch. Weigand, "Economic Machine Adjustment in the Case of Product Screening", *Statistical Papers* 33, pp 171-184, 1992.

24. Costa, A. F. B., "Joint Economic Design of $\bar{x}$ and $R$ Control Charts for Processes Subject to Two Independent Assignable Causes", *IIE Transactions* 25, pp 27-33, 1993.

25. Del Castillo, E. and D. C. Montgomery, "Optimal Design of Control Charts for Monitoring Short Production Runs", *Economic Quality Control* 8, pp 225-240, 1993.

26. Duncan, A. J., "The Economic Design of $\bar{x}$ Control Charts Used to Maintain Current Control of a Process", *Journal of the American Statistical Association* 51, pp 228-242, 1956.

27. Frahm, P., "$\bar{x}$ - Control Charts and Age Replacement Policies", *Economic Quality Control* 7, pp 85-96, 1992.

28. Frahm, P., "Alterserneuerung und Blockerneuerung unter Einbeziehung von Stichprobenkontrollen", *Dr.-Thesis*, Würzburg, 1994.

29. Ho, C. and K.E. Case, "Economic Design of Control Charts: A Literature Review for 1981-1991", *Journal of Quality Technology* 26, pp 39-53, 1994.

30. Hryniewicz, O., "The Economic Design of a Certain Class of Control Charts: A General Approach", *Technical Reports of the Würzburg Research Group on Quality Control*, 11, 1988.

31. Hryniewicz, O., "A Simple and Generally Applicable Approximation Technique for the Determination of the Economic Design of Control Charts", *Technical Reports of the Würzburg Research Group on Quality Control*, 15, 1988.

32. Hryniewicz, O., "Economic Design of Attribute Control Charts Based on Double Sampling Plans", *Technical Reports of the Würzburg Research Group on Quality Control*, 17, 1989.

33. Hryniewicz, O., "The Performance of Differently Designed $p$-Control Charts in the Presence of Shifts of Unexpected Size", *Economic Quality Control* 4, pp 7-18, 1989.

34. Keats, J. B. and J. R. Simpson, "Comparison of $\bar{x}$ and the CUSUM Control Charts in an Economic Model", *Economic Quality Control* 9, pp 203-220, 1994.

35. Kurc, K., "The Performance of Differently Designed $\bar{x}$ Control Charts in the Presence of a Shift of Unexpected Size", *Economic Quality Control* 6, pp 3-15, 1991.

36. Lorenzen, T. J. and L.C. Vance, "The Economic Design of Control Charts: A Unified Approach", *Technometrics* 28, pp 3-10, 1986.

37. McWilliams, T. P., "Economic Control Chart Designs and the In-control Time Distribution: A Sensitivity Analysis", *Journal of Quality Technology* 21, pp 103-110, 1989.

38. McWilliams, T. P., "Economic, Statistical, and Economic-Statistical $\bar{x}$ Chart Designs", *Journal of Quality Technology* 26, pp 227-238, 1994.

39. Montgomery, D. C., "The Economic Design of Control Charts: A Review and Literature Survey", *Journal of Quality Technology* 12, pp 75-87, 1980.

40. Montgomery, D. C., "The Economic Design of an $\bar{x}$ Control Chart", *Journal of Quality Technology* 14, pp 40-43, 1982.

41. Montgomery, D.C., "*Introduction to Statistical Quality Control*", 2nd ed, New York: Wiley, 1991.

42. Montgomery, D.C., "The Use of Statistical Process Control and Design of Experiments in Product and Process Improvement", *IIE Transactions* 24, pp 4-17, 1992.

43. Montgomery, D.C., J.C.C. Torng, J.K. Cochran, and F.P. Lawrence, "Statistically Constrained Economic Design of the EMWA Control Chart", *Journal of Quality Technology* 27, pp 250-256, 1995.

44. Moskowitz, H. and R. Plante, "Effect of Quality Loss Function on the Economic Design of $\bar{x}$ Process Control Charts", *European Journal of Operational Research* 72, pp 333-349, 1994.

45. Mukherjee, S. P. and A. Mandal, "Economic Design of $\bar{x}$-Chart Revisited", *Calcutta Statistical Bulletin* 43, pp 109-118, 1993.

46. Nandi, S. N., "Economic Design of *np*-Control Charts When There is a Multiplicity of Assignable Causes", *IAPQR Transactions* 17, pp. 1-9, 1992.

47. Rahim, M. A., "Economically Optimal Determination of the Parameters of $\bar{x}$-Charts with Warning Limits When Quality Characteristics Are Non-Normally Distributed", *Engineering Optimization* 7, pp 289-301, 1984.

48. Rahim, M. A., "Economic Model of $\bar{x}$ Chart Under Non-Normality and Measurement Errors", *Computers and Operations Research* 12, pp 291-299, 1985.

49. Rahim, M. A., "Determination of Optimal Design Parameters of Joint $\bar{x}$ and $R$ Charts", *Journal of Quality Technology* 21, pp 65-70, 1989.

50. Rahim, M. A., "Economic Design of $\bar{x}$ Control Charts Assuming Weibull In-Control Times", *Journal of Quality Technology* 25, pp 296-305, 1993.

51. Rahim, M. A. and P. K. Banerjee, "A Generalized Model for the Economic Design of $\bar{x}$-Control Charts for Production Systems with Increasing Failure Rate and Early Replacement", *Naval Research Logistics* 40, pp 787-809, 1993.

52. Rahim, M. A., R. S. Lashkari, and P. K. Banerjee, "Joint Economic Design of Mean and Variance Control Charts", *Engineering Optimization* 14, pp 65-78, 1988.

53. Roy, S. D. and S. Kakoty, "An Economic Design of CUSUM Control Chart for Controlling the Mean of a Normal Distribution", *IAPQR Transactions* 18, pp 29-42, 1993.

54. Saniga, E. M., "Joint Economically Optimal Design of $\bar{x}$ and R Control Charts", *Management Science* 24, pp 420-431, 1977.

55. Saniga, E. M., "Joint Economic Design of $\bar{x}$ and $R$ Control Charts with Alternative Process Models", *AIIE Transactions* 11, pp 254-260, 1979.

56. Saniga, E. M., "Heuristic Economical Design for $\bar{x}$ and $R$ Control Charts", *Frontiers in Statistical Quality Control* 3, pp 220-228, 1987.

57. Saniga, E. M., "Economic Statistical Control-Chart Designs with an Application to $\bar{x}$ and $R$ Charts", *Technometrics* 31, pp 313-320, 1989.

58. Saniga, E.M., D.J. Davis, and T.P. McWilliams, "Economic, Statistical, and Economic-Statistical Design of Attribute Charts", *Journal of Quality Technology* 27, pp 56-73, 1995.

59. Saniga, E. M. and D. C. Montgomery, "Economically Quality Control Policies for a Single Cause System", *AIIE Transactions* 13, pp 258-264, 1981.

60. Saniga, E. M. and T. P. McWilliams, "Economic, Statistical, and Economic-Statistical Design of Attribute Charts", *Journal of Quality Technology* 27, pp 56-73, 1995.

61. Svoboda, L., "Economic Design of Control Charts: A Review and Literature Survey (1979-1989)", *In: Statistical Process Control in Manufacturing.* Eds. J.B. Keats and D.C. Montgomery. New York: Marcel Dekker, 1991.

62. Tagaras, G., "Economic $\bar{x}$ Charts with Asymmetric Control Limits", *Journal of Quality Technology* 21, pp 147-154, 1989.

63. Tagaras, G., "Power Approximation in the Economic Design of Control Charts", *Naval Research Logistic Quarterly* 36, pp 639-654, 1989.

64. Tagaras, G., "Economic Design of Time-Varying and Adaptive Control Charts", in K. S. Al-Sultan and M. A. Rahim (Editors), *Optimization in Quality Control*, Kluwer Academic Publishers, 1997.

65. Tagaras, G. and H. L. Lee, "Economic Design of Control Charts with Different Control Limits for Different Assignable Causes", *Management Science* 34, pp 1347-1366, 1988

66. Taylor, H. M., "The Economic Design of Cumulative Sum Control Charts", *Technometrics* 10, pp 479-488, 1968.

67. Vaughan, T. S. and Peters, M. H., "Economic Design of Fraction Nonconforming Control Charts with Multiple State Changes", *Journal of Quality Technology* 23, pp 32-43, 1991.

68. Vance, L. C., "A Bibliography of Statistical Quality Control Chart Techniques", 1970-1980, *Journal of Quality Technology* 15, pp 59-62, 1983.

69. Weigand, Ch., "A New Approach for Optimal Control of a Production Process", *Economic Quality Control* 7, pp 225-251, 1992.

70. Woodall, W. H., "The Statistical Design of Quality Control Charts", *The Statistician* 34, pp 155-160, 1985.

71. Woodall, W. H., "The Design of CUSUM Quality Control Charts", *Journal of Quality Technology* 18, pp 99-102, 1986.

72. Woodall, W. H., "Weakness of the Economic Design of Control Charts", *Technometrics* 28, pp 408-409, 1986.

73. Woodall, W. H., "Conflicts Between Deming's Philosophy and the Economic Design of Control Charts", *Frontiers in Statistical Quality Control* 3, pp 242-248, 1987.

74. Woodall, W. H., and F. W. Faltin, "An Overview and Perspective on Control Charting", *in: Statistical Applications in Process Control and Experimental Design.* Eds. J.B. Keats and D.C. Montgomery. New York: Marcel Dekker, 1995.

**References on Sampling Plans**

1. Alidaee, B., "On Optimal Ordering Policy of a Sequential Model", *Journal of Optimization Theory and Applications*, 83(1), pp 199-205, 1994.

2. Bennett, G.K., K.E. Case and J.W. Schmidt, "The economic effects of inspector error on attribute sampling plans", *Naval Research Logistics Quarterly* 21, pp 431-443, 1974.

3. Britney, R., "Optimal Screening Plans for Nonserial Production Systems", *Management Science* 18, pp 550-559, 1972.

4. Chakravarty, A.K., and A. Shtub, "Strategic Allocation of Inspection Effort In A Serial, Multi-Product Production Systems", *IIE Transactions* 19, pp 13-22, 1987.

5. Cozzolino, J.M., "Optimal Burn-In Testing of Repairable Equipment", *Naval Research Logistics Quarterly*, 17, pp 167-182, 1970.

6. Duffuaa, S. O., "Impact of Inspection Errors on Performance measure of a complete inspection plan", *International Journal of Production Research*, 34(7), pp 2035-2049, 1996.

7. Duffuaa, S. O., and Al-Najjar "An Optimal complete Inspection Plan for critical Multicharacteristic components", Journal of Operational Research Society, 1995, pp 930-942.

8. Duffuaa, S. O., and I. Nadeem, "A Complete Inspection Plan for Dependent Multicharacteristic Critical Components" *International Journal of Production Research*, 32(8), pp 1897-1907, 1994.

9. Duffuaa, S.O. and A. Raouf, "A cost minimization model for dependent multicharacteristic inspection", *Proceedings IXth International Conference on Prod. and Exhibition*, Cincinati, Ohio, pp 738-746, 1987.

10. Duffuaa, S.O., and A.Raouf, "A cost minimization model for dependent multicharacteristics components", *Proceedings of the IXth ICPR*, pp 738-746, 1989.

11. Duffuaa, S.O., and A. Raouf, "Mathematical optimization models for multicharacteristics repeat inspection", *Journal of Applied Mathematial Modelling*, 13, pp 408-412, 1989.

12. Duffuaa, S.O., and A. Raouf, "An optimal sequence in multicharacteristic inspection", *Journal of Optimization Theory and Applications*, 67(1), pp 79-86, 1990.

13. Hui, Y.V., "Economic Design of Complete Inspection for Bivariate Products", *International Journal of Production Research*, 28, pp 259-265, 1990.

14. Hui, Y.V., "Economic Design of a Complete Inspection Plan with Feedback Control", *International Journal of Production Research*, 29, pp 2151-2158, 1991.

15. Jain, J.K., "A model for determining the optimal number of inspections minimizing inspection cost", *Unpublished M.Sc. Thesis*, University of Windsor, Ontario, Canada, 1977.

16. Kim, S.B., and D. Bai, "Economic Design of One-Sided Screening Procedures Based on a Correlated Variable With All Parameters Unknown", *Metrika*, 39, pp 85-93, 1992a.

17. Kim, S.B., and D. Bai, "Economic Screening Procedures in Logistic and Normal Model", *Naval Research Logistics* 37, pp 263-280, 1992b.

18. Lee, H.L., "On the Optimality of a Simplified Multicharacteristic Component Inspection Models", *IIE Transactions* 20, pp 392-298, 1988.

19. Lee, H.L. and M. Rosenblatt, "Optimal Inspection and Ordering Policies for Products with Imperfect Quality", *IIE Transactions* 17, pp 284-289, 1985

20. Lo, Y. and K. Tang, "Economic design of multicharacteristic models for a three-class screening procedure", *International Journal of Production research* 28, pp 2341-2351, 1990.

21. Melloy, N.J., "Determining the Optimal Process Mean and Screening Limits for Packages Subject to Compliance Testing", *Journal of Quality Technology* 23, pp 318-323, 1991.

22. Park, J.S., M. H. Peters, and K. Tang, "Optimal Inspection Policy in Sequential Screening", *Management Science* 37, pp 1058-1061, 1991.

23. Raouf, A., J.K.P. Jain, and P.T. Sathe, "A Cost Minimization Model for Multicharacteristic Component Inspection", *IIE Transactions* 15, pp 187-194, 1983.

24. Raz, T., "A Survey of Models for Allocating Inspection Effort in Multistage Production Systems", *Journal of Quality Technology*, 18, pp 239-247, 1986.

25. Schneider, H., C. O'Cinneide, and K. Tang, "Optimal Control of a Production Process Subject to AOQL Constraint", *Naval Research Logistics* 35, pp 383-395, 1988.

26. Schneider, H., K. Tang, and C.O'Cinneide, "Optimal Control of a Production Process Subject to Random Deterioration", *Operations Research* 38, pp 1116-1122, 1990.

27. Stewart, L.T. and J.D. Johnson, "Determining Optimum Burn-In and Replacement Times Using Bayesian Decision Theory", *IEEE Transactions on Reliability*, 21, pp 170-175, 1972.

28. Stewart, R., D. Montgomery, and R. Hikes, "Choice of Double Sampling Plans Based on Priority distribution and Costs", *AIIE Transactions*, 10(1), pp 19-30, 1978.

29. Tang, K. and J. Lo, "Determination of the Process Mean when Inspection Is Based on a Correlated Variable", *IIE Transactions* 25, pp 66-72, 1993.

30. Tang, K. and H. Schneider, "The Economic Effects of Inspection Error on a Complete Inspection Plan", *IIE Transactions* 19, pp 421-428, 1987.

31. Tang, K. and H. Schneider, "Selection of the Optimal Inspection Precision Level for a Complete Inspection Plan", *Journal of Quality Technology* 20, pp 153-156, 1988.

32. Tang, K. and H. Schneider, "Cost Effectiveness of Using a Correlated Variable in a Complete Inspection Plan When Inspection Error is Present", *Naval Research Logistics* 37, pp 893-904, 1990.

**References on Taguchi's Quality Control Models**

1. Barker, T.B., "Quality Engineering by Design: Taguchi's Philosophy", *Quality Progress*, December pp 32-42, 1986.

2. Box, G.E.P. "Discussion, Off-line, Quality Control Parameter Design and the Taguchi Method", *Journal of Quality Technology*, 17(4), pp 189-190, 1985.

3. Burgam, P.M., "Design of Experiments-the Taguchi Way", *Manufacturing Engineering* 94(5), pp 44-47, 1985.

4. Gunter, B., "A perspective on the Taguchi Methods", *Quality Assurance*, 13(3), pp 81-87, 1987.

5. Kacker, R.N., and A.C. Shoemaker, "Robust design: A Cost Effective Method for Improving Manufacturing Processs", *AT&T Technical Journal*, 65(2), 1986.

6. Kolarik, W.J., *Creating Quality: Concepts, Systems, Strategies, and Tools*, McGraw Hill, NY, 1995.

7. Pignatiello, J.R., "An Overview of the Strategy and Tactics of Taguchi", *IIE Transactions*, September 1988.

8. Ross, P.G., *Taguchi Techniques for Quality Engineering*, Englewood Cliffs, NJ, Prentice Hall, 1989.

9. Sullivan, L.P., "Reducing Variability: A New Approach to Quality", *Quality Progress*, 17(7), pp 15-21, 1984.

10. Sullivan, L.P., "The Power of Taguchi Methods", *Quality Assurance*, 13(3), pp 88-90, 1987.

11. Taguchi, G., "Off-line and On-line Quality Control Systems", *Proceedings of the International Conference on Quality Control*, B4, pp 1-5, 1978.

12. Taguchi, G., *Introduction to Quality Engineering*, Tokyo: Asian Productivity Center, 1986.

13. Taguchi, G., E.A. Elsayed, and T. Hsiang, *Quality Engineering in Production Systems*, McGraw Hill, 1989.

14. Taguchi, G., and Y. Wu, *Introduction to Off-line Quality Control*, Tokyo: Central Japan Quality Control Association, 1979.

**References on Economic Models Combining Quality, Production and Maintenance**


1. Ben Daya, M., "Integrated Production Maintenance and Quality Model Using Imperfect Maintenance Concept", Working paper, Systems Engg. Dept., KFUPM, Dhahran, Saudi Arabia, 1996.
2. Ben Daya, M., and M. A. Rahim, "Effect of Maintenance on the Economic Design of $\bar{x}$-charts", Working paper, Systems Engg. Dept., KFUPM, Dhahran, Saudi Arabia, 1996.
3. Ben Daya, M., and S.O. Duffuaa, "Maintenance and Quality: the Missing Link", Journal of Quality in Maintenance Engg., 1(1), pp 20-26, 1995.
4. Cheng, T.C.E., "An Economic Order Quantity Model with Flexibility and Reliability Considerations", *European Journal of Operational Research*, 39, pp 174-179, 1989.
5. Cheng, T.C.E., "An Economic Order Quantity Model with Demand-Dependent Unit Production Cost and Imperfect Production Process", *IIE Trans.*, 23(1), pp 23-28, 1991a.
6. Cheng, T.C.E., "EPQ with Process Capability and Quality Assurance Considerations", *Operational Research Society*, 42(8), pp 718-720, 1991b.
7. Chiu, H.N. and B.S. Huang, "The Economic Design of $\bar{x}$-control Charts Under Preventive Maintenance Policy", *Journal of Quality in Maintenance Engg.*, 13(1), pp 61-71, 1996.
8. Goyal, S. K., A. Gunasekaran, "Effect of dynamic process quality control on the economics of production", *International Journal of Operations and Production Management*, 10, pp 69-77, 1990.
9. Goyal, S. K., A. Gunasekaran, T. Martikainen, and P. Yli-Olli, "Integrating Production and Quality Control Policies: A Survey", *European Journal of Operational Research*, 69, pp 1-13, 1993.
10. Gunnasekaran, A., A.R. Korukonda, I. Virtanen, and P. Yli-Olli, "Optimal Investment and lot-sizing policies for improved productivity and quality," *International Journal of Production Research*, 33(1), pp 261-278, 1995.
11. Huang, B., H. Chiu, "The Quality Management of the Imperfect Production Process Under Two Monitoring Policies", *Int. J. of Quality and Reliability Management*, 12(3), pp 19-31, 1995.

12. Khouja, M., and A. Mehrez, "Economic Production Lot Size Model With Variable Production Rate and Imperfect Quality",*J. of Operational Research Society*, 45(12), pp 1405-1417, 1994.

13. Lio, M.J., S.T. Tseng and T.M.Lin, "The Effects of Inspection Errors to the Imperfect EMQ Model", *IIE Trans*, 26(2), pp 42-51, 1994.

14. Lorenzen, T.J., and L.C. Vance, "The Economic Design of Control Charts: A Unified Approach", *Technometrics*, 28, pp 3-10, 1986.

15. Makhdoum, M.A.A., *Integrated Production, Quality, and Maintenance Models Under Various Preventive Maintenance Polilcies*, MS Thesis, King Fahd University of Petroleum and Minerals, Dhahran, Saudi Arabia, 1996.

16. Montgomery, D.C., and R.G. Heikes, "Process Failure Mechanism and Optimal Design of Fraction Defective Control Charts", *AIIE Trans.*, 8, 1976.

17. Peters, M.H., H. Schneider and K. Tang, "Joint Determination of Optimal Inventory and Quality Control Policy", *The Institute of Management Science*, 34(8), pp 991-1004, 1988.

18. Porteus, E.L., "Optimal Lot Sizing, Process Quality Improvement and Setup Cost Reduction", *Operations Research*, 34, pp 137-44, 1986.

19. Porteus, E.L., "The Impact of Inspection Delay on Process and Inspection lot Sizing", *The Institute of Management Science*, pp 999-1007, 1990.

20. Rahim, M.A., "Joint Determination of Production Quantity, Inspection Schedule and Control Chart Design", *IIE Trans.*, 26(6), pp 2-11, 1994.

21. Rosenblatt, M.J., and H.L.Lee, "Economic Production Cycles With Imperfect Production Process", *IIE Trans.*, 18, pp 48-55, 1986.

22. Yum, B. J., and E. D. McDowell, "The Optimal Allocation of Inspection Effort in a Class of Nonserial Production Systems", *AIIE Transactions*, 13, pp 285-293, 1981.

23. Yum, B. J., and E. D. McDowell, "Optimal Inspection Policies in a Serial Production System Including Scrap Rework and Repair: An MILP Approach", *International Journal of Production Research*, 25, 1451-1460, 1987.

2

# SOME CONTEMPORARY APPROACHES TO OPTIMIZATION MODELS IN PROCESS CONTROL

**M. A. Rahim[1] and K. S. Al-Sultan[2]**

[1]*Faculty of Administration,*
*University of New Brunswick,*
*Fredericton, N.B., E3B 5A3,*
*Canada.*

[2]*Systems Engineering Department,*
*King Fahd University of Petroleum and Minerals,*
*Dhahran 31261,*
*Saudi Arabia.*

## ABSTRACT

This article provides an overview of recent work on optimization models in quality control, joint control of production quantity and quality, economic selection of target means and optimal determination of production runs. A brief description is provided for each of the problems that are addressed in this article. The important and interesting findings are highlighted and some new directions for further research are outlined.

**Key Words:** optimization models, quality control, economic design, control charts, target mean, deteriorating processes.

## 1 OPTIMIZATION MODELS IN QUALITY CONTROL DESIGN

### Problem Description

Statistical process control is used to measure the performance of a process. A process is said to be operating in statistical control when the only source of variation is a natural cause. The process must be brought back into statistical control by detecting and eliminating assignable causes of variation. Control charts are widely used to maintain statistical control of a manufacturing process which is subject to assignable causes. These causes induce shifts in process parameters (mean or standard deviation or both). To use a control chart, the user must specify the sample size ($n$), the sampling interval ($h$) and the control limit coefficient ($k$). The design of control charts with respect to economic criteria has been a subject of interest during the last four decades. Duncan (1956) first proposed the economic design of $\overline{x}$-control charts to control normal process means under the assumption that the occurrence time of the assignable cause has an exponential distribution (i.e., Markovian shock). In Duncan's approach to Markovian shock models, the length of the sampling intervals is kept constant, and that the occurrence time of the single assignable cause has an exponential distribution. This assumption has been widely used in subsequent work on the subject. Literature surveys of related work were presented by Gibra (1975), Montgomery (1980), Vance (1983), Svoboda (1991), and Ho and Case (1994). Under the same assumption of the Markovian shock model, Lorenzen and Vance (1986) developed a unified model which is flexible enough to apply to a situation where production continues/ceases during the search for assignable causes.

## 1.1 Economic Design of $\overline{x}$-Control Charts Assuming Weibull In-Control Times

Studies involving non-Markovian models were made by Baker (1971), Montgomery and Heikes (1976), Heikes et al.(1974), and, Banerjee and Rahim (1987). However, in these studies on non-Markovian models, with the exception of Banerjee and Rahim (1987), the length of the sampling intervals was specified and not a part of the decision variables. Utilizing the renewal theory approach, Banerjee and Rahim (1987) derived economic models for some non-Markovian processes. However, the issue of non-uniform sampling schemes had not been addressed until Banerjee and Rahim (1988) showed that increasing the frequency of sampling with the age of the system yields a lower operational cost per hour for a Weibull distributed shock model. Further, Rahim (1993) provided a computer program that determines the economically optimal design parameters of the chart. The optimal value of the design parameters, sample size, sampling intervals and control limit coefficient are determined by minimizing the expected cost per unit time. The length of the sampling intervals is chosen to maintain a constant integrated hazard rate over each sampling interval. The program is written based on Banerjee and Rahim (1988). It compared three cases: (1) A Weibull shock model with a variable sampling interval scheme, (2) a Weibull shock model under a uniform sampling interval, and (3) an exponential shock model under a constant sampling interval. Based on the expected loss-cost, the result of case 1 is found to be most economical compared to those of both case 2 and case 3. There has been growing interest among practitioners and academicians about the application of this readily available program (McWilliams (1994)). This program may be useful to other areas of research, such as in maintenance and replacement problems (Montgomery (1991, 1992), Porteus and Angelus (1996), Tagaras (1997), Yang and Makis (1997)). The effects of Weibull process failure mechanisms on economic design of charts are also studied by McWilliams (1989), Parkhideh and Case (1989), Chung and Lin (1993) and Moskowitz et al.(1994).

## 1.2 Economic Design of $\overline{x}$-Control Charts Assuming Gamma Distribution In-Control Times

The Weibull distribution has been widely applied to many non-Markovian process failure mechanisms. However, there are many other probability distributions that are useful in the fields of reliability and quality control engineering. One such distribution is gamma which has a number of important applications (for example, Tadikamalla (1979)). Furthermore, the renewal functions for Weibull cases do not converge to their asymptotic expressions as fast as gamma cases. The lifetime distribution of batteries was found to be adequately represented by a gamma distribution rather than a Weibull distribution (Soland (1968)). Rahim (1997) developed an economic model assuming gamma distributed in-control times. This work is motivated by the idea of perfect switching of repairable equipment adherent to statistical process control. The problem can be viewed as a combination of inspection policy and control policy. There are five inspection policies (A, B, C, D, and E) presented. These are described in Chapter 5 of this book. The paper presents an interesting practical application of techniques involved that practitioners may find useful.

The main difference from the paper of Banerjee and Rahim (1988) is the assumption of the possibility of age dependent repair before failure. That is, whether or not additional economies can be achieved by introducing the notion of preventive replacement. The residual life beyond a certain age for systems involving increasing hazard rate shock models will be rather short. Consequently, frequent sampling will be necessary after the system attains a certain age. This, in turn, may increase the operational cost as a result of frequent sampling. Therefore, it is conceivable that terminating a production cycle at any time beyond this age might yield additional economies. The question of replacement before failure does not arise for the Markovian shock model because of its memory-less property. Following the similar approach of Banerjee and Rahim (1988), the focus of this study is to propose a manner in which the frequency of sampling is to be regulated, while taking into account the underlying probability distribution of the in-control duration. The criterion for choosing the sampling plan is that the expected cost per hour of operation should be minimum and the length of the sampling intervals is to be chosen in such a way as to maintain a constant integrated hazard over each sampling interval.

## 1.3 Economic Design of $\overline{x}$-Control Charts Under Truncated Double Exponential Shock Models

Collani et al.(1992) provided an economic model for determining the two design parameters, namely the inspection interval and the maximum length of renewal cycle length. His paper deals with the economic design of periodic inspection and renewal policies for production processes subject to wear-out phenomena.

## 1.4 A Generalized Model for the Economic Design of Control Charts

Again, this work is an extension of the previous work of Banerjee and Rahim (1988). Rahim and Banerjee (1993) developed a model for the economic design of control charts, where a general distribution of in-control periods having an increasing failure rate is assumed and the possibility of age-dependent salvage value of the equipment is introduced. Truncated and non-truncated probability models are chosen. It is shown that substantial economic benefits can be achieved by adopting a non-uniform inspection scheme and by truncating a production cycle when it attains a certain age. The work has been well received by others, for example, Porteus and Angelus (1996) and Tagaras (1997). Surtihadi and Raghavachari (1994) provided an economic design of charts for general in-control distributions. Pignatiello and Tsai (1988) provided a method for determining the design parameters when cost parameters are not precisely known.

## 1.5 A Dynamic Economic Model for $\overline{x}$-Control Chart Design

The models of the above sections treated only the sampling interval as a time-varying parameter. Recently, Parkhideh and Case (1989) considered the economic design of dynamic $\overline{x}$-control charts in which the design parameters ($n, h$ and $k$) are varying over time. In addition, Parkhideh and Case (1989) provided recurrence expressions for each of the decision variables. However, there were six decision variables in their design methodology, which makes it very complicated to fix the maximum average hourly net income obtained from the process. Most recently, Ohta and Rahim (1996) proposed an alternative and simplified design methodology for dynamic $\overline{x}$-control charts. The optimal values are obtained by imposing the following constraints. The optimal sampling interval, $h_i$, is chosen such that the integrated hazard rate over each sampling interval is constant. The optimal sample size, $n_i$ is chosen such that the relative sample size per unit time during each sampling interval is constant. Analogously, the optimal control limit coefficient, $k_i$, is chosen so that the power of the control chart remains constant over each sampling interval. The process failure mechanism is assumed to follow a Weibull shock model and the product quality characteristic is considered to be normal. Computational experience indicates that the proposed dynamic non-uniform control chart design is much simpler and provides a lower cost than that of Parkhideh and Case's dynamic model (1989) and seems superior to Duncan's static model. A dynamic programming approach to the economic design of $\overline{x}$-control charts is proposed by Tagaras (1994) for the modelling and cost minimization of statistical process control. Tagaras (1994) also provided a dynamic and control chart for finite production runs. Olorunniwo (1992) provided a partially dynamic $\overline{x}$-control chart.

## 1.6 Economic Design of $\overline{x}$ and R Charts

The $\overline{x}$-control chart tells us whether changes have occurred in the process mean. This might be due to tool wear, or a gradual increase in temperature. The R-chart values indicate that a gain or loss in uniformity has occurred. Such change might be due to worn out bearings, a loose tool part, vibration, or an erratic flow of lubrications to a machine. The two types of charts go hand in hand when monitoring variable quality characteristics.

Considerable attention in recent years has also been devoted to the economic design of $\overline{x}$ and R control charts (for example, Saniga (1977, 1979, 1989, 1991), Jones and Case (1981), Saniga and Montgomery (1981), Rahim (1989) and

Costa (1993)). The growing interest may be understood by considering the much greater power (in a statistical sense) of joint $\overline{x}$ and R charts over $\overline{x}$ and R charts alone. One can imagine many examples for the application of the joint and R control charts. Shewhart (1931) reported, if the output variable involves some physical property of a material, then there exists some relationship between mean value of the product quality and its variance. The work of Arcelus and Banerjee (1987), Arcelus et al.(1982), Bisgaard et al.(1984) and Rahim and Lashkari (1984) indicate that the process variance tends to increase with an increase in the process mean as tools wear out. A typical situation arises owing to vibration and/or pressure, whereby accumulation inside a spray nozzle may puff up, resulting in a positive shift in the process mean and the process variance. Changes in the process variance have also been experienced in chemical processes, oil refineries, etc., (Castillo (1996)). Montgomery (1991) provided a typical example for joint $\overline{x}$ and R charts. A manufacturer produces non-returnable glass bottles for packaging a carbonated soft drink beverage. The wall thickness of the bottles is an important quality characteristic. If the wall is too thin, internal pressures generated during filling will cause the bottle to burst. The manufacturer has used $\overline{x}$ and R control charts for process surveillance. Saniga (1979) and, Saniga and Montgomery (1981) developed models for the joint economically optimal design of $\overline{x}$ and R control charts for processes subject to a single assignable cause. In both studies, it was assumed that the occurrence of the assignable cause results in a simultaneous shift in the process mean and process variance. A computer program for the optimal economic design of joint $\overline{x}$ and R charts, based on the economic model of Saniga and Montgomery (1981) was written by Rahim (1989).

Saniga (1979) in his initial model, assumed that the process is subject to two assignable causes. The occurrence of the first assignable cause results in a shift in the process mean while the occurrence of the second assignable cause results in a shift of the process variance. However, he assumes that the occurrence of an assignable cause of one kind blocks the occurrence of an assignable cause of the other. Consequently, it does not permit both process parameters to be out-of-control at the same time. This assumption may not be applicable in many industrial situations. Relaxing this assumption, Costa (1993) began on the more realistic premise that the occurrence of an assignable cause of one kind does not necessarily prevent the occurrence of the other kind of assignable cause kind.

In all these above models, however, it is assumed that the occurrence times of the assignable causes are described by exponential distributions. This is the underlying reason for the standard suggestion that the sampling interval be constant. Although the assumption of an exponential process failure mechanism simplifies the model, it may not be appropriate for some processes which deteriorate with time.

Costa and Rahim (1996) consider the problem of a continuous production process whose mean and variance are simultaneously monitored by an $\bar{x}$-chart and R-chart, respectively. The paper combines two existing process control models for the joint economic design of $\bar{x}$ and R charts: (i) two assignable causes are allowed to occur independently according to exponential distributions and may be present simultaneously by Costa (1993); (ii) Banerjee and Rahim (1988) developed charts under the assumption of Weibull distribution for the occurrence times, where the sampling interval is variable and monotonically decreasing. The product variable quality characteristic is assumed to be normally distributed and the process is subject to two independent assignable causes (such as: tool wear-out, overheating or vibration). One changes the process mean and the other process variance. The occurrence of one kind of assignable causes does not preclude the occurrence of the other kind. The occurrence times of the assignable causes are described by a Weibull distribution having increasing failure rates. A cost model is developed for determining the design parameters of joint $\bar{x}$ and R control charts. A non-uniform decreasing sampling interval scheme is adopted to incorporate the effects of process deterioration and a two-step search procedure is employed to determine the economically optimum design parameters. Finally, a sensitivity analysis of the model with respect to Weibull distribution parameters is performed, some new results are derived and some interesting findings are observed. Consideration of a general distribution may be an avenue for future research (Surtihad and Raghavachari (1994)).

## 1.7 Dynamic Joint Optimization Models in Quality Control and Maintenance

Fine and Porteus (1987) provided a very interesting paper on dynamic process improvement. Porteus and Angelus (1996) outline the opportunities for improved statistical process control. Tagaras (1995) developed a dynamic control chart for fine production runs. The link between maintenance and quality, although not completely missing, is not adequately addressed in the literature (Ben-Daya and Duffuaa (1995)). A common feature of the models that are addressed in the above sections is the assumption that the process shifts from an in-control state to the out-of-control state according to some probability distribution and the shift is detected by inspection. The process is restored to the in-control state at a fixed cost. The new dimension that one may like to bring to the modelling of this problem is to include the maintenance effort as a decision variable and that is to be optimized. Also, maintenance effort has an impact on the equipment failure rate distribution, which needs to be modelled. The effect of maintenance on quality has been discussed by Taguchi et al.(1989), Tagaras (1988) and Ben-Daya and Rahim (1996) and Rahim et al.(1996).

## 1.8 Constrained Optimization Models in Quality Control

McWilliams (1997) provided an approach to determine the parameters by placing constraints on ARLs, false alarm probability and power of the chart in detecting a shift. These models are described in this book in Chapter 6. McWilliams (1992) also provided an economic model with cycle duration constraints. More recently, Castillo et al.(1996) proposed a model that also considered statistical constraints in the performance of the resulting design. Collani (1997) viewed over both economic and statistical approaches and provided a simplified approach for the determination of the economic design of a control chart and this described in Chapter 3 of this book.

## 1.9 Economic Design of Time-Varying and Adaptive Control Charts

In economic design of time-varying control charts, the design parameters are changed in a predetermined fashion as the production process goes on. On the other hand, in adaptive control charts, some of the parameters are changed during production based on the sample information. Tagaras (1997) presents economic design of time-varying and adaptive control charts in this book in Chapter 4. The paper concludes by summarizing the findings so far and proposing further areas of future research.

## 1.10 Statistical Process Control and Taguchi Method

Benton (1991) showed how the Taguchi method complements many of the features of statistical process control. Nayebpur and Woodall (1993) developed an economic model under a geometric process failure mechanism and a comparison was made between their model and Taguchi's methods. Tagaras (1994) presented an economic model for the selection of acceptance sampling plans by variables under the assumption of quadratic quality cost. Elsayed and Chen (1993) provided an economic design of control charts using quadratic loss function.

# 2 JOINT OPTIMIZATION MODELS FOR PRODUCTION QUALITY AND QUANTITY

**Problem Description**

A common feature of the problems stated in Section 1 was mainly the quality of the product. There is no doubt the importance of quality control has been growing faster than ever before. In many industrial situations, however, the quality and quantity of the products are both equally important. The major thrust of this study was to develop an integrated model for production quantity and quality control for a class of deteriorating process. The research carried out and, the results are very significant in handling various problems related to inspection policy, maintenance schedules, inventory control and quality control.

Makhdoum (1996) reported that there are two different approaches for integrating quality control and economic production quantity. The first approach computes the expected percentage of defective items in production systems, while the latter monitors the nonconformities using control charts (see for example, Rosenblatt and Lee (1986), Porteus (1986, 1990), Liou et al.(1994), Cheng (1989, 1991a, 1991b) Khouja and Mehrez (1994), Peters et al.(1988)).

Goyal et al.(1993) has presented an excellent survey on integrating production and quality policies. They suggested that the available literature on the integrated production design issue can be classified as follows:

1. Models based on rework and lot-sizing (Gupta and Chakraborty (1984), Tayi and Ballou (1988)).
2. Models based on inspections and lot-sizing (Ballou and Pazer (1982), Barad (1990), Eppen and Hurst (1974), Yum and McDowell (1987)).

3. Models based on process control and lot-sizing policies (Cheng (1989, 1991), Tapiero et al.(1987), Tapiero and Hsu (1987)).
4. Models based on transfer and production lot-sizing policies (Goyal (1976), Goyal (1978), Goyal and Szendrovits (1986), Szendrovits and Drezrer (1980)).

A comprehensive survey of models integrating production and quality control can be found in Goyal et al.(1993) and Makhdoum (1996). The next section presents some models that report new directions for further research.

## 2.1 Joint Determination of Production Quantity, Inspection Schedule and Control Chart Design

Rahim (1994) presents a model for jointly determining an economic production quantity, inspection schedule and control chart design of an imperfect production process. The work has opened up a new avenue for further research. Traditionally, quality control and inventory control problems have been viewed as two isolated problems. From the producer's point of view, however, the economic quality and the economic quantity are equally important. The objective of the work has been to develop an integrated model for both the inventory and quality control problems. This work has gained importance in industries, particularly in production and manufacturing environments. In his recent work, Tagaras (1997) pointed out that this work enriches the previous model by integrating it with an economic production quantity model.

## 2.2 Generalized Model for Joint Determination of Production Run, Inspection Schedules, and Control Chart Design

Rahim and Ben Daya (1996) generalized the previous model stated in Section 2.1, integrating a quality and inventory control problem. In this model a more realistic assumption was introduced concerning the stopping of the machine during in-control phases because of false alarms. These types of situations occur more frequently in the pipeline, textile, pulp and paper, and ceramic industries.

## 2.3 Integrated Model for the Joint Optimization of Maintenance Level and Control Chart Design

Statistical process control and preventive maintenance have been treated separately in the past. Tagaras (1988) presents an economic model that incorporates both process control and maintenance policy. However, Markovian deterioration assumption was made. Ben Daya and Rahim (1996) developed an integrated model for the joint optimization of the maintenance level and economic design of control chart design under non-Markovian deterioration assumption. This is done for a deteriorating process where the in-control period follows a general probability distribution with increasing hazard rates. In the proposed model, preventive maintenance (PM) activities reduce the shift rate to the out-of-control state proportional to the PM level. Armstrong and Atkins (1996) developed joint optimization of maintenance and inventory policies for a simple system. Tseng (1996) provided optimal preventive maintenance policies for a deteriorating production system. It may be worthwhile to incorporate the quality control aspect into both Armstrong and Atkin's (1996) and Tseng's (1996) works.

## 2.4 A Generalized Integrated Model for Quality, Quantity, and Inspection Schedules for a Deteriorating Production Process Having Deteriorating Product Items

In all these above studies, the effect of the deteriorating process has been well considered. However, the effect of deteriorating items has been ignored. The effect of deteriorating items on raw materials is important in many inventory systems in a way that cannot be ignored. Deteriorating items include food products, pharmaceutical products, photographic films, radioactive substance, volatile liquids, etc. In recent decades, researchers have shown a great deal of interest in studying this project, Rahim and Ben Daya (1996) studied the simultaneous effects of both deteriorating product items and deteriorating production process on economic production quantity, inspection schedules and economic design of control charts. Deterioration times for both product and process are assumed to follow two parameter Weibull distributions. The product quality characteristic is assumed to be normally distributed. Sensitivity analysis of all the input factors is carried out over an adequate range.

# 3 ECONOMIC SELECTION OF PROCESS PARAMETERS

**Problem Description**

In Sections 1 and 2, it was assumed that the process parameters were specified. However, in some manufacturing situations, production cost and selling price are functions of the process parameter(s) (mean or variance or both), and lower and/or upper specifications limits.

## 3.1 Economic Selection of Target Mean

The traditional solution to this problem consists of developing an economic model that determines the optimal target mean for variables quality characteristics (see for example, Hunter and Kartha (1977), Golhar and Pollock (1988), Carlsson (1984), Carlsson (1989), Schmidt and Pfeifer (1991), Boucher and Jafari (1991), Melloy (1991), Dodson (1993) and others).

Considerable attention has been paid to the study of the container filling problem by several researchers. Springer (1951) developed a method for determining the most economic position of a process mean. In doing so, he considered a manufacturing situation where both upper and lower specifications limits are of interest and that the financial loss due to producing a product above the upper specification limit would not necessarily be equal to the loss when producing one below the lower specification limit.

A simple method was suggested for determining the optimum target mean which minimized the total cost. The distribution of the product quality characteristic was assumed to be normal or gamma. Bettes (1962) considered a similar type of problem with a given lower specification limit, however, he assumed an arbitrary upper specification limit. Moreover, he assumed undersized and oversized items were reprocessed at a fixed cost. Nelson (1979) provided a nomograph for Springer's analytical solution. Hunter and Kartha (1977) investigated the optimization of a target mean when a lower specification limit was fixed. Their stated problem was as follows: each item is inspected to determine whether it satisfies a lower specification limit. If it does, then that item is sold at the regular price. If it does not, the item is sold at a reduced price in a secondary market. Thus, the customer is compensated for poor quality but does not pay for excessive good quality. A net income function, consisting of income from the accepted items, give-away cost of material in excess of the lower specification, and income from the rejected items, was developed. They provided a simple procedure for obtaining the optimal process mean.

Bisgard et al.(1984) extended Hunter and Kartha's (1977) work to include selecting the most favorable quality characteristic distribution for the product. Industrial examples of normal, lognormal and poisson distributions were provided. Carlsson (1984) modified the work of Hunter and Kartha (1977) to include both fixed and variable costs. He derived a revenue function whereby the customer paid extra for good quality and was compensated for poor quality. Both Huter and Kartha (1977) and Carlsson (1984) assumed that rejected products were sold in a secondary market. Golhar (1987), however, studied the case in which the rejected product was recycled, i.e., the under-filled items are emptied and refilled, so that it could be sold in the primary market. Golhar and Pollock (1988) extended this work to include an upper specification limit and provided solutions for determining both optimal process mean and upper specification limit. Schmidt and Pfeifer (1991) considered the problem of economic selection of the target mean and the upper limit for the canning problem with limited capacity. In what may be considered a direct extension of Springer's (1951) work, Montogmery (1991) suggested the general problem of a normal distributed quality characteristic in which the profit function was related to conformance to upper and lower specifications. The profit function was asymmetric. That is, the profit obtained by a product below the lower specification limit was different from that obtained from a product above the upper specification limit.

Economic selection of the mean and upper limit for a container-filling process under capacity constraints is studied by Liu et al.(1997) and is reported in Chapter 7 of this book. Optimal target value using multiple criteria economic selection models are investigated by Carlsson (1989) and also described in Chapter 8 of this book.

## 3.2 Inspection Plan Models

All of the above models assume that 100% inspection is used. However, there are some models which assume that some sampling plan is used. Carlsson (1989) considered the case of acceptance sampling where the rejection criterion is based on the sample mean. Boucher and Jafari (1991) considered the case in which the reject criterion is based on the number of nonconforming units in the sample.

Al-Sultan (1994) extended Boucher and Jafari's (1991) model to the case of two machines in series. He used a rejection criterion which is based on the number of nonconforming units in the sample. Products can either be rejected after the first or the second machine. Rejected items are sold at different reduced prices. Pulak and Al-Sultan (1996a) also extended Boucher and Jafari's (1991) model to the case where rectifying inspection is used.

## 3.3 Models for more than one Quality Characteristic

All of the above papers considered the case when there is only one quality characteristic of interest, and it is a variable characteristic. However,there are situations where there is more than one quality characteristic of interest, and some of them are attributes, while others might be variables. Arcelus and Rahim (1990a) have considered joint determination of optimum variable and attribute target means where items are acceptable if they meet the specifications for both types of quality characteristics at the same time; otherwise, the items are sold as scrap at reduced prices. Arcelus and Rahim (1991,1994) developed an economic item by item plan that simultaneously determines the most profitable target mean value for both single variable and single attribute quality characteristic. Arcelus and Rahim (1990b) have also considered a lot by lot sampling plan, where the lot is acceptable if the number of rejected items does not exceed a predetermined number. The objective is to select a setting that will maximize the expected income per lot. Carlsson (1992) has considered the quality section of a two-dimensional process level under single acceptance sampling by variables and developed a model for it. Elsayed and Chen (1993) have considered the problem for multiple characteristic products, and employed the Taguchi's loss function approach to determine the optimal level settings.

## 3.4 Models with Nonconstant Variance

A common assumption in the above studies is that process variability is constant. Schmidt and Pfeifer (1989) considered Golbar's (1987) model, and investigated cost saving due to variance reduction. This was the first attempt to include the effect of variance on the total expected cost. Golhar and Pollock (1992) developed a procedure for studying the cost savings due to variance reduction in their model (Golhar and Pollock (1988)). Al-Sultan and Pulak (1996) used the same procedure proposed by Golhar and Pollock (1992) to study the effect of variance reduction on the cost of their model (Pulak and Al-Sultan (1996a)).

The optimal strategy for the producer for the so-called "filling problem" or "canning problem" has always been to focus the process mean as the primary decision variable. All the previously developed models with the exception of the above three attempts, however, overlooked a very important factor: that is, variability associated with filling operations. The problem of simultaneously selecting the most economical target mean and variance for a continuous production process was studied by Warren et al.(1996). The process involves the filling of containers. Initially, they consider the problem of finding the optimal target mean under the assumption that variance is known. Then an economic model for the selection of a target variance is developed using both customer and producer viewpoints, which are assumed to be independent of the product quality characteristic distributions. Since a variance is always linked to the level of mean value, it is hard to distinguish a control factor which affects only the variance. As such, this assumption is reasonable. Defining $\sigma_i^2$ as the variance at the intersection of producer and customer quality cost, $\sigma^2$ as the variance at which the minimum of the total cost is attained, one can see there exist three possible outcomes. (i) $\sigma^2$=$\sigma_i^2$, (ii) $\sigma^2 < \sigma_i^2$ and $\sigma^2 > \sigma_i^2$. All of these cases were addressed mathematically and several results were derived. Generally, a customer's quality cost increases while the producer's quality cost decreases, with the increase in process variability. Further research on an integrated model for selecting both target mean and variance will be worth investigating. Of course, it should be kept in mind, while the process is operated in on-target with a minimum variance it will produce a minimum number of non-conforming products.

## 3.5 Uniformity of Production Versus Conformance to Specifications

Arcelus and Rahim (1996) presented a generalized model for controlling both conformance to specifications and uniformity of production. The rationale behind this approach lied in the need to account for two sets of costs. One set deals with the cost related to the uniformity-of-production objective. The other set deals with the cost related to the expected profit by equating the cost of conformance to specifications and uniformity criteria. This approach upholds the modern concept of Taguchi's loss-function which states that any deviation from the target mean incurs an economic loss, even if the value of the quality characteristic lies within the specification limits. The economics of quality improvement suggest implication for future development and practical applications (Arcelus (1997)).

## 3.6 Solution Methods and Programs

There have been many attempts to simplify solving the models in the literature. For example, Nelson (1978) has developed a graphical solution for Hunter and Kartha's (1977) model. Nelson (1979) developed a nomograph for Springer's (1951) model. Some programs have been written for solving some of the above models. Golhar (1988) developed a FORTRAN Program to implement a procedure suggested by Golhar and Pollock (1988) which computes the optimal target mean and upper limit for the fill. Pulak and Al-Sultan (1996b) developed a FORTRAN package which solves ten of the above models. They have also given some suggestions for good initial solutions for the solution procedure.

# 4 OPTIMAL CONTROL MODELS FOR DETERIORATING PROCESSES

### Problem Description

As the production process operates overtime, the process parameters are subject to shift and/or drift due to tool-wear that may cause the quality of the output to deteriorate. Some degree of deterioration may be tolerated at a cost. However, it may be less costly to intervene by overhauling, adjusting or resetting the production process after a specified production run. Basically, this is one of process control problems where mean shifts gradually rather than shifting instantaneously by a fixed amount. Next, we present various models developed for the above described problem.

## 4.1 Models with Linear Trend

The problem of controlling production processes which are subject to tool-wear was studied by Hall and Eilon (1963) under the assumption that the process mean changes while the process variance remains constant throughout the production period.

Taha (1966) proposed a methodology for determining the optimal cycle length for a cutting tool with the assumption that the tool wears out with time, which causes the production of defective items. Gibra (1967) proposed models for determining the optimal production run for both stable and unstable processes.Smith and Vemuganti (1968) extended the model of Taha (1966) to include the initial mean setting and the rate of wear of the tool as parameters.

Arcelus and Banerjee (1985) generalized the Bisgaard et al. (1984) model to include a linear shift in the mean. Rahim and Lashkari (1985) relaxed the assumption of constant variances and developed a model for determining the length of the production run where both the mean and variance are changing. Rahim and Raouf (1988) extended the work further by considering a process having multilevel tool wear. Other related work can also be found in Kamat (1976), Arcelus et al.(1985), Quesenberry (1988) and Pugh (1988).

## 4.2 Models with Random Drift

Rahim and Banerjee (1988) determined the production run for a process with random linear drift. They assumed that the time at which the process starts drifting follows an exponential probability distribution model. Al-Sultan and Al-Fawzan (1996a) investigated the saving due to a variance reduction in the above model. Al-Sultan and Al-Fawzan (1996b) extended further the above model for the case of two specification limits. Al-Sultan and Al-Fawzan (1996c) also extended Rahim and Banerjee's (1988) model for the case of multistage production systems with deteriorating processes.

Schneider et al.(1990) proposed a model for determining the optimal initial setting of the process mean and a lower point where the process mean has to be adjusted when it reaches it. Kubat and Lam (1992) assumed that the deterioration is modelled by a Wiener (Brownian) process with a positive drift. The process is adjusted when the deteriorating mean reaches an action limit which is a decision variable. Other related work can also be found in Pate-Cornel et al.(1987).

## 4.3 Other Related Models

Gibra (1974) assumed that the drift in the process mean is nonlinear. Drezner and Wesolowsky (1989) extended the work of Gibra (1967) by using the quadratic loss function for modelling the problem. Jeang and Yang (1992) generalized the work of Drezner and Wesolowsky (1989) and developed a model in which they considered the trend to be a monotone nonlinear function and their model finds the initial mean setting and the optimal tool replacement.

Rahim (1996) considers the problem of joint determination of optimal production run and production rates for a process having a multilevel tool wear. In the past, consideration of production rates has not been taken into account in determining optimal production run. However, production rates in many ways affect the tool life time and inventory policies. The proposed economic model combines both tool replacement cost, the preventive maintenance cost, the product quality loss and the inventory cost. Effects of preventive maintenance is also studied. A comparative study with and without preventive maintenance would be of considerable interest and practical value. A typical application of these models can be found in a metal cutting process, drilling process or grinding process, where wear occurs at both cutting edges and pads.

Other related models can be found in Arcelus et al.(1982), Arcelus and Banerjee (1985) and Arcelus and Banerjee (1987). See also an extensive survey by Al-Fawzan and Al-Sultan (1996).

# 5 CONCLUDING REMARKS

This paper attempts to overview briefly some of the recent contributions related to optimization in quality control. Many papers are simply mentioned, while others are discussed extensively. Although, the presentation of the different contributions is not fairly balanced, the paper addressed some important and interesting issues in optimization in quality and process control. It also provided a few new directions for further research. A complete review of the recent contributions is hardly feasible. However, this chapter serves as a readily available reference for the use of future researchers.

**Acknowledgements**
The authors are grateful to professor George Tagaras for his constructive criticism on the first draft of the paper. His valuable suggestions are incorporated and some of his remarks are also included in the concluding remarks of the paper. The authors are also thankful to Professor Olle Carlsson, Professor Hiroshi Ohta and Professor A.F.B Costa for their comments.

# REFERENCES

[1] Al-Fawzan, M.A. and K.S. Al-Sultan, "The Optimal Control of A Production Process Subject to Drift and Shift in The Process Mean: A Survey", *20th International Conference on Computers and Industrial Engineering, Korea,1*, October 6-9, pp 961-964, 1996.

[2] Al-Sultan, K.S., "An Algorithm for the Determination of the Optimal Target Values for Two Machines in Series with Quality Sampling Plans", *International Journal of Production Research* 12(1), pp 37-45, 1994.

[3] Al-Sultan, K.S. and M.A. Al-Fawzan, "Variance Reduction in a Process with random linear drift", accepted for Publication in *International Journal of Production Research*, 1996a.

[4] Al-Sultan, K.S. and M.A. Al-Fawzan, "An Extension of Rahim and Banerjee's Model for a Process with Upper and Lower Specification Limits", submitted, 1996b.

[5] Al-Sultan, K.S. and M.A. Al-Fawzan, "Determination of the Optimal Process Means and Production Cycles for Multi-stage Production Systems Subject to Process Deterioration", accepted for publication in *Production Planning and Control*, 1996c.

[6] Al-Sultan, K.S. and M.F.S. Pulak, "Process Improvement by Variance Reduction for a Single Filling Operation with Rectifying Inspection", Accepted for publication in *Production Planning and Control*, 1996.

[7] Al-Sultan, K.S. and M.A. Rahim, "Economic Selection of Process Parameters: A Literature Survey", Working Paper, Department of Systems Engineering, King Fahd University of Petroleum and Minerals, 1994.

[8] Arcelus, F.J., "Uniformity of Production Vs. Conformance to Specifications in the Canning Problem", *Optimization in Quality Control*, edited by K.S. Al-Sultan and M.A. Rahim, Kluwer Academic Publishers, 1997.

[9] Arcelus, F.J., P.K. Banerjee, "Selection of the Most Economical Production Plan in a Tool-wear Process", *Technometrics*, 27(4), pp 433-437, 1985.

[10] Arcelus, F.J., P.K. Banerjee, "Optimal Production Plan in a Tool Wear Process with Rewards for Acceptable, Undersized and Oversized Parts", *Engineering Costs and Production Economics*, 11, pp 13-19, 1987.

[11] Arcelus, F.J., M.A. Rahim, "Optimal Process Levels for the Joint Control of Variables and Attributes", *European Journal of Operational Research* 45, pp 224-230, 1990a.

[12] Arcelus, F.J., and M.A. Rahim, "Optimal Settings for Variable-Attribute Quality Control Problem", *Journal of Chinese Institute of Industrial Engineering*, 7(2), pp 57-62, 1990b.

[13] Arcelus, F. J. and M.A. Rahim, "Joint Determination of Optimum Variable and Attribute Target Means", *Naval Research Logistics*, 38(6), pp 851-864, 1991.

[14] Arcelus, F. J. and M.A. Rahim, "Simultaneous Economic Selection of a Variable and an Attribute Target Mean", *Journal of Quality Technology*, 26(2), pp 125-133, 1994.

[15] Arcelus, F. J. and M.A. Rahim, "Reducing Performance Variation in the Canning Problem", accepted in *European Journal of Operational Research*, 1996.

[16] Arcelus, F.J., P.K. Banerjee, and R. Chandra, "Optimal Production Run for a Normally Distributed Quality characteristic Exhibiting Non-Negative Shifts in Process Mean and Variance", *IIE Transactions* 14, pp 90-98, 1982.

[17] Arcelus, F.J., P.K. Banerjee, and R. Chandra, "The Optimal Schedule to Produce a given Number of Acceptable Parts with a Specified Confidence Level", *International Journal of Production Research*, 23(1), pp 185-196, 1985.

[18] Armstrong, M. J. and D. R. Atkins, "Joint Optimization of Maintenance and Inventory Policies for a Simple System", *IIE-Transactions*, 28(5), pp 415-424, 1996.

[19] Baker, K. R., "Two Models in the Economic Design of an $\bar{x}$-Control Chart", *AIIE Transactions*, 3, pp 257-263, 1971.

[20] Ballou, D.P., and H.L. Pazer, "The impact of Inspector Fallibility on The Inspection Policy in Serial Production Systems", *Management Science*, 28, pp 387-399, 1982.

[21] Banerjee, P.K. and M.A. Rahim, "Economic Design of $\bar{x}$- Control Charts: A Renewal Theory Approach", *Engineering Optimization*, 12(1), pp 63-73, 1987.

[22] Banerjee, P.K. and M.A. Rahim, "Economic Design of $\bar{x}$- Control Charts Under Weibull Shock Models," *Technometrics*, 30(44), pp 407-414, 1988.

[23] Baker, K.R., "Two Process Models in the Economic Design of a $\bar{x}$-Chart", *AIIE Transactions*, 3, pp 257-263, 1971.

[24] Barad, M., "A Break-even Level Approach to Location of Inspection Stations in A Multi-Stage Production Process", *International Journal of Production Research*, 28, pp 29-45, 1990.

[25] Ben Daya, M., "Integrated Production Maintenance and Quality Model using Imperfect Maintenance Concept", Working paper, Systems Engineering Department, KFUPM-Dhahran, Saudi Arabia, 1996.

[26] Ben-Daya, M., and S.O. Duffuaa, "Maintenance and Quality: The Missing Link", *Journal of Quality in Maintenance Engineering*, 1(1), pp 20-26, 1995.

[27] Ben-Daya, M. and M. A. Rahim "Effect of Maintenance on the Economic Design of Control Chart", Working Paper #96-033, Faculty of Administration, University of New Brunswick, Fredericton, Canada, 1996.

[28] Benton, W.C., "Statistical Process Control and the Taguchi Method: A Comparative Evaluation", *International Journal of Production Research*, 29(9), pp 1761-1770, 1991.

[29] Bettes, D.C., "Finding an Optimal Target Value in Relation to a Fixed Lower Limit and an Arbitrary Upper Limit", *Applied Statistics*, pp 202-210, 1962.

[30] Bisgaard, S., W.G. Hunter, and L. Pallesen, "Economic Selection of Quality of Manufactured Product", *Technometrics* 26, pp 9-18, 1984.

[31] Boucher, T. O. and M. A. Jafari, "The Optimal Target Value for Single Filling Operations With Quality Sampling Plans", Journal of Quality Technology, 23(1), pp 44-47, 1991.

[32] Carlsson, O., "Determining the Most Profitable Process Level for a Production Process Under Different Sales Conditions", *Journal of Quality Technology*, 16(1), pp 44-49, 1984.

[33] Carlsson, O., "Economic Selection of a Process Level Under Acceptance Sampling by Variables", Engineering Costs and Production Economics, 16, pp 69-78, 1989.

[34] Carlsson, O., "Quality Selection of a Two-Dimensional Process Level Under Single Acceptance Sampling by Variables", *International Journal of Production Economy*, 27, pp 43-56, 1992.

[35] Castillo, E. D., "Multiresponse Process Optimization via Constrained Confidence Region", *Journal of Quality Technology*, 28, pp 61-80, 1996.

[36] Castillo, E. D., P. Markin, and D. C. Montgomery, "Multiple-Criteria Optimal Design of Control Charts", *IIE Transactions*, 28(6), pp 463-474, 1996.

[37] Fine, C.H. and E.L. Porteus, "Dynamic Process Improvement", Working Paper # 1952-87, Alfred P. Sloan School of Management, Massachusetts Institute of Technology, Cambridge, Massachusetts, U.S.A., 1987.

[38] Cheng, T.C.E., "An Economic Production Quantity Model with Flexibility and Reliability Considerations", *European Journal of Operation Research*, 39, pp 174-179, 1989.

[39] Cheng, T.C.E., "EPQ with Process Capability and Quality Assurance Considerations", *Journal of the Operational Research Society*, 42, pp 713-720, 1991a.

[40] Cheng, T.C.E., "An Economic Order Quantity Model with Demand-Dependent Unit Production Cost and Imperfect Production Processes", *IIE Transactions*, 23(1), pp 23-28, 1991b.

[41] Chiu, H.N. and B.S. Huang, "The Economic Design of $\overline{x}$ and S2 Control Charts with Preventive Maintenance and Increasing Hazard Rate", *Journal of Quality in Maintenance Engineering*, 1(4), pp 17-40, 1995.

[42] Chiu, H.N. and B.S. Huang, "The Economic Design of $\bar{x}$-Control Charts under Preventive Maintenance Policy", *Journal of Quality in Maintenance Engineering*, 13(1), pp 61-71, 1996.

[43] Chung,, K. J. and C. N. Lin., "The Economic Design of Dynamic-Control Charts under Weibull shock Models", *International Journal of Quality and Reliability*, 10(8), pp 41-56, 1993.

[44] Collani, V. E., P. Frahm, and P. Gabriel, "Economic Inspection and Renewal Policies in Case of Unperfect Renewals", *Economic Quality Control* 7, pp 195-212, 1992.

[45] Collani, V. E., "Determination of the Economic Design of Control Charts Simplified", *Optimization in Quality Control*, edited by K. S. Al-Sultan and M. A. Rahim, Kluwer Academic Publishers, 1997.

[46] Costa, A. F.B., "Joint Economic Design of $\bar{x}$ and R Control Charts for Process Subject to Two Independent Assignable causes", *AIIE Transactions* 25, pp 27-33, 1993.

[47] Costa, A. F.B. and M. A. Rahim, "Economic Design of $\bar{x}$ and R Control Charts Under Weibull Shock Models", Working Paper # 96-034, Faculty of Administration, University of New Brunswick, Fredericton, Canada, 1996.

[48] Dodson, B. L., "Determining the Optimal Target Value for a Process with Upper and Lower Specifications", *Quality Engineering*, 5(3), pp 393-402, 1993.

[49] Drezner, Z. and G.O. Wesolowsky, "Optimal Control of A Linear Trend Process with Quadratic Loss", *IIE Transactions*, 21(1), pp 66-72, 1989.

[50] Duncan, A. J., "The Economic Design of $\bar{x}$-Charts Used to Maintain Current Control of a Process", *Journal of the American Statistical Association*, 11, pp 228-242, 1956.

[51] Elsayed, E. and A. Chen, "An Economic Design of $\bar{x}$-Control Charts using Quadratic Loss Function", *International Journal of Production Research*, 32(4), pp 873-887, 1994.

[52] Eppen, G.D., and E.G. Hurst, "Optimal Allocation of Inspection Stations in A Multi Stage Production Processes", *Management Science*, 10, pp 1194-1200, 1974.

[53] Fine, C.H. and E.L. Porteus, "Dynamic Process Improvement", Working Paper #1952-87, Alfred P. Sloan School of Management, Massachusetts Institute of Technology, Cambridge, Masachusettes, U.S.A., 1987.

[54] Gibra, I. N., "Optimal Control of Processes Subject to Linear Trends", *Journal of Industrial Engineering*, 18(1), pp 34-41, 1967.

[55] Gibra, I. N., "Economically Optimal Determination of the Parameters of $\bar{x}$-Control Charts." *Management Sciences*, 17, pp 635-646, 1971.

[56] Gibra, I. N., "Optimal Production Runs of Processes Subject to Systematic Trends", *International Journal of Production Research*, 12(4), pp 511-517, 1974.

[57] Gibra, I. N., "Recent Development in Control Charts Techniques", *Journal of Quality Technology*, 7, pp 183-192, 1975.

[58] Golhar, D. Y., "Determination of the Best Mean Contents for a Canning Problem", *Journal of Quality Technology*, 19(2), pp 82-84, 1987.

[59] Golhar, D. Y., "Computation of the Optimal Process mean and the Upper Control Limit for a Canning Problem", *Journal of Quality Technology*, 20(3), pp 193-195,1988.

[60] Golhar, D.Y. and S.M. Pollock, "Determination of the Optimal Process Mean and the Upper Limit for a Canning Problem", *Journal of Quality Technology*, 9(4), pp 176-181, 1988.

[61] Golhar, D.Y. and S.M. Pollock, "Cost Savings Due to Variance Reduction in a Canning Process", *IIE Transactions*, 24(1), pp 89-92, 1992.

[62] Goyal, S.K., "Note on: Manufacturing Cycle Time Determination for A Multi-Stage Economic Production Quantity Model", *Management Science*, 23, pp 332-333, 1976.

[63] Goyal, S.K., "Economic Batch Quantity in A Multi-Stage Production System", *International Journal of Production Research*, 16, pp 267-273, 1978.

[64] Goyal, S.K., and A.Z. Szendrovits, "A Constant Lot Size Model With Equal and Unequal Sizes Batch Shipments Between Production Stages", *Engineering Costs and Production Economics*, 10, pp 203-210, 1986.

[65] Goyal, S.K., A. Gunasekaran, T. Martikainen and P. Yli-Olli "Integrating Production and Quality Control Policies: Assumers", *European Journal of Operational Research*, 69, pp 1-13, 1993.

[66] Gupta, T.I. and S. Chakraborty, "Looping in a Multi-Stage Production System", *International Journal of Production Research*, 22, pp 299-311, 1984.

[67] Hall, R. I. and S. Eilon, "Controlling Production Processes which are Subject to Linear Trend", *Operational Research Quarterly*, 14(3), pp 179-189, 1963.

[68] Heikes, R. G., D. C. Montgomery, And J. Y. H. Yeung, "Alternative Process Models in Economic Design of T Control Charts", *AIIE Transactions*, 6, pp 55-61, 1974.

[69] Ho, C. and K. E. Case, "Economic Design of Control Charts: A Literature Review for 1981-1991", *Journal of Quality Technology* 26, pp 39-53, 1994.

[70] Hunter, W. G. and C. P. Kartha, "Determining the Most Profitable Target Value for a Production Process", *Journal of Quality Technology*, 9(4), pp 176-181, 1977.

[71] Jeang, A. and K. Yang, "Optimal Tool Replacement with Nondecreasing Tool Wear", *International Journal of Production Research*, 30(2), pp 299-314, 1992.

[72] Jones, L.L. and K.E. Case, "Economic Design of $\overline{x}$ and R Control chart", *AIIE Transactions*, 13, pp 182-195, 1981.

[73] Kamat, S.J., "A Smoothed Bayes Control Procedure for The Control of a Variable Quality Characteristic with Linear Shift", *Journal of Quality Technology*, 8(2), pp 98-104, 1976.

[74] Khouja, M and A. Mehrez, "Economic Production Lost Size Model with Variable Production Rate and Imperfect Quality", *Journal of Operational Research Society*, 45(12), pp 1405-1417, 1994.

[75] Koo, T. Y. and K. E. Case, "Economic Design of $\overline{x}$-Control Charts for use in Monitoring Continuous Flow Process", *International Journal of Production Research*, 28, pp 2001-2011, 1990.

[76] Kubat, P. and C.Y.T. Lam, "Optimal monitoring strategies for slowly deteriorating repairable systems", *IEEE Transactions on Communications*, 40(4), pp 661-665, 1992.

[77] Lee, H.L., K. Moinzadeh, and G. Tagaras, "A Model for Continuous Production Control with Warning Signals to Fault Occurrences", *Journal of Operational Research Society*, 37(5), pp 515-523. 1986.

[78] Lee, H.L., and M.J. Rooseblatt, "Economic Design and Control of Monitoring Mechanisms in Automated Production Systems", *IIE Transactions*, 20(2), pp 201-208, 1988.

[79] Liu, J., K. Tang, and Y. K. Chun, "Economic Selection of the Mean and Upper Limit for a Container-Filling Process Under Capacity Constraints", *Optimization in Quality Control*, edited by K. S. Al-Sultan and M. A. Rahim, Kluwer Academic Publishers, 1997.

[80] Liou, M.J., S.T. Tseng, and T.M. Lin "The Effects of Inspection Errors to the Imperfect EQM Model", *IIE Transaction*, 26(2), pp 42-51, 1994.

[81] Lorenzen, T. I. and L. C. Vance, "The Economic Design of Control Charts: A Unified Approach", *Technometrics*, 28, pp 3-10, 1986.

[82] Makhdoum, M.A.A., "Integrated Production, Quality, and Maintenance Models under Various Preventive maintenance Policies", Unpublished MS thesis, King Fahd University of Petroleum and Minerals, Dhahran, Saudi Arabia, 1996.

[83] McWilliams, T. P., "Economic Control Chart Designs and the In-Control Time Distribution: A Sensitivity Study", *Journal of Quality Technology*, 21, pp 103-110, 1989.

[84] McWilliams, T. P., "Economic Control Models with Cycle Duration Constraints", Economic Quality Control, 7, pp 164-194, 1992.

[85] McWilliams, T. P., "Economic, Statistical, and Economic Statistical $\bar{x}$-Chart Design", *Journal of Quality Technology*, 26(3), pp 227-238, 1994.

[86] McWilliams, T. P., "Constrained Optimization Models for Determining Economic Control Chart Parameters", *Optimization in Quality Control*, edited by K. S. Al-Sultan and M. A. Rahim, Kluwer Academic Publishers, 1997.

[87] Melloy, B.J., "Determining the Optimal Process Mean and Screening Limits for Packages subject Compliance Testing", *Journal of Quality Technology*, 23(4), pp 318-323, 1991.

[88] Misra, R., "Optimum Production Lot Size Model for a System with Deteriorating Inventory," *International Journal of Production Research*, 13(5), pp 459-505, 1975.

[89] Montgomery, D. C., "The Economic Design of Control Charts: A Review and Literature Survey." *Journal of Quality Technology*, 12, pp 75-87, 1980.

[90] Montgomery, D. C., "*Introduction to Statistical Quality Control*". Second Edition, John Wiley and Sons, New York, pp 428-429, 1991.

[91] Montgomery, D. C., "The Use of Statistical Process Control and Design of Experiments in Product and Process Improvement". *AIIE Transactions* 24(5), pp 4-17, 1992.

[92] Montgomery, D. C. and R. G. Heikes, "Process Failure Mechanisms and Optimal Design of Fraction Defective Control Charts." *IIE Transactions,* 24, pp 4-17, 1976.

[93] Moskowitz, H., R. Plante, and Y.H. Chun "Effect of Process Failure Mechanisms on Economic $\bar{x}$-Control Charts", *AIIE Transactions,* 26(6), pp 12-21, 1994.

[94] Nayebpour, M.R. and W.H. Woodall, "An Analysis of Taguchi's On-Line Quality Monitoring Procedures for Attributes", *Technometrics,* 35(1), pp 53-60, 1993.

[95] Nelson, L. S., "Best Target Value for a Production Process", *Journal of Quality Technology,* 10(2), pp 88-89, 1978.

[96] Nelson, L. S., "Nomograph for Setting Process to Minimize Scrap Cost", 11(1), pp 48-49, 1979.

[97] Ohta, H. and M. A. Rahim, "A Dynamic Economic Model for an $\bar{x}$-Control Chart Design". Working Paper #96-40, Faculty of Administration, University of New Brunswick, Canada, 1996.

[98] Olorunniwo, F.O., "Economic Design of a Partially dynamic $\bar{x}$-Control Chart",*IJQRM,* 6(3), pp 24-37. 1992.

[99] Parkhideh, B. and K.E. Case, "The Economic Design of a Dynamic $\bar{x}$-Control Chart," *IIE Transactions,* 21(4), pp 313-323, 1989.

[100] Pate-Cornel, M.E., H.L. Lee, and G. Tagaras, "Warnings of Malfunction: The Decision to Inspect and Maintain Production Processes on Schedule or on Demand", *Management Science,* 33(10), pp 1277-1290, 1987.

[101] Peters, M.H., H. Schneider, and K. Tang "Joint Determination of Optimal Inventory and Quality Control Policy", *The Institute of Management Science,* 34(8), pp 991-1004, 1988.

[102] Pignatiello, J.J., and A. Tsai, "Optimal Economic Design of a $\bar{x}$-Control Charts", IIE-Transaction, 21(4), pp 313-323, 1988.

[103] Porteus, E.L., "Optimal Lot Sizing, Process Quality Improvement and Setup Cost Reduction", *Operations Research,* 34, pp 137-44, 1986.

[104] Porteus, E.L., "The impact of inspection delay on process and inspection lot sizing", *The Institute of Management Science*, pp 999-1007, 1990.

[105] Porteus, E.L. and Angelus, A. "Opportunities for Improved Statistical Process Control," accepted in *Management Science*, 1996.

[106] Pugh, G.A., "An Algorithm for Economically Setting A Uniformly-Shifting Process", *Computers and Industrial Engineering*, 14(3), pp 237-240, 1988.

[107] Pulak, M.F.S. and K.S. Al-Sultan, "On the Optimum Targeting for a Single Filling Operation with Rectifying Inspection", *Omega*, to appear, 1996a.

[108] Pulak, M.F.S. and K.S. Al-Sultan, "A Computer Package for Process Mean Targeting", *Journal of Quality Technology*, to appear, 1996b.

[109] Quesenberry, C.P., "An SPC Approach to Compensating a Tool-Wear Process", *Journal of Quality Technology*, 20(4), pp 220-229, 1988.

[110] Rahim, M. A., "Determination of Optimal Design Parameters of Joint and R Charts.",*Journal of Quality Technology* 21, pp 21-70, 1989.

[111] Rahim, M. A., "Economic Design of Control Charts Assuming Weibull Distribution In-Control Times", *Journal of Quality Technology*, 25, pp 296-305, 1993.

[112] Rahim, M. A., "Joint Determination of Production Quantity, Inspection Schedule, and Control Chart Design", *AIIE Transactions*, 26, pp 2-11, 1994.

[113] Rahim, M.A., "Joint Determination of Optimal Production Run and Production Rates for a Process having Multilevel Tool Wear", Working Paper, Faculty of Administration, University of new Brunswick, Fredericton, Canada, 1996.

[114] Rahim, M. A., "Economically Optimal Design of $\overline{x}$-Control Charts Assuming Gamma Distributed In-Control Times", *Optimization in Quality Control*, edited by K. S. Al-Sultan and M. A. Rahim, Kluwer Academic Publishers, 1997.

[115] Rahim, M. A. and P. S. Banerjee, "Optimal Production Run for A Process with Random Linear Drift", *Omega*, 16(4), pp 347-351, 1988.

[116] Rahim, M. A. and P.K. Banerjee, "A Generalized Model for the Economic Design of Control Charts for Production Systems with Increasing Failure Rate and Early Replacement." *Naval Research Logistics* 40, pp 787-809, 1993.

[117] Rahim, M. A. and M. Ben-Daya, "A Generalized Economic Model for Joint Determination of Production Run, Inspection Schedule, and Control Chart Design". Working Paper #96-032, Faculty of Administration, University of New Brunswick, Fredericton, Canada, 1996.

[118] Rahim, M. A., H. Ohta and M. Ben-Daya, "An Integrated Dynamic Joint Optimization Model for Controlling both Quality and Maintenance Policies", Working Paper #96-042, Faculty of Administration, University of New Brunswick, Fredericton, Canada, 1996.

[119] Rahim, M. A. and M. Ben-Daya, "Joint Optimization of Production Quantity, Inspection Policy and Quality Control for an Imperfect Process with Deteriorating Inventory Systems", Working Paper #96-041, Faculty of Administration, University of New Brunswick, Fredericton, Canada, 1996.

[120] Rahim, M. A. and R. S. Lashkari, "Optimal Decision Rules for Determining the Length of Production Run", *Computers and Industrial Engineering* 9, pp 195-202, 1984.

[121] Rahim, M.A. and A. Raouf, "Optimal Production Run for a Process Having Multilevel Tool Wear," *International Journal of Systems Science*, 9(1), pp 139-149, 1988.

[122] Rosenblatt, M.J. and H.L. Lee, "Economic Production Cycles with Imperfect Production Processes", *IEE Transactions*, 18, pp 48-55, 1986.

[123] Saniga, E. M., "Joint Economically Optimal Design of $\bar{x}$-Control and R Control Charts", *Management Science*, 24, pp 420-431, 1977.

[124] Saniga, E. M., "Joint Economic Design of $\bar{x}$-Control and $R$ Control Charts with Alternate Process Models", *AIIE Transactions*, 11, pp 254-260, 1979.

[125] Saniga, E. M. and D. C. Montgomery, "Economical Quality Control Policies for a Single Cause System", *AIIE Transaction*, 13, pp 258-264, 1981.

[126] Saniga, E. M., "Economic Statistical Control Chart Design with an Application to R Charts", *Technometrics*, 31, pp 313-320, 1989.

[127] Saniga, E. M., "Joint Statistical Design of $\bar{x}$-Control and $R$ Control Charts", *Journal of Quality Technology*, 23, pp 156-162, 1991.

[128] Schmidt, R. L. and P. E. Pfeifer, "An Economic Evaluation of Improvements in Process Capability for a Single-Level Canning Problem", *Journal of Quality Technology*, 21(1), pp 16-19, 1989.

[129] Schmidt, R. L. and P. E. Pfeifer, "Economic Selection of the Mean and Upper Limit for a Canning Problem With Limited Capacity", *Journal of Quality Technology*, 23(4), pp 312-317, 1991.

[130] Schneider H., K. Tang. and C. O'Cinneide, "Optimal Control of a Production Process Subject to Random Deterioration", *Operations Research*, 38(6), pp 1116-1122, 1990.

[131] Shewhart W. A., "*Economic Control of Quality of Manufactured Products*". D Van Nostrand Company, Inc., Princeton, N.J., 1931.

[132] Smith B.E., and R.R. Vemuganti, "A learning Model for Processes with Tool Wear", *Technometrics*, 10(2), pp 379-387, 1968.

[133] Springer, C. H., "A Method for Determining the Most Economic Position of a Process Mean", *Industrial Quality Control*, July, 36-39, 1951.

[134] Soland, R. M., "Availability of Renewal Functions for Gamma and Weibull Distributions with Increasing Hazard Rate", *Operational Research*, 16, pp 536-543, 1968.

[135] Stack, N.D., and R. Wild, "The Nature, Performance and Operating Characteristics of Serial Production Systems", *International Journal of Production Research*, 32, pp 2287-2302, 1980.

[136] Surtihad, J. and M. Raghavachari, "Exact Economic Design of Charts for General Time In-Control Distributions",*International Journal of Production Research* 32, pp 2287-2302, 1994.

[137] Svoboda, L., "Economic Design of Control Charts: A Review and Literature Survey (1979-1989)", *Statistical Process Control in Manufacturing*, edited by J. B. Keats and D. C. Montgomery, Marcel Dekker, New York, NY, pp 331-330, 1991.

[138] Szendrovits, A.Z. and Z. Drezner, "Optimizing Multi-Stage Production with Constant Lot Size and Varying Numbers of Batches",*OMEGA*, 8, pp 623-629, 1980.

[139] Tadikamalla, P. R., "An Inspection on Policy for the Gamma Failure Distributions",*Journal of Operational Society*, 30, pp 77-80, 1979.

[140] Taha, H.A., "A Policy for Determining the Optimal Cycle Length for a Cutting Tool", *Journal of Industrial Engineering*, 17(3), pp 157-162, 1966.

[141] Taguchi, G., E.A. Elsayed, and T. Hsing, "*Quality Engineering in Production Systems*", McGraw Hill, New York, NY., 1989.

[142] Tagaras, G., "An Integrated Cost Model for the Joint Optimization of Process Control and Maintenance", *Journal of Operational Research Society*, 39(8), pp 757-766, 1988.

[143] Tagaras, G., "A Dynamic Programming Approach to the Economic Design of $\bar{x}$-Control Charts", *IIE Transactions*, 26(3), pp 48-56, 1994.

[144] Tagaras, G., "Dynamic Control Charts for Finite Production Runs", European *Journal of Operations Research* to appear, 1995.

[145] Tagaras, G., "Economic Decision of Time-Varying and Adaptive Control Charts", *Optimization in Quality Control*, edited by K. S. Al-Sultan and M. A. Rahim, Kluwer Academic Publishers, 1997.

[146] Tapiero, C.S., P.H. Ritchken,and A. Reisman, "Reliability, Pricing and Quality Control", *European Journal of Operational Research*, 31, pp 37-45, 1987.

[147] Tapiero, C.S., and L.F. Hsu, "Quality Control of the M/MI/I Queue", *International Journal of Production Research*, 25, pp 447-455, 1987.

[148] Tayi, G.K., and D.P. Ballou, "An Integrated Production-Inventory Model with Reprocessing and Inspection", *International Journal of Production Research*, 26, pp 1299-1315, 1988.

[149] Taylor, H. M., "Statistical Control of a Gaussian Process", *Technometrics*, 9, pp 29-42, 1967.

[150] Tseng, S. T., "Optimal Preventive Maintenance Policy for Deteriorating Production Systems". *IIE-Transactions*, 25(8), pp 687-694, 1996.

[151] Vance, L. C., "Bibliography of Quality Control Chart Techniques". *Journal of Quality Technology*, 15, pp 59-62, 1983.

[152] Warren, G., A. Rahim, and J. Bhadury, "Simultaneous Economic Selection of Target Mean and Variance", Working Paper #96-017, Faculty of Administration, UNB, Fredericton, Canada, 1996.

[153] Wiklund, S., "Adjustment Strategies When Using Shewart Charts". Economic Quality Control, *Journal and News Letter for Quality and Reliability*, 8, pp 3-21, 1993.

[154] Yang, J. and V. Makis, "Optimal Control and Monitoring of Deteriorating Production Process," to appear in *Optimization in Quality Control*, edited by K.S. Al-Sultan and M.A. Rahim, Kluwer Academic Publishers, 1997.

[155] Yum, B.J., and E.D. McDowell, "Optimal Inspection Policies in a Serial Production System Including Scrap Rework and Repair: An MILP approach", *International Journal of Production Research*, 25, pp 1451-1460, 1987.

# PART II

## ECONOMIC DESIGN OF CONTROL CHARTS

# 3

# DETERMINATION OF THE ECONOMIC DESIGN OF CONTROL CHARTS SIMPLIFIED

**E. v. Collani**

*Institut für Angewandte Mathematik und Statistik,*
*Universität Würzburg,*
*Sanderring 2,*
*D-97070 Würzburg,*
*Germany.*

## ABSTRACT

Control charts are widely used in industry for monitoring manufacturing processes. In spite of the fact that a wrongly selected control chart design may cause considerable losses, industry refrains from using an economic design which guarantees to some extent a design adapted to the given technical and economic conditions.

One of the many reasons for this surprising fact is the complicated structure of the objective function used for determining the economic design. Generally, a large number of different input parameters makes optimization cumbersome and allows only simple control chart policies.

A simplified approach is developed by means of which

1. the treatment of the problem in a more general setting becomes possible,
2. the number of input parameters explicitly entering the objective function is reduced considerably, and
3. the optimization procedure is separated into two steps referring to the decision procedure, and the sampling interval, respectively.

Besides the more technical advantages of the simplified approach, it also provides some interesting insight into the relevant interrelationship between input parameters and design parameters. Finally, this chapter is not intended to be a concluding investigation, but rather aims to point out the need for and show the direction of further research.

**Key words:** control chart, economic design, monitoring policy

# 1 INTRODUCTION

Since Duncan's pioneering paper on the economic design of control charts in 1956, the relevant literature has grown considerably. There are useful review papers by Montgomery (1982), Svoboda (1991), and Ho and Case (1994). Recent developments include:

- As a reaction to Woodall's critique (1986, 1987) the so-called *economic-statistical design* of control charts was introduced, particularly by Saniga (1989, 1995).
- Many efforts have been made during the last few years to simplify the model, especially with respect to the economic parameters involved (Collani (1988, 1989)).
- Moreover, various extensions concerning the process in question have been investigated (Collani (1994), Castillo and Montgomory (1993)), and more sophisticated decision procedures besides the classical Shewhart control chart have been considered by many authors (Keats and Simpson (1994), Montgomory et al.(1995)).

Although there have been important theoretical advances, economic control procedures are almost unknown in industrial practice. Undoubtedly, one important reason is the fact that there are no *economic standards* for statistical process control. But before any standard can be developed, the procedures for determining the economic design of control charts must necessarily be further simplified. There will be no standard for a procedure requiring to specify ten or even more different input parameters.

In this chapter a simplified model is developed, which leads to a dramatic decrease in the number of input parameters, and readily allows various extensions of the existing methodology, and thereby, hopefully, will enable the development of economic standards for statistical process control.

## 1.1 Wear-Out and Sudden Shocks

Among others, quality of industrial production processes may suffer from wear-out phenomena continuously growing over time, and from sudden shocks occurring at a random time. Either of them affects process quality and thus process yield. Consequently, counteractions are taken aimed at removing signs of wear before they may result in a decrease of the process yield, and detecting shocks which have occurred before their impact has become serious.

Generally, wear-out phenomena are compensated by

- continuous process maintenance, including routine inspections and repairs/ renewals,

and shocks are detected by

- monitoring the process.

Here, we deal primarily with the design of monitoring policies assuming that any sign of wear is removed in due time by a continuous maintenance policy.

Frequently, the problem with randomly occurring shocks is that they cannot be readily recognized, but that they have to be discovered, e.g., by expensive checks of the process itself. Thus, in extreme situations it may even happen that monitoring the process by means of process checks is more expensive than not detecting shocks until the next regular maintenance action. Alternatively to process checks, one may monitor the quality of the output being affected indirectly by any shock, and release an alarm whenever a deterioration of quality is observed. Thus, a two-step problem has to be solved: first a monitoring strategy has to be selected, and second the design of the policy has to be determined by means of which the losses resulting from shocks may be controlled efficiently.

According to common understanding, a sampling policy is a very appropriate means for solving the above described problem. A sampling policy doesn't observe the process state directly, but the quality of the output. If the quality turns out to be too bad, a decision is made in favor of a process intervention additionally to those performed according to the continuous maintenance.

Any sampling policy is given by three sequences:

- a sequence of timepoints $\{t_i\}_{i=1,2,\cdots}$, where $t_i$ gives the time when the *ith* sample has to be drawn from the production process determining the *sampling intensity*,
- a sequence of sample sizes $\{n_i\}_{i=1,2,\cdots}$, where $n_i$ is the sample size of the *ith* sample given by $\{X_1^{(t_i)}, \cdots, X_n^{(t_i)}\}$ determining the *sampling accuracy*, and
- a sequence of decision functions $\{\gamma_i\}_{i=1,2,\cdots}$, where $\gamma_i$ is the decision function to be applied at time $t_i$ , with

$$\gamma_i\left(X_1^{(t_1)}, \cdots, X_n^{(t_1)}, \cdots, X_1^{(t_i)}, \cdots, X_n^{(t_i)}\right) = \begin{cases} 0 & : \text{ no intervention} \\ 1 & : \text{ intervention} \end{cases}$$

Very often sampling policies in industrial practice exhibit the following properties:

- $t_{i+1} - t_i = h$ (fixed sampling interval)
- $n_i = n$ (fixed sample size)
- $\gamma_i = \gamma(X_1^{(t_i)}, \cdots, X_n^{(t_i)})$ (independent decisions)

For the remainder we assume that the sampling policy has a fixed sampling interval $h$. Moreover, we start with the investigation of the case *fixed sample size* and *independent decisions*.

# 2 THE PROCESS MODEL

## 2.1 The Output Model

It is assumed that the quality of an item produced is determined by exactly one observable characteristic denoted by $X$. There are no assumptions made concerning the distribution of $X$. It might be a measurable characteristic usually modeled by a normal distribution, or a dichotomous attribute, like conforming or nonconforming modelled by the binomial distribution, or $X$ might describe the number of nonconformities on an item usually leading to the Poisson distribution. The process output can be described by means of the sequence

$$\{X_i\}_{i \in IN} \tag{3.1}$$

Let the number of $X_i$ generated per time unit of operation be constant. This number is called production speed, and is denoted by $v$:

$$v := \text{production speed}. \tag{3.2}$$

Next, it is assumed that the state of the process can be described by some properties of the random variables $X_i$, e.g., the expectation of $X_i$ (called the process mean) or the variance of $X_i$ (called the process variability). Hence, the process state at time $t = i/v$ is characterized by the distribution function adopted by the random variables $X_i$.

Moreover, it is assumed that there are only two distinct process states, the in–control state, called State I, and the out–of–control state, called State II. When operating in State I, any intervention (repair, renewal, etc.) concerning the process decreases the process yield and therefore should be avoided. When operating in State II, an appropriate process intervention increases the process yield. Thus, we look for a policy which

- guarantees only few and cheap actions when the process is operating in-control, and
- decides in due time for an appropriate process intervention, whenever the process has entered State II.

State I and State II are described by some distributional properties of $X$, hence we define:

**Definition**

The process is operating at time $t = i/v$ in

- State I, if the random variable $X_i$ is distributed according to $F_I(x)$, and
- State II, if the random variable $X_i$ is distributed according to $F_{II}(x)$.

Finally, it is assumed that the random variables $X_i$ are conditionally independent, conditioned under $F_I(x)$ or $F_{II}(x)$, respectively.

## 2.2 The Transition Model

The transition behaviour of the process is based on the following assumptions:

- The process starts in–control, i.e., in State I.
- After some random time of operation a shock occurs and the process changes to the out–of–control state, i.e., State II.
- When operating in State II, a change to State I can only be accomplished by means of a process intervention.

Thus the process is run in the following way: it is started in-control. After some time a shock leads to an assignable cause and the process enters State II. After detection by means of a monitoring policy, the assignable cause is removed and State I restored. Hence, the process operates alternating in State I and State II, respectively. Any wear-out phenomena occurring in time are compensated by continuous maintenance not specified in detail.

Obviously, the lengths of the out-of-control periods depend solely on the monitoring policy and are investigated later on. The lengths of the consecutive in-control periods depend on the process in question and on the process interventions performed.

Let $\tau_0$ denote the in-control period after the start of the process with

$$P(\tau_0 \le t) = \begin{cases} 0 & \text{if } t < 0 \\ F_0(t) & \text{if } t \ge 0 \end{cases} \tag{3.3}$$

Generally, there are several different possibilities for removing an assignable cause. The differences refer to the "degree of repair" which is expressed by the distribution function of the length of the subsequent in-control period.

Let $K \in \mathbb{N}$ denote the number of different types of process corrections, and $t_s$ the "age" of the process at arrival of the shock. The random time of operating in-control following a process correction after detection of State II depends on

- the type $k \in K$ of the process correction, and

- on the "process age" $t_s$ at shock arrival.

Denoting by $\tau_{t_s}^{(k)}, t_s \geq 0, 1 \leq k \leq K$ the time of operating in-control following a process correction of type $k$ performed at "process age" $t_s$, we have:

$$P(\tau_{t_s}^{(k)} \leq t) = \begin{cases} 0 & \text{if } t < 0 \\ F_{t_s}^{(k)}(t) & \text{if } t \geq 0 \end{cases} \tag{3.4}$$

If

$$F_{t_s}^{(k_0)}(t) = F_0(t) \tag{3.5}$$

for some $k_0 \in \{1, \cdots, K\}$ and any $t_s > 0$ then the intervention $k_0$ restores the starting conditions, and therefore it is called *process renewal.*

It is assumed that the first derivative of $F_{t_s}^{(k)}$ exists for any $k, 1 \leq k \leq K$ and $t_s \geq 0$:

$$f_{t_s}^{(k)}(t) := \frac{d}{dt} F_{t_s}^{(k)}(t). \tag{3.6}$$

As an illustration, consider a process of age $t_s$,when a shock occurs and terminates the actual in-control period. Very often the following two types of process interventions are investigated in literature:

1. A renewal removing the assignable cause and additionally restoring the starting conditions of the process. In this case the length of the subsequent in-control period has according to (3.5) the distribution function $F_0(t)$.
2. A minimal repair which removes the assignable cause, but doesn't change the "age of the process", i.e., after a minimal repair the age of the process is the same as at the arrival time of the shock $t_s$, hence the subsequent in-control period is distributed according to $F_{t_s}(t)$, or

   $$F_{t_s}(t) = P(\tau_0 \leq t + t_s | \tau_0 > t_s)$$

   i.e., $F_{t_s}(t)$ is given by the conditional distribution of $\tau_0$ under the condition that a shock has not occurred for a period of length $t_s$.

Having assumed that the process is continuously maintained and thus signs of wear-out are removed before they can affect the process, the following approximation with respect to the distribution of the lengths of the in-control periods seems to be justified:

**Assumption 1**

Any removal of an assignable cause following a shock restores the starting conditions of the process. Hence:

$$F_{t_s}^{(k)}(t) = F_0(t) \tag{3.7}$$

for $1 \leq k \leq K$ and any $t_s$.

Consequently, when operating in State II we need to consider in the remainder only one type of intervention, which is a *process renewal.*

# 3 THE SAMPLING POLICY

The problem is to determine a monitoring policy in order to decide on possible process interventions. For investigating sampling policies, we have to specify a sample.

**Sample**

For any $j \geq n$ and $n \geq 1$ the subsequence $\{X_{j-n-1}, \ldots, X_j\}$ of the process $\{X_i\}_{i=1,2,\cdots}$ is called a sample of size $n$ at time $t = j/v$.

In order to simplify the problem considerably we make the following assumption with respect to a sample:

**Assumption 2**

Let $\{X_{j-n-1}, \ldots, X_j\}$ be a sample, then $\{X_{j-n-1}, \ldots, X_j\}$ are independent, identically distributed random variables.

Assumption 2 is justified whenever a transition from State I to State II during sampling occurs only with small probability, which holds if the sample size $n$ is small compared with the expected number of items produced during an in-control period $E[\tau_0]v$.

Taking a sample for process monitoring doesn't touch the process itself, but leads to a decision on actions having direct effects on the process, which are called process interventions and which are defined here in the following way:

**Process Intervention**

An action after which the process operates in State I with probability 1 is called *process intervention.*

Accordingly, there are two types of process interventions to be considered:

1. A process intervention when the process is actually operating in State I, merely reveals this fact (e.g., by a process inspection) with probability 1.
2. A process intervention when the process is actually operating in State II, reveals this fact with probability 1, and additionally transfers the process back to State I (e.g., by a repair) again with probability 1.

Now we are in the position to define a simple periodic sampling policy:

**Sampling Policy**

A simple periodic sampling policy is given by three quantities $(h, n, \gamma)$, and the following rule: Take every $h$ hours of operation a random sample of size $n$ with $0 < n < hv$ from the process, and decide by means of the decision function $\gamma$ on the basis of the sampling observation upon the actual state of the process. In case the decision is in favor of State II (i.e., the out-of-control state) a process intervention is undertaken. Otherwise the process is left alone.

The pair $(n, \gamma)$ is called ***sampling plan.***

More formally, a sampling policy is given by an element of the set

$$\mathcal{S} = \left\{(h, n, \gamma) | h \in \mathbb{R}^+, n \in \mathbb{N}, n < hv, \gamma \in \Gamma\right\}$$

where $\Gamma$ is the set of admissible decision functions.

For theoretical and practical reasons it is often desirable to consider an *extended* set of monitoring policies denoted by $\overline{\mathcal{S}}$ instead of $\mathcal{S}$:

$$\overline{\mathcal{S}} = \left\{(h, n, \gamma) | h \in \overline{\mathbb{R}_0^+}, n \in \mathbb{N}_0, n \leq hv, \gamma \in \overline{\Gamma}\right\}$$

where $\overline{\Gamma} := \Gamma \cup \gamma^{(0)} \cup \gamma^{(1)}$ with

$$\gamma^{(0)} = 0 \tag{3.8}$$
$$\gamma^{(1)} = 1 \tag{3.9}$$

i.e., applying $\gamma^{(0)}$ means to decide always against an intervention, and applying $\gamma^{(1)}$ means to decide always in favor of an intervention.

$\overline{\mathcal{S}}$ includes five different types of monitoring policies:

1. $\mathcal{S}_I = \{(h, n, \gamma) | h = 0, n = 0, \gamma^{(1)}\}$
   Interpretation: Perform at each time point an intervention.
2. $\mathcal{S}_{II} = \{(h, n, \gamma) | h = \infty, 0 \leq n < \infty, \gamma \in \overline{\Gamma}\} \cup \{(h, n, \gamma) | 0 < h < \infty, n = 0, \gamma = \gamma^{(0)}\}$
   Interpretation: Dispense with monitoring and interventions.
3. $\mathcal{S}_{III} = \{(h, n, \gamma) | 0 < h < \infty, n = 0, \gamma = \gamma^{(1)}\}$
   Interpretation: Dispense with sampling actions, perform at each time point $\nu \cdot h$, $\nu = 1, 2, ...$, an intervention.

4. $\mathcal{S}_{IV} = \{(h, n, \gamma) | 0 < h < \infty, 1 \leq n < hv, \gamma \in \Gamma\}$

   Interpretation: The process is monitored by means of sampling. An intervention is performed if the decision based on the sampling observations is in favor of State II.

5. $\mathcal{S}_{V} = \{(h, n, \gamma) | 0 < h < \infty, n = hv, \gamma \in \Gamma\}$

   Interpretation: The process is monitored by screening, i.e., each item is tested. After any subsequences of $n$ items tested a decision is made based on the screening observations. An intervention is performed if the decision based on the screening observations is in favor of State II.

It is easily seen that each of the five subsets of $\overline{\mathcal{S}}$ may be appropriate in some real-world situations:

- $\mathcal{S}_I$ is appropriate in the following situation:

  Consider the case that State I just means that the process is operating and State II means that the process has stopped and that the state of the process is known at each timepoint $t$. This case can be described by

$$X_i = \begin{cases} 1 & \text{if the process is operating at time } t = i/v \\ 0 & \text{if the process has stopped } t = i/v \end{cases}$$

  with the following degenerated distributions:

$$P(X_i = 1 | \text{State I}) = 1$$

$$P(X_i = 0 | \text{State II}) = 1$$

- $\mathcal{S}_{II}$ is appropriate, if any process intervention isn't rewarding. This holds e.g., in the case of rather frequent continuous maintenance actions, by means of which any departure from State I is discovered in due time.

- $\mathcal{S}_{III}$ is appropriate, if an intervention is cheap compared with sampling.

- $\mathcal{S}_{IV}$ is appropriate, if interventions are expensive compared with sampling.

- $\mathcal{S}_{V}$ is appropriate, if sampling is cheap compared with the costs of operating out-of-control, or screening is performed by an automatic testing device.

We call $(h, n, \gamma) \in \mathcal{S}_{IV} = \mathcal{S}$ a proper sampling policy. Here, we investigate primarily the case that a proper sampling policy shall be used, but as shown e.g., in Collani (1987, 1989) the different strategies can be investigated simultaneously.

**Remark:** Frequently, in maintenance and replacement theory State II of the process is called failure indicating that process operation comes to a stop after the process has entered the out-of-control state. In statistical quality control the model of failure is extended in the following way: it is assumed that the process does not stop after a transition to State II, but continues to operate, with the only difference that after a "failure" has occurred the process deteriorates and runs on a lower level of efficiency described by the distribution function $F_{II}(x)$. Consequently a process "failure" is called "assignable cause" or "special cause". The first term goes back to Walter A. Shewhart, and the second term was introduced by W. Edwards Deming, one of the greatest protagonists of quality of our times.

## 3.1 The Decision Function

One of the major problems arising when dealing with sampling policies as given above is the selection of an appropriate decision function based on a statistical test of the hypotheses:

$H_0$ : $F_I(x)$ is the actual process distribution function, i.e., the process is operating in State I.
$H_1$ : $F_{II}(x)$ is the actual process distribution function, i.e., the process is operating in State II.

For this situation we consider decision functions which adopt only the two values 0 and 1 with the following meaning:

$$\gamma_{(\ell_1,\ldots,\ell_m)} = \begin{cases} 0 & : \text{ no intervention, i.e., decision for State I} \\ 1 & : \text{ intervention, i.e., decision for State II} \end{cases} \tag{3.10}$$

where $(\ell_1, \cdots, \ell_m)$ are the design parameters of the decision function.

Evidently there are two possible wrong decisions:

1. A decision for State II although the process is operating in State I, called error of Type I, and
2. A decision for State I although the process is operating in State II, called error of Type II.

In order to describe the statistical properties of a given decision function, two classes are generally distinguished. Those where the actual decision depends exclusively on the outcome of the actual sample, and those where the actual decision depends not only on the outcome of the actual sample but also on the outcome of past samples.

For the first class the so-called *error probabilities* are used for characterizing the properties of a decision function:

| | |
|---|---|
| Probability of Type I error: | $\alpha := P(\text{Type I error}) = P(\gamma = 1 \mid \text{State I})$ |
| Probability of Type II error: | $\beta := P(\text{Type II error}) = P(\gamma = 0 \mid \text{State II})$ |

The error probabilities depend generally on the design parameters $(\ell_1, \cdots, \ell_m)$ and on the sample size $n$. Whenever this dependence shall be exhibited either of the following notations will be used:

$$\alpha = \alpha_{(n,\ell_1,\cdots,\ell_m)} \quad \text{or} \quad \alpha = \alpha_{(n,\gamma)} \tag{3.11}$$

$$\beta = \beta_{(n,\ell_1,\cdots,\ell_m)} \quad \text{or} \quad \beta = \beta_{(n,\gamma)} \tag{3.12}$$

For the second class of decision functions the above defined error probabilities depend on the outcome of past samples, and therefore are not constant for the different decision points. Hence, it does not make sense to use them in order to compare different decision functions. Instead of the error probabilities, the so-called average run lengths (ARL) are used, i.e., the expected number of decisions until the first alarm under the condition that the process state hasn't changed.

**Definition**

Consider a process operating in one state only, and let $\gamma_i$ denote the $i$-th decision and $RL$ be defined by

$$RL := \min\{m|\gamma_m = 1, \gamma_i = 0 \text{ for } i < m\}.$$

Then $ARL := E[RL]$ is called the average run length of $\gamma$ for the process state in question.

For the model considered here, we have two states, State I and State II, respectively, and therefore a decision function is characterized by the two corresponding average run lengths:

| | | | |
|---|---|---|---|
| Average Run Length in State I: | $ARL(I)$ | := | E[$RL$\| State I] |
| Average Run Length in State II: | $ARL(II)$ | := | E[$RL$\| State II] |

Just as the error probabilities, the average run lengths depend generally on the design parameters $(\ell_1, \cdots, \ell_m)$ and on the sample size $n$. Thus, whenever it is necessary to exhibit this dependence either of the following notations will be used:

$$ARL(I) = ARL(I)_{(n,\ell_1,\cdots,\ell_m)} \text{ or } ARL(I) = ARL(I)_{(n,\gamma)} \quad (3.13)$$

$$ARL(II) = ARL(II)_{(n,\ell_1,\cdots,\ell_m)} \text{ or } ARL(II) = ARL(II)_{(n,\gamma)} \quad (3.14)$$

Note that after each alarm, i.e., decision for State II, the process starts operating in State I, because either it was a wrong decision or a subsequent intervention has transferred the process again to State I.

There are two problems concerning the decision function $\gamma$. The first refers to the selection of the set $\Gamma$ of admissible decision functions, and the second refers to the determination of appropriate design parameters $(\ell_1, \cdots, \ell_m)$ for $\gamma \in \Gamma$. The first problem has not been dealt with seriously in literature. For the second one, there are two competing and controversial approaches (Woodall (1986)):

- the *statistical approach*, which determines $(\ell_1, \cdots, \ell_m)$ to meet some given requirements with respect to the statistical properties of $\gamma$, and

- the *economic approach*, which determines $(\ell_1, \cdots, \ell_m)$ so that the process meets in a certain sense optimally its overall economic objective.

In order to apply the second approach, the relevant economic aspects of the situation in question have to be modelled.

# 4 THE ECONOMIC MODEL

The aim is to determine an appropriate control policy $(h, n, \gamma)$ for a given production process which takes into account all relevant economic parameters. Hence, these economic parameters have to be specified. There are two groups of parameters to be considered: one is related to the process itself, whose only task is to produce items meeting some quality requirements. The second group is related to the monitoring actions and interventions to be performed according to the policy used.

For the sake of simplicity we assume that the items produced are classified in *conforming* and *nonconforming* items - i.e., we assume that there are specifications. Let $Sp$ with $Sp \subset I\!R$ denote the specification interval:

$$X \quad \text{conforming} \quad \Leftrightarrow \quad X \in Sp \tag{3.15}$$

then from an economic point of view the items produced are completely described by:

- $g^+$ = average profit derived from a conforming item,
- $g^-$ = average profit derived from a nonconforming item.

To have a meaningful situation, it is assumed that $g^+ > g^-$ holds.

The expected profit per item when operating in State I denoted by $g_I$, and in State II denoted by $g_{II}$, respectively, characterizes the process in question. Let $G$ be the random profit for an item produced, $p_I$ the nonconforming probability in State I, and $p_{II}$ the nonconforming probability in State II, then:

- $g_I = E[G|\text{State I}] = g^+(1 - p_I) + g^- p_I$

- $g_{II} = E[G|\text{State II}] = g^+(1 - p_{II}) + g^- p_{II}$

with

$$p_I = \int_{X \notin S_P} dF_I(x) \qquad \text{and} \qquad p_{II} = \int_{X \notin S_P} dF_{II}(x)$$

According to our assumptions there are a maximum of 3 different actions to be performed using a periodic control policy $(h, n, \gamma)$. One monitoring action (sampling) and two types of process interventions, called in the remainder inspection and renewal. Hence, we need at least three different quantities for describing the economic effects of these actions. Two of these actions, namely sampling or inspection, i.e., intervention after a false alarm, have no impact on the process itself, and therefore they only cost money and are completely described by their respective costs:

- $a^*n$ = cost for taking and evaluating a sample of size $n$
  $a^*$ is called *sampling cost.*
- $e^*$ = average cost of performing an inspection when the process is operating in State I
  $e^*$ is called *inspection cost.*

Note that an intervention following a Type I error can generally be looked upon as a search for the assignable cause, and therefore it is called inspection. The inspection cost $e^*$ includes all costs related to the intervention: the actual costs as well as any costs caused e.g., by a shut-down of the process.

It is assumed here that either sampling or inspection cost are positive, although the cases $a^* = 0$ or $e^* = 0$ are not at all uninteresting:

- $a^* = 0$ can be used to describe the case where the process is continuously monitored e.g., by an automatic testing device, i.e., when a screening policy $(\gamma \in \mathcal{S}_V)$ is applied.
- $e^* = 0$, on the other hand, describes the case, where at each time point the actual state of the process is known at zero costs. This case happens if any shock causes an immediately observable process failure and leads to a policy $\gamma \in \mathcal{S}_I$.

Moreover, it seems to be quite clear that if $e^*$ is not considerably larger than $a^*$, then it should be better to dispense with sampling and to use a policy $\gamma \in \mathcal{S}_{III}$ instead.

An intervention after a true alarm is a process renewal with the immediate consequences that the process is brought back into State I, and that it will continue operating in State I for the random time $\tau_0$. During the time $\tau_0$ of operation the profit derived from the process will be increased compared with the profit immediately before the process renewal. The surplus in profit minus the costs of one process renewal describes its economic property sufficiently well. Its expectation denoted by $b^*$ is called *average benefit per renewal*:

- $b^*$ = average benefit per renewal

We obtain:

$$b^* = (g_I - g_{II})E[\tau_0]v - r^* \tag{3.16}$$

with $r^*$ := average cost of a process renewal and

$$E[\tau_0] = \int_0^\infty \overline{F}_0(t)dt \qquad \text{with} \quad \overline{F}_0(t) = 1 - F_0(t) \tag{3.17}$$

To get a meaningful problem it is assumed here that the average benefit per renewal is positive:

$$b^* > 0 \tag{3.18}$$

Of course, the opposite may happen:

- In a case where the average surplus derived from a renewal is less than the cost of a renewal, it happens that $b^* \leq 0$. In such a case, interventions (and consequently also monitoring actions) should not be performed as they don't improve the process yield. Hence, a policy $\gamma \in \mathcal{S}_{II}$ should be used.

The economic parameters $a^*$, $e^*$ and $b^*$ are called *economic key parameters.* It should be mentioned that they are derived from a possibly large number of different primary economic parameters. It follows that each set of economic key parameters represents a whole family of different economic situations.

# 5 THE OBJECTIVE FUNCTION

The economic approach in statistical process control is based on an objective function which reflects the overall aim of having the process run, i.e., to make profit. Profit is defined as the difference between yield and expenses. To check whether all relevant economic parameters have been specified so far, we figure out that the yield is completely considered by $g^+$ and $g^-$. The process expenses when the process is operating (i.e., producing) are taken into account again by the parameters $g^+$ and $g^-$, and when it is shut down (for performing an inspection or renewal, respectively) they are taken into account by $e^*$ and $r^*$. The expenses caused by the control policy $(h, n, \gamma)$ are taken into account by $a^*$ (sampling), $e^*$ (inspection) and $r^*$ (renewal). Thus we may conclude that the approach used here is based on a complete economic model.

The time between two successive process renewals is called a *renewal cycle.* The renewal cycles divide the process with respect to time in stochastically equivalent and independent parts. For each renewal cycle we define the following quantities:

| | | |
|---|---|---|
| $A_I$ | = | number of monitoring actions (= samples) while the process is operating in State I |
| $A_F$ | = | number of process inspections, i.e., false alarms |
| $A_{II}$ | = | number of monitoring actions (= samples) while the process is operating in State II |

Let $T$ denote the time of operation during one renewal cycle. Then we have:

$$T = (A_I + A_{II})h \tag{3.19}$$

with expectation

$$E[T] = (E[A_I + A_{II}])h. \tag{3.20}$$

It makes sense to take into account the discrete structure of the process by measuring the length of a renewal cycle by the number of items $X_i$ produced instead of the time of operation $T$:

$N$ = number of items produced during one renewal cycle,

Then

$$E[N] = E[T]v = E[A_I + A_{II}]hv. \tag{3.21}$$

Economically, a renewal cycle is fully described by the corresponding income denoted by $I$ or its expectation given by:

$$\begin{aligned} E[I] &= g_I E[\tau_0]v + g_{II}(E[T] - E[\tau_0])v - r^* - e^* E[A_F] - a^* n E[A_I + A_{II}] \\ &= g_{II}(E[A_I + A_{II}])hv + b^* - e^* E[A_F] - a^* n E[A_I + A_{II}] \end{aligned} \tag{3.22}$$

where

1. The first term in (3.22) gives the average income if the process would operate all the time in State II.
2. The second term adds the average benefit due to the fact that a renewal was performed at the beginning of the renewal cycle.

Thus, the sum of the two first terms represents the average yield derived from the items produced reduced by the average cost of one renewal.

3. The third term gives the expected expenses for unnecessary interventions after false alarms.
4. Finally, the last term represents the total expected sampling costs.

The long run profit per item produced denoted by $\Pi^*$, which is equivalent to the long run profit per time unit of operation, is defined as the quotient of the expected income per renewal cycle $E[I]$ and the expected number of items produced during one renewal cycle $E[N]$:

$$\begin{aligned}\Pi^*(h,n,\gamma) &= \frac{E[I]}{E[N]} = \frac{b^* - e^* E[A_F]}{E[A_I + A_{II}]hv} - \frac{a^* n}{hv} + g_{II} \\ &= \frac{1}{hv}\left\{\frac{b^* - e^* E[A_F]}{E[A_I + A_{II}]} - a^* n\right\} + g_{II} \qquad (3.23)\end{aligned}$$

With (3.23) a suitable objective function has been derived, allowing to define an optimal monitoring policy:

**Definition**

A monitoring policy $(h^*, n^*, \gamma^*) \in \mathcal{S}$ is called $\Pi^*$-optimal in the class $\Gamma$ of admissible decision functions and with respect to the long run profit per item, if

$$\Pi^*(h^*, n^*, \gamma^*) \geq \Pi(h, n, \gamma) \qquad \text{for any} \quad (h, n, \gamma) \in \mathcal{S}$$

Note that due to the restriction $(h, n, \gamma) \in \mathcal{S}$ it might happen that there is no $\Pi^*$-optimal monitoring policy (Collani (1987)).

The optimization problem given by the above definition is not a trivial one. Therefore, the remainder is mainly devoted to attempts for simplifying this problem.

The following linear transformation of $\Pi^*(h, n, \gamma)$ leads to a first simplification:

$$\Pi(h,n,\gamma) = \frac{\left(\Pi^*(h,n,\gamma) - g_{II}\right) v E[\tau_0]}{e^*} \qquad (3.24)$$

$$= \frac{E[\tau_0]}{h}\left\{\frac{\frac{b^*}{e^*} - E[A_F]}{E[A_I + A_{II}]} - \frac{a^*}{e^*}n\right\}$$

$$= \frac{E[\tau_0]}{h}\left\{\frac{b - E[A_F]}{E[A_I + A_{II}]} - an\right\} \tag{3.25}$$

where $b := \frac{b^*}{e^*}$ and $a := \frac{a^*}{e^*}$

The linear transformation (3.24) used for obtaining $\Pi(h, n, \gamma)$ has an obvious interpretation:

- by subtracting $g_{II}$ the overall profit per unit is replaced by the surplus per unit due to the control policy,
- by multiplying with $v$ the surplus per unit is replaced by the surplus per hour,
- the multiplication by $E[\tau_0]$ means that the arbitrary time unit ***hour*** is replaced by the time unit *expected length of an in-control period* being in some sense characteristic to the process,
- finally, the division by the positive constant $e^*$ enables the introduction of the two relative parameters

  1. $b$ = relative average benefit per renewal, and
  2. $a$ = relative sampling cost

  thus reducing the number of economic parameters explicitly entering the objective function by one, and showing that not the *key economic parameters* itself, but their ratios are decisive for the economic design of a process monitoring policy.

One of the problems to be solved is the determination of the expectations $E[A_I], E[A_F]$ and $E[A_{II}]$. The first one doesn't depend on the sampling plan $(n, \gamma)$ and is given by:

- $$E[A_I] = \sum_{j=1}^{\infty} \overline{F}_0(jh) \tag{3.26}$$

The two other ones depend heavily on the sampling plan $(n, \gamma)$ and are dealt with later on.

Before starting a detailed discussion of the objective function, we note the following two features of $\Pi(h, n, \gamma)$:

- Generality:
  So far no distributional assumptions with respect to the quality characteristic $X$ or the length of an in-control period $\tau_0$ have been made. The only restrictions refer to the sampling interval and the sample size being assumed to be fixed.

- Simplicity:
  There are only *two* economic parameters entering explicitly $\Pi(h, n, \gamma)$, namely the relative sampling cost $a$ and the relative benefit per renewal $b$.

# 6 INDEPENDENT DECISIONS

Independent decisions are predominant at least in industrial practice, where Shewhart-type procedures are widely used. Therefore, we concentrate our investigations at this stage on the case of independent decisions.

It follows that there are constant error probabilities $\alpha$ and $\beta$, and hence

- $$E[A_F] = \alpha \cdot E[A_I] \tag{3.27}$$
- $$E[A_{II}] = \frac{1}{1-\beta} \tag{3.28}$$

Inserting (3.27) and (3.28) into the objective function (3.25) yields:

$$\Pi(h, n, \gamma) = \frac{E[\tau_0]}{h} \left\{ \frac{b - \alpha \cdot E[A_I]}{E[A_I] + \frac{1}{1-\beta}} - an \right\} \tag{3.29}$$

where $\alpha$ and $\beta$ are the error probabilities, which generally depend on

- some distributional parameters of $F_I(x)$ and $F_{II}(x)$,
- the sample size $n$, and
- the design parameters $(\ell_1, \cdots, \ell_m)$ of the decision function $\gamma$.

## 6.1 First Simplification

As shown in Frahn (1994) the expected number of monitoring actions during an in-control period can be approximated under very general conditions sufficiently well by $E_1[A_I]$:

$$E_1[A_I] = \frac{E[\tau_0]}{h} - \frac{1}{2} \tag{3.30}$$

Inserting (3.30) into (3.29) results in a first approximation of the objective function (3.25):

$$\Pi_1(h,n,\gamma) = \frac{E[\tau_0]}{h}\left\{\frac{\left(b+\frac{a}{2}\right) - \alpha \cdot \frac{E[\tau_0]}{h}}{\frac{E[\tau_0]}{h} + \left(\frac{1}{1-\beta} - \frac{1}{2}\right)} - an\right\} \tag{3.31}$$

The objective function $\Pi_1$, which as will be seen later simplifies particularly the problem of calculating the optimal sampling interval, motivates the definition of $\Pi_1$-optimality:

**Definition**

A monitoring policy $(h_1^*, n_1^*, \gamma_1^*) \in \mathcal{S}$ is called $\Pi_1$-optimal if

$$\Pi_1(h_1^*, n_1^*, \gamma_1^*) \geq \Pi_1(h,n,\gamma) \qquad \text{for any} \quad (h,n,\gamma) \in \mathcal{S}$$

If the $\Pi_1$-optimal sample size $n_1^*$ and the $\Pi_1$-optimal decision function $\gamma_1^*$ are known, the $\Pi_1$-optimal sampling interval $h_1^*$ is easily obtained as the solution of

$$\left.\frac{\partial \Pi_1(h,n,\gamma)}{\partial h}\right|_{n=n_1^*,\gamma_1=\gamma_1^*} = 0$$

Elementary calculations yield the following explicit expression for the $\Pi_1$-optimal sampling interval $h_1^*$:

$$h_1^* = 2\,\frac{1-\beta^*}{1+\beta^*}\,\frac{E[\tau_0]}{\sqrt{\frac{2[b(1-\beta^*)+\alpha^*]}{(an_1^*+\alpha^*)(1+\beta^*)}}-1} \tag{3.32}$$

where $\alpha^* = \alpha_{n_1^*,\gamma_1^*}$ and $\beta^* = \beta_{n_1^*,\gamma_1^*}$ denote the error probabilities obtained for the $\Pi_1$-optimal sample size $n_1^*$ and the $\Pi_1$- optimal decision function $\gamma_1^*$.

Besides the explicit formula for the optimal sampling interval, the objective function $\Pi_1(h,n,\gamma)$ illustrates an important feature of the optimal sampling interval $h^*$, which was already observed in Arnold and Collani (1989), namely that $h^*$ depends rather on the expectation $E[\tau_0]$ than on the whole shape of the distribution function $F_0(t)$ of $\tau_0$.

## 6.2 Second Simplification

As a consequence of the above observation, we may approximate the distribution of $\tau_0$ by the exponential distribution with expectation equal to $E[\tau_0]$; i.e., instead of $F_0(t)$ we use the distribution function $\hat{F}(t)$ given by:

$$\hat{F}(t) = \begin{cases} 0 & : \quad t < 0 \\ 1 - e^{\frac{t}{E[\tau_0]}} & : \quad t \geq 0 \end{cases} \tag{3.33}$$

Note that in most papers on the economic design of control charts the assumption $F_0(t) = \hat{F}(t)$ is made from the very beginning.

With (3.33) a second approximation of $E[A_I]$ besides (3.30) is obtained:

$$E_2[A_I] = \frac{1}{e^{\frac{h}{E[\tau_0]}} - 1} \tag{3.34}$$

Using $E_2[A_I]$ results in a second approximation $\Pi_2(h, n, \gamma)$ of the objective function (3.29):

$$\Pi_2(h, n, \gamma) = \frac{E[\tau_0]}{h} \left\{ \frac{b\left(e^{\frac{h}{E[\tau_0]}} - 1\right) - \alpha}{e^{\frac{h}{E[\tau_0]}} - \beta} \cdot (1 - \beta) - an \right\} \tag{3.35}$$

With $\Pi_2(h, n, \gamma)$ we define:

**Definition**

A monitoring policy $(h_2^*, n_2^*, \gamma_2^*) \in \mathcal{S}$ is called $\Pi_2$-optimal if

$$\Pi_2(h_2^*, n_2^*, \gamma_2^*) \geq \Pi_2(h, n, \gamma) \qquad \text{for any} \quad (h, n, \gamma) \in \mathcal{S}$$

Comparing the two approximations $\Pi_1(h, n, \gamma)$ and $\Pi_2(h, n, \gamma)$, we notice that the former leads to an explicit formula for the optimal sampling interval, but the latter coincides with the exact objective function in the important case of exponentially distributed times between consecutive shocks. This is the reason why the following considerations are based on $\Pi_2(h, n, \gamma)$.

Up to now, we have simplified the determination of an approximate $\Pi^*$-optimal sampling interval $h^*$. Unfortunately, the determination of $n^*$ and $\gamma^*$ may constitute a more difficult numerical problem. In the following section it is shown how this problem can be simplified, too.

## 6.3 Third Simplification

We start with stating that generally the $\Pi^*$-optimal sampling interval $h^*$ divided by the expected length of an in-control period $E[\tau_0]$ is a small number, i.e.,

$$\frac{h^*}{E[\tau_0]} \ll 1. \tag{3.36}$$

This observation will be used in the following to derive a third simplification.

The problem is to determine a monitoring policy $(h, n, \gamma) \in \mathcal{S}$ where $\gamma$ depends on its design parameters $(\ell_1, \cdots, \ell_m)$ and the sample size $n$. Define the set $\mathcal{U}$ by

$$\mathcal{U} = \left\{(h, n, \ell_1, \cdots, \ell_m) | (h, n, \gamma_{(\ell_1, \cdots, \ell_m)}) \in \mathcal{S}\right\} \tag{3.37}$$

In order to avoid the necessity of discrete optimization (at least with respect to $n$, and often also with respect to some of the decision design parameters) and to further simplify the objective function by reducing the number of input parameters explicitly entering the objective function, we introduce generalized variables:

$$(x, y, z_1, \cdots, z_m) \in \mathbb{R}^{m+2} \tag{3.38}$$

defined by means of a suitably chosen transformation $W$ on $\mathcal{U}$:

$$\begin{pmatrix} x \\ y \\ z_1 \\ \cdots \\ z_m \end{pmatrix} = \begin{pmatrix} \frac{h}{E[\tau_0]} \\ w(n, \ell_1, \cdots, \ell_m) \\ w_1(n, \ell_1, \cdots, \ell_m) \\ \cdots \\ w_m(n, \ell_1, \cdots, \ell_m) \end{pmatrix} \tag{3.39}$$

with the inverse $W^{-1}$:

$$\begin{pmatrix} h \\ n \\ \ell_1 \\ \cdots \\ \ell_m \end{pmatrix} = \begin{pmatrix} xE[\tau_0] \\ s(y, z_1, \cdots, z_m) \\ s_1(y, z_1, \cdots, z_m) \\ \cdots \\ s_m(y, z_1, \cdots, z_m) \end{pmatrix} \tag{3.40}$$

Let

$$\mathcal{U}_w = W(\mathcal{U}) \quad \text{be the image set of } \mathcal{U} \text{ with respect to } W \tag{3.41}$$

and

$$\mathcal{U}_w^{(0)} \supset \mathcal{U}_w \quad \text{a suitably chosen open inclusive set of } \mathcal{U}_w \tag{3.42}$$

with the Jacobian determinant $\frac{\partial(xE[\tau_0],s,s_1,\cdots,s_m)}{\partial(x,y,z_1,\cdots,z_m)} \neq 0$ in $\mathcal{U}_w^{(0)}$.

The elements $u_w^{(0)} \in \mathcal{U}_w^{(0)}$ are called *generalized monitoring policies.* It should be noted that generally $W^{-1}(u_w^{(0)}) \notin \mathcal{U}$, i.e., generally the image of an $u_w^{(0)}$, is not a proper monitoring policy.

By means of (3.40) the error probabilities $\alpha_{(n,\ell_1,\cdots,\ell_m)}$ and $\beta_{(n,\ell_1,\cdots,\ell_m)}$ are defined on $\mathcal{U}_w$ as functions of the generalized variables, and denoted by $\alpha_{(y,z_1,\cdots,z_m)}$ and $\beta_{(y,z_1,\cdots,z_m)}$ for any $(y, z_1, \cdots, z_m) \in U_w$. If the extensions of $\alpha_{(y,z_1,\cdots,z_m)}$ and $\beta_{(y,z_1,\cdots,z_m)}$ on $\mathcal{U}_w^{(0)}$ exist, they are called *generalized error probabilities.* It is assumed:

**Assumption 3**

The *generalized error probabilities* $\alpha = \alpha_{(y,z_1,\cdots,z_m)}$ and $\beta = \beta_{(y,z_1,\cdots,z_m)}$

- exist, and
- are differentiable in $\mathcal{U}_w^{(0)}$ with respect to each of their arguments.

With (3.40) and Assumption 3, a third version of the objective function is defined on the open set $\mathcal{U}_w^{(0)}$ of generalized monitoring policies.

$$\Pi_3(x, y, z_1, \cdots, z_m) = \frac{1}{x}\left\{\frac{b\,(e^x - 1) - \alpha}{e^x - \beta} \cdot (1 - \beta) - as\right\} \tag{3.43}$$

with $\alpha = \alpha_{(y,z_1,\cdots,z_m)}$ , $\beta = \beta_{(y,z_1,\cdots,z_m)}$ and $s = s(y, z_1, \cdots, z_m)$.

Evidently, $\Pi_3$ coincides with $\Pi_2$ in each point. The latter is defined in the following sense:

$\Pi_3(x, y, z_1, \cdots, z_m) = \Pi_2(h, n, \gamma)$ for any pair $(x, y, , z_1, \cdots, z_m) \in \mathcal{U}_w$ and $(h, n, \ell_1, \cdots, \ell_m) \in \mathcal{U}$ with

$$\begin{aligned} x &= \frac{h}{E[\tau_0]}, \quad \text{and} \\ y &= s(n, \ell_1, \cdots, \ell_m), z_i = s_i(n, \ell_1, \cdots, \ell_m), i = 1, \cdots, m \end{aligned}$$

Therefore, $\Pi_3$ may be used to obtain an approximately $\Pi_2$-optimal control policy. But at first we have to define a procedure assigning a proper monitoring policy to each generalized monitoring policy.

For a given generalized monitoring policy $u_w^{(0)} \in \mathcal{U}_w^{(0)}$ define

$$\hat{u}_w = \arg\min_{u_w \in \mathcal{U}_w} ||u_w - u_w^{(0)}|| \tag{3.44}$$

where $|| \cdot ||$ denotes the Euclidean norm.

$\hat{u}_w$ is called *restricted generalized monitoring policy.* For any restricted generalized monitoring policy $\hat{u}_w$, we have

$$W^{-1}(\hat{u}_w) = u \in \mathcal{U} \tag{3.45}$$

thus $\hat{u}_w$ determines a proper monitoring policy.

**Definition**

A generalized monitoring policy ${}^*u_w^{(0)} \in \mathcal{U}_w^{(0)}$ is called $\Pi_3$-optimal if

$$\Pi_3\left({}^*u_w^{(0)}\right) \geq \Pi_3\left(u_w^{(0)}\right) \qquad \text{for any} \quad u_w^{(0)} \in \mathcal{U}_w^{(0)}$$

and

**Definition**

Let ${}^*u_w^{(0)}$ be a $\Pi_3$-optimal generalized monitoring policy and ${}^*\hat{u}_w$ the corresponding restricted generalized monitoring policy, then $(h_3^*, n_3^*, \gamma_3^*) \in \mathcal{S}$ determined by ${}^*u = W^{(-1)}\left({}^*\hat{u}_w\right)$ is called a $\Pi_3$-optimal monitoring policy.

According to Assumption 3, the generalized error probabilities $\alpha = \alpha_{y,z_1,\cdots,z_m}$ and $\beta = \beta_{y,z_1,\cdots,z_m}$ are differentiable in $\mathcal{U}_w^{(0)}$.

Thus, necessary conditions for a $\Pi_3$-optimal generalized monitoring policy $(x^*, y^*, z_1^*, \cdots, z_m^*) \in \mathcal{U}_w^{(0)}$ are given by:

$$\left.\frac{\partial \Pi_3(x, y, z_1, \cdots, z_m)}{\partial x}\right|_{x=x^*,y=y^*,z_1=z_1^*,\cdots,z_m=z_m^*} = 0, \qquad (3.46)$$

$$\left.\frac{\partial \Pi_3(x, y, z_1, \cdots, z_m)}{\partial y}\right|_{x=x^*,y=y^*,z_1=z_1^*,\cdots,z_m=z_m^*} = 0, \qquad (3.47)$$

$$\left.\frac{\partial \Pi_3(x, y, z_1, \cdots, z_m)}{\partial z_i}\right|_{x=x^*,y=y^*,z_1=z_1^*,\cdots,z_m=z_m^*} = 0 \qquad (3.48)$$

$$\text{for} \quad i = 1, \cdots, m$$

From (3.46), (3.47), and (3.48) the following system of equations for $x^*, y^*$ and $z_i^*, i = 1, \cdots, m$ is obtained:

$$b\frac{(e^x - 1)(e^x - \beta) - x\, e^x(1 - \beta)}{(e^x - \beta)} = as\frac{e^x - \beta}{1 - \beta} + \alpha\left(1 + \frac{x\, e^x}{e^x - \beta}\right) \qquad (3.49)$$

$$a(e^x - \beta)s_y + (1 - \beta)\alpha_y + \frac{b\,(e^x - 1)^2}{e^x - \beta}\beta_y - \alpha\frac{e^x - 1}{e^x - \beta}\beta_y = 0 \qquad (3.50)$$

$$a(e^x - \beta)s_{z_i} + (1 - \beta)\alpha_{z_i} + \frac{b\,(e^x - 1)^2}{e^x - \beta}\beta_{z_i} - \alpha\frac{e^x - 1}{e^x - \beta}\beta_{z_i} = 0 \qquad (3.51)$$

for $i = 1, \cdots, m$

where the abbreviation $u_v = \frac{\partial u}{\partial v}$ is used.

(3.49), (3.50) and (3.51) stand for $m + 2$ quite complicated equations, and solving them constitutes in general a difficult problem. Therefore, a further simplification would be desirable. In a first step (3.49) is utilized to eliminate the relative benefit $b$ in (3.50) and (3.51).

As can be verified by simple calculations (3.49) is equivalent to:

$$b\frac{(e^x-1)^2}{e^x-\beta} = \frac{(e^x-1)^2}{(e^x-1)(e^x-\beta)-xe^x(1-\beta)} \cdot \left[as\frac{e^x-\beta}{1-\beta}+\alpha\left(1+\frac{xe^x}{e^x-\beta}\right)\right] \tag{3.52}$$

Inserting (3.52) into (3.50) and (3.51) leads to a system of equations for $y^*$ and $z_i^*, i=1,\cdots,m$ which doesn't contain any more explicitly the relative benefit per renewal $b$.

$$\frac{(e^x-1)^2}{(e^x-1)(e^x-\beta)-xe^x(1-\beta)}\left[as\frac{e^x-\beta}{1-\beta}+\alpha\left(1+\frac{xe^x}{e^x-\beta}\right)\right]\beta_y + a(e^x-\beta)s_y+(1-\beta)\alpha_y-\alpha\frac{e^x-1}{e^x-\beta}\beta_y = 0 \tag{3.53}$$

$$\frac{(e^x-1)^2}{(e^x-1)(e^x-\beta)-xe^x(1-\beta)}\left[as\frac{e^x-\beta}{1-\beta}+\alpha\left(1+\frac{xe^x}{e^x-\beta}\right)\right]\beta_{z_i} + a(e^x-\beta)s_{z_i}+(1-\beta)\alpha_{z_i}-\alpha\frac{e^x-1}{e^x-\beta}\beta_{z_i} = 0 \tag{3.54}$$

for $i=1,\cdots,m$

As already noticed, $\frac{h^*}{E[\tau_0]}$ is a small quantity, and therefore so is $x^*$. Hence, we may let $x$ tend to zero in (3.53) and (3.54) and solve the limiting equations for an approximation.

The limits for $x \to 0$ of the left hand sides of (3.53) and (3.54), respectively, are readily obtained by *de l'Hospital's Rule* resulting in the following system of differential equations for approximate values $\tilde{y}^*$ and $\tilde{z}_i^*$ of $y^*$ and $z_i^*$, respectively:

$$(as_y+\alpha_y)(1-\beta)+\frac{2}{1+\beta}(as+\alpha)\beta_y = 0 \tag{3.55}$$

$$(as_{z_i} + \alpha_{z_i})(1-\beta) + \frac{2}{1+\beta}(as+\alpha)\,\beta_{z_i} = 0 \text{ for } i = 1, \cdots, m \quad (3.56)$$

Using (3.55) and (3.56) instead of (3.50) and (3.51) has two decisive advantages:

- The optimization with respect to $y$ and $z_i, i = 1, \cdots, m$, i.e., the determination of a $\Pi_3$-optimal generalized sampling plan $(y^*, z_1^*, \cdots, z_m^*)$, is performed independently of the variable $x$, i.e., the generalized sampling interval.
- (3.55) and (3.56) contain only one economic parameter, namely the relative sampling cost $a$.

**Definition**

Let $(\tilde{y}^*, \tilde{z}_1^*, \cdots, \tilde{z}_m^*)$ be a generalized sampling plan obtained as solution of (3.55) and (3.56), and let $(\hat{y}^*, \hat{z}_1^*, \cdots, \hat{z}_m^*)$ be the corresponding restricted generalized sampling plan, then

$$\begin{aligned} \hat{n}^* &= s(\hat{y}^*, \hat{z}_1^*, \cdots, \hat{z}_m^*) \\ \hat{\ell}_1^* &= s_1(\hat{y}^*, \hat{z}_1^*, \cdots, \hat{z}_m^*) \\ \cdot\cdot &\;\; \cdot\cdot \quad \cdots \\ \hat{\ell}_m^* &= s_m(\hat{y}^*, \hat{z}_1^*, \cdots, \hat{z}_m^*) \end{aligned} \quad (3.57)$$

$$(3.58)$$

is called an asymptotic $\Pi_3$-optimal sampling plan.

In this section it was assumed that the generalized error probabilities exist and are differentiable. For this case an explicit optimization algorithm and an important example are given in the following sections.

## 6.4 Differentiable Error Probabilities

Now we are in a position to formulate a relative simple algorithm for determining an approximately $\Pi^*$-monitoring policy for the case that

- $\mathcal{U} = \left\{(h, n, \ell_1, \cdots, \ell_m) | h \in \mathbb{R}^+, n \in \mathbb{N}, (\ell_1, \cdots, \ell_m) \in (\mathbb{R}^+)^m\right\}$, and

- the extensions of $\alpha_{(n,\ell_1,\cdots,\ell_m)}$ and $\beta_{(n,\ell_1,\cdots,\ell_m)}$ on $(I\!R^+)^{m+1}$ are differentiable with respect to each of their arguments.

Then, Assumption 3 is met by defining the generalized variables of the sampling plan in (3.39) by

$$
\begin{aligned}
y &= w(n,\ell_1,\cdots,\ell_m) = n \\
z_1 &= w_1(n,\ell_1,\cdots,\ell_m) = \ell_1 \\
&\cdots \\
z_m &= w_m(n,\ell_1,\cdots,\ell_m) = \ell_m
\end{aligned}
\tag{3.59}
$$

and setting

$$\mathcal{U}_w^{(0)} = (I\!R^+)^{m+2}$$

With (3.59) the following general algorithm is obtained:

**Algorithm A1**

Let $\gamma = \gamma_{(\ell_1,\cdots,\ell_m)}$ be a decision function with differentiable error probabilities $\alpha$ and $\beta$. Then an approximately $\Pi^*$-optimal monitoring policy $(\hat{h}^*, \hat{n}^*, \hat{\gamma}^*)$ for given

1. *economic key parameters* $a^*$, $e^*$ and $b^*$
2. mean time between shocks $E[\tau_0]$
3. distributional parameters completely describing $F_I$ and $F_{II}$

is obtained in 4 steps:

- **Step 1:** Calculate
  - the relative sampling cost $a = \frac{a^*}{e^*}$
  - the relative benefit per renewal $b = \frac{b^*}{e^*}$
- **Step 2:** Determine a solution $(\tilde{y}^*, \tilde{z}_1^*, \cdots, \tilde{z}_m^*)$ of the system of equations:

$$(a+\alpha_y)(1-\beta)+\frac{2}{1+\beta}(ay+\alpha)\,\beta_y=0 \qquad (3.60)$$

$$\alpha_{z_i}(1-\beta)+\frac{2}{1+\beta}(ay+\alpha)\,\beta_{z_i}=0 \qquad (3.61)$$

$$\text{for} \quad i=1,\cdots,m$$

- **Step 3:** Set

$$\begin{aligned}\hat{n}^* &= \text{closest positive integer to} \quad \tilde{y}^* \\ \hat{\ell}_i^* &= \tilde{z}_i^* \quad \text{for} \quad i=1,\cdots,m\end{aligned}$$

- **Step 4:** Determine the approximately optimal sampling interval:

$$\hat{h}^* = 2\,\frac{1-\hat{\beta}^*}{1+\hat{\beta}^*}\,\frac{E[\tau_0]}{\sqrt{\frac{2[b(1-\hat{\beta}^*)+\hat{\alpha}^*]}{(a\hat{n}^*+\hat{\alpha}^*)(1+\hat{\beta}^*)}-1}}$$

where $\hat{\alpha}^* = \alpha_{\hat{n}^*,\hat{\ell}_1^*,\cdots,\hat{\ell}_m^*}$ and $\hat{\beta}^* = \beta_{\hat{n}^*,\hat{\ell}_1^*,\cdots,\hat{\ell}_m^*}$.

## 6.5 The Economic Design of $\bar{x}$ - Charts

In the previous sections we have developed a rather general framework for determining the approximate economic design of monitoring policies. In this section we take the well-known Shewhart $\bar{x}$-charts as an example to illustrate the use of the methodology.

It is assumed that there is only one, normally distributed quality characteristic $X$ with constant variance $\sigma^2$. With $X_i$ being the quality characteristic of the $i$-th item produced, the production process can be modelled by the sequence:

$$\{X_i\}_{i=1,2,\ldots} \qquad \text{with} \quad X_i \sim N\left(\mu_i,\sigma^2\right) \qquad (3.62)$$

Assuming the two-sided case, the expectation of the quality characteristic $X$ may adopt three values, $\mu_0$, the target value, and $\mu_0 \pm \delta$, where $\delta > 0$ is called the shift parameter. The process starts on target, i.e., with expectation $\mu = \mu_0$. After the occurrence of an assignable cause, the expectation of the quality characteristic of the item produced is either $\mu_0 + \delta$ or $\mu_0 - \delta$.

The process is said to operate at time $t = \frac{i}{v}$ in

$$\begin{array}{lll} \text{State I} & \text{iff} & E[X_i] = \mu_0 \\ \text{State II} & \text{iff} & E[X_i] = \mu_0 \pm \delta\sigma \end{array} \tag{3.63}$$

The best independent decision function for this situation is based on the Gauss–test and will be denoted by $\gamma_G$. For given sample size $n \in \mathbb{N}$ the decision function $\gamma_G$ is completely determined by the alarm limit $k \in \mathbb{R}^+$, i.e., for $\gamma_G$ we have $m = 1$ and $\ell_1 = k$. Let $(X_{j-n+1}, \cdots, X_j)$ be the random sample at time $t = \frac{j}{v}$, then

$$\gamma_G(X_{j-n+1}, \cdots, X_j) = \begin{cases} 0 & \text{if } \left|\frac{\bar{X}-\mu_0}{\sigma}\sqrt{n}\right| < k \\ 1 & \text{otherwise.} \end{cases} \tag{3.64}$$

Following the usual notation, a monitoring policy with decision function $\gamma_G$ is given by the triple $(h, n, k)$. For given shift parameter $\delta$ the probabilities of wrong decisions when using $(h, n, k)$ are easily obtained:

- Probability of false alarm denoted by $\alpha$ (Type I error):

$$\alpha = 2\Phi(-k) \tag{3.65}$$

- Probability of not detecting a shift of size $\delta$ denoted by $\beta$ (Type II error):

$$\beta = \Phi\left(k - \delta\sqrt{n}\right) - \Phi\left(-k - \delta\sqrt{n}\right) \tag{3.66}$$

Inserting the error probabilities (3.65) and (3.66) into the objective function (3.35) yields:

$$\Pi_2(h,n,k) = \frac{E[\tau_0]}{h}\left\{\frac{b\left(e^{\frac{h}{E[\tau_0]}}-1\right)-2\Phi(-k)}{e^{\frac{h}{E[\tau_0]}}-\Phi\left(k-\delta\sqrt{n}\right)+\Phi\left(-k-\delta\sqrt{n}\right)}\cdot\right.$$
$$\left.\left[\left(\Phi\left(-k+\delta\sqrt{n}\right)+\Phi\left(-k-\delta\sqrt{n}\right)\right]-an\right\} \qquad (3.67)$$

The special form of error probabilities in the case of $\gamma_G$ suggest introducing the following generalized variables:

$$x = \frac{E[\tau_0]}{h}$$
$$y = \delta\sqrt{n}$$
$$z = k$$

resulting in the following objective function $\Pi_3$:

$$\Pi_3(x,y,z) = \frac{1}{x}\left\{\frac{b\left(e^{x}-1\right)-2\Phi(-z)}{e^{x}-\Phi(z-y)+\Phi(-z-y)}\cdot\right.$$
$$\left.\left(\Phi(-z+y)+\Phi(-z-y)\right)-a_0y^2\right\} \qquad (3.68)$$

where $a_0 = \frac{a}{\delta^2}$.

We have

- the relation between $\Pi_3$ and $\Pi_2$ is given by

$$\Pi_3(x,y,z) = \Pi_2(h,n,k) \quad \text{for any pair} \quad (x,y,z) \quad \text{and} \quad (h,n,k)$$
$$x = \frac{h}{E[\tau_0]},\ y = \delta\sqrt{n},\ z = k \qquad (3.69)$$

- $\Pi_3(x,y,z)$ is differentiable in $(\mathbb{R}^+)^3$ with respect to each of the three variables $x, y$ and $z$ .

Note that no distributional parameter at all (neither from $F_I(x)$, $F_{II}(x)$ nor from $F_0(t)$) enters explicitly $\Pi_3(x,y,z)$, but only the two parameters $b$ and $a_0$.

Let $(x^*, y^*, z^*)$ be a $\Pi_3$-optimal solution, then according to the previous section, an asymptotic $\Pi_3$-optimal sampling plan $(\tilde{y}^*, \tilde{z}^*)$ can be obtained as simultaneous solutions of the following two equations:

$$2a_0 y + \frac{2}{1+\beta}\left(a_0 y^2 + \alpha\right)\beta_y = 0 \tag{3.70}$$

$$(1-\beta)\,\alpha_z + \frac{2}{1+\beta}\left(a_0 y^2 + \alpha\right)\beta_z = 0 \tag{3.71}$$

with

$$\alpha = 2\Phi(-z), \qquad \beta = \Phi(z-y) - \Phi(-z-y) \quad \text{and}$$

$$\alpha_z = \frac{\partial \alpha}{\partial z} = -2\phi(z), \quad \beta_y = \frac{\partial \beta}{\partial y} = -\phi(z-y) + \phi(z+y)$$

$$\beta_z = \frac{\partial \beta}{\partial z} = \phi(z-y) + \phi(z+y)$$

where $\phi$ denotes the density function of the standardized normal distribution. Thus we obtain the following algorithm.

## *An Algorithm for Economic $\overline{x}$-Charts*

In the case of two-sided $\overline{x}$-charts for a normally distributed quality characteristic an approximately $\Pi^*$-optimal $\overline{x}$-chart may be obtained by Algorithm A1.1:

**Algorithm A1.1**

Let $\gamma_G$ be the decision function based on the two-sided Gauss-test. Then an approximately $\Pi^*$-optimal monitoring policy $(\hat{h}^*, \hat{n}^*, \hat{k}^*)$ for given

1. *economic key parameters* $a^*$, $e^*$ and $b^*$
2. mean time between shocks $E[\tau_0]$
3. shift parameter $\delta$

is obtained in 4 steps:

- **Step 1:** Set

$$a_0 = \frac{a^*}{e^* \delta^2}$$

$$b = \frac{b^*}{e^*}.$$

- **Step 2:** Determine a solution $(\hat{y}^*, \hat{z}^*)$ of the two equations (3.70) and (3.71).
- **Step 3:** Set

$$\hat{n}_1^* = \text{ nearest positive integer to } \left(\frac{\hat{y}^*}{\delta}\right)^2 \quad (3.72)$$

$$\hat{k}_1^* = \hat{z}^* \quad (3.73)$$

- **Step 4:** Determine the approximately optimal sampling interval:

$$\hat{h}_1^* = 2\,\frac{1-\hat{\beta}^*}{1+\hat{\beta}^*}\,\frac{E[\tau_0]}{\sqrt{\frac{2[b(1-\hat{\beta}^*)+\hat{\alpha}^*]}{(a\hat{n}_1^*+\hat{\alpha}^*)(1+\hat{\beta}^*)}} - 1)} \quad (3.74)$$

where $\hat{\alpha}^* = 2\Phi(-\hat{k}_1^*)$ and $\hat{\beta}^* = \Phi(\hat{k}_1^* - \delta\sqrt{\hat{n}_1^*}) - \Phi(-\hat{k}_1^* - \delta\sqrt{\hat{n}_1^*})$

Besides the *economic key parameters* it is necessary to know numerically the distributional parameters $\delta$, i.e., the shift parameter, and $E[\tau_0]$, i.e., the expected length of an in-control period.

## *Accuracy of the $\overline{X}$-Algorithm*

In Collani (1989), there are nomograms for determining the approximate economic design of $\overline{x}$-charts, which are based on the algorithms given in the preceding section and which are used here for demonstrating its accuracy.

| $\delta$ | $a$ | $b$ | $n^*$ | $k^*$ | $\hat{n}^*$ | $\hat{k}^*$ | $\delta$ | $a$ | $b$ | $n^*$ | $k^*$ | $\hat{n}^*$ | $\hat{k}^*$ |
|---|---|---|---|---|---|---|---|---|---|---|---|---|---|
| 0.5 | 0.0001 | 10 | 80 | 3.278 | 80 | 3.28 | 1.5 | 0.0001 | 10 | 12 | 3.872 | 12 | 3.87 |
| | | 50 | 80 | 3.283 | 80 | 3.28 | | | 50 | 12 | 3.873 | 12 | 3.87 |
| | | 500 | 80 | 3.285 | 80 | 3.28 | | | 500 | 12 | 3.874 | 12 | 3.87 |
| | 0.001 | 10 | 51 | 2.534 | 52 | 2.56 | | 0.001 | 10 | 9 | 3.261 | 9 | 3.26 |
| | | 50 | 52 | 2.555 | 52 | 2.56 | | | 50 | 9 | 3.266 | 9 | 3.26 |
| | | 500 | 53 | 2.57 | 52 | 2.56 | | | 500 | 9 | 3.268 | 9 | 3.26 |
| | 0.01 | 10 | 22 | 1.525 | 25 | 1.65 | | 0.01 | 10 | 6 | 2.527 | 6 | 2.54 |
| | | 50 | 24 | 1.605 | 25 | 1.65 | | | 50 | 6 | 2.542 | 6 | 2.54 |
| | | 500 | 25 | 1.644 | 25 | 1.65 | | | 500 | 6 | 2.550 | 6 | 2.54 |
| 1.0 | 0.0001 | 10 | 24 | 3.654 | 24 | 3.66 | 2.0 | 0.0001 | 10 | 7 | 3.999 | 7 | 4.00 |
| | | 50 | 24 | 3.656 | 24 | 3.66 | | | 50 | 7 | 4.000 | 7 | 4.00 |
| | | 500 | 24 | 3.657 | 24 | 3.66 | | | 500 | 7 | 4.000 | 7 | 4.00 |
| | 0.001 | 10 | 17 | 2.999 | 17 | 3.01 | | 0.001 | 10 | 5 | 3.381 | 5 | 3.40 |
| | | 50 | 17 | 3.005 | 17 | 3.01 | | | 50 | 5 | 3.384 | 5 | 3.40 |
| | | 500 | 17 | 3.009 | 17 | 3.01 | | | 500 | 5 | 3.386 | 5 | 3.40 |
| | 0.01 | 10 | 10 | 2.183 | 10 | 2.23 | | 0.01 | 10 | 4 | 2.750 | 4 | 2.73 |
| | | 50 | 10 | 2.208 | 10 | 2.23 | | | 50 | 4 | 2.761 | 4 | 2.73 |
| | | 500 | 10 | 2.221 | 10 | 2.23 | | | 500 | 4 | 2.766 | 4 | 2.73 |

**Table 1** Exact and approximate economic sampling plans for $\bar{x}$-charts.

In Table 1 the exact and the approximate designs of sampling plans for an $\bar{x}$-chart are listed covering a wide range of the relevant input parameters. It can be seen that from the viewpoint of industrial practice there is no difference between the approximate and the exact economic design.

## *Weigand's Solution*

The determination of $(\hat{h}_1^*, \hat{n}_1^*, \hat{k}_1^*)$ includes the solution of a system of differential equations, and therefore a computer is necessary. Of course, it would be desirable to have a simpler solution algorithm which could be used on shop floor level without a PC. The following approximation, due to Weigand (1992), is based on a graphical solution of (3.70) and (3.71) given in Collani (1989). The resulting approximately optimal monitoring policy is denoted by $(\hat{h}_w^*, \hat{n}_w^*, \hat{k}_w^*)$ and shall illustrate the possibility of developing closed form solutions:

**Weigand's Algorithm:**

Let $\gamma_G$ be the decision function based on the two-sided Gauss-test. Then an approximately $\Pi^*$-optimal monitoring policy $(\hat{h}_w^*, \hat{n}_w^*, \hat{k}_w^*)$ for given *economic key parameters* $a^*$, $e^*$ and $b^*$ and distributional parameters $\delta$ and $E[\tau_0]$ is obtained in 5 steps:

- Step 1: Set

$$a_0 = \frac{a^*}{e^* \delta^2}$$

$$b = \frac{b^*}{e^*}.$$

- Step 2: If $a_0 < 0.1$ and $b > 1$, then calculate

$$y = 0.019 + \sqrt{-2.93722 \cdot ln(a_0) - 3.2255}$$

Set

$$\hat{n}_w^* \quad = \quad \text{closest positive integer to} \quad \left(\frac{y}{\delta}\right)^2$$

- Step 3: Calculate

$$z = -0.9 + \sqrt{0.798442 \cdot \delta^2 \hat{n}_w^* - 0.03034 \cdot \delta \sqrt{\hat{n}_w^*} + 1.57566}$$

Set

$$\hat{k}_w^* \quad = \quad z$$

- Step 4: Calculate

$$\alpha = 2\Phi(-\hat{k}_w^*)$$
$$\beta = \Phi(\hat{k}_w^* - \delta\sqrt{\hat{n}_w^*}) - \Phi(-\hat{k}_w^* - \delta\sqrt{\hat{n}_w^*})$$

- Step 5: Set

$$\hat{h}_w^* = E[\tau] \cdot \frac{\alpha + a_0\delta^2\hat{n}_w^* + \sqrt{(\alpha + a_0\delta^2\hat{n}_w^*)\frac{b(1-\beta)+\alpha}{1-0.5(1-\beta)}}}{b + 0.5\alpha - a_0\delta^2\hat{n}_w^*\left(\frac{1}{1-\beta} - 0.5\right)}$$

The expression for $\hat{h}_w^*$ which is not equal to (3.32) illustrates the fact that there are a multitude of different expressions for an approximately optimal sampling interval derived in literature.

## 6.6 Non-Differentiable Error Probabilities

In this section the results obtained so far are extended to the case that the derivatives of the generalized error probabilities do not exist or are too cumbersome to work with.

### Example

*np* - control chart

For illustrating the difficulties which may arise take the so-called *P*-chart used in the case that there is not a measurable quality characteristic but each item produced can only be classified either in conforming, $X = 0$, or nonconforming, $X = 1$. In this case we have:

- process distribution: $X \sim \mathrm{Bi}(1, p)$
- process states:
  State I: $E[X] = p_I$
  State II: $E[X] = p_{II}$
  with $0 \le p_I < p_{II} \le 1$
- test statistic: $TS = \sum_{i=1}^{n} X_i$
- decision function:

$$\gamma = \begin{cases} 0 & TS \le c \\ 1 & TS > c \end{cases}$$

  with $0 \le c < n$, where $n$ is the sample size and $c$ is called acceptance number.
- error probabilities:

$$\alpha_{n,c} = \mathrm{P(alarm|State\ I)} = 1 - \sum_{m=0}^{c} \binom{n}{m} p_I^m (1 - p_I)^{n-m}$$

$$\beta_{n,c} = \mathrm{P(no\ alarm|State\ II)} = \sum_{m=0}^{c} \binom{n}{m} p_{II}^m (1 - p_{II})^{n-m}$$

Here we have $n \in \mathbb{N}$ and $c \in \mathbb{N}_0$ and the extension of the error probabilities to an open inclusive set of $\mathcal{U}$ do not exist without more ado. One could, of course, utilize the relation between the Binomial and the F-distribution, leading to the following representation of the error probabilities:

$$\alpha_{n,c} = \int_0^{\frac{n-c}{c+1}\frac{p_I}{1-p_I}} f_{2(c+1),2(n-c)}(y)dy \tag{3.75}$$

$$\beta_{n,c} = 1 - \int_0^{\frac{n-c}{c+1}\frac{p_{II}}{1-p_{II}}} f_{2(c+1),2(n-c)}(y)dy \tag{3.76}$$

with

$$f_{\nu_1,\nu_2}(y) = \frac{\Gamma\left(\frac{\nu_1+\nu_2}{2}\right)}{\Gamma\left(\frac{\nu_1}{2}\right)\Gamma\left(\frac{\nu_2}{2}\right)} \nu_1^{\frac{\nu_1}{2}} \nu_2^{\frac{\nu_2}{2}} y^{\frac{\nu_1}{2}-1}(\nu_1 + \nu_2 y)^{-\frac{\nu_1+\nu_2}{2}} \quad \text{for} \quad y \geq 0$$

With (3.75) and (3.76) it is in principle possible to define differentiable generalized error probabilities, but the problem gets rather complicated and a simpler method would be desirable.

For arriving at a simpler and more general method, the system of differential equations (3.55), (3.56) is used to derive an approximate objective function for the sample size and the decision function, which is also valid for the general case.

It is easily verified that the system of differential equations (3.55) and (3.56) is equivalent to

$$\frac{\frac{\partial}{\partial y}(as+\alpha)}{as+\alpha} + \frac{\frac{\partial}{\partial y}\left(\frac{2}{1-\beta}-1\right)}{\frac{2}{1-\beta}-1} = 0 \tag{3.77}$$

$$\frac{\frac{\partial}{\partial z_i}(as+\alpha)}{as+\alpha} + \frac{\frac{\partial}{\partial z_i}\left(\frac{2}{1-\beta}-1\right)}{\frac{2}{1-\beta}-1} = 0 \quad \text{for} \quad i = 1, \cdots, m \tag{3.78}$$

By integrating (3.77) and (3.78), it is readily seen that solving the system (3.55) and (3.56) for $\tilde{y}^*$ and $\tilde{z}_i^*, i = 1, \cdots, m$, respectively, is equivalent to finding a relative minimum $\tilde{y}^*$ and $\tilde{z}_i^*, i = 1, \cdots, m$, of the function $\tilde{L}(y, z_1, \cdots, z_m)$:

$$\tilde{L}(y, z_1, \cdots, z_m) = \left(\frac{1}{1-\beta} - \frac{1}{2}\right) \cdot (as + \alpha) \tag{3.79}$$

with

$$\alpha = \alpha(y, z_1, \cdots, z_m), \qquad \beta = \beta(y, z_1, \cdots, z_m) \qquad s = s(y, z_1, \cdots, z_m)$$

Restricting the search on the domain $\mathcal{U}_w$ instead of $\mathcal{U}_w^{(0)}$ and by applying the transformation (3.39), the function $L$ in the original design parameters $(n, \ell_1, \cdots, \ell_m)$ is obtained:

$$L(n, \ell_1, \cdots, \ell_m) = \left(\frac{1}{1-\beta} - \frac{1}{2}\right) \cdot (an + \alpha) \tag{3.80}$$

with $\alpha = \alpha_{(n,\ell_1,\cdots,\ell_m)}$ and $\beta = \beta_{(n,\ell_1,\cdots,\ell_m)}$ being the probabilities of an error of Type I and Type II, respectively. The loss function (3.80) was already derived in Hryniewicz (1988) in a slightly different setting.

The interpretation of (3.80) is straightforward:

- The first factor gives the average run length in STATE II corrected by $-\frac{1}{2}$ - see (3.30) - which can be looked upon as the average time per renewal cycle operating out-of-control, which is proportional to the corresponding loss.
- The second factor is the average relative cost for one monitoring action when operating in-control, i.e., the sum of costs for sampling and false alarm measured in units of the average cost caused by a false alarm $e^*$.
- Consequently, $L$ will be called in the remainder a *loss function.*

Note that (3.80) reflects the interesting fact that the optimal economic sample size and the optimal decision function are more or less independent not only of the average benefit derived from an intervention when operating out-of-control but also of the average length of an in-control period. They are determined rather by the monitoring costs given by $an$ and $E[A_F]$, and the time of being out-of-control determined by $E[A_{II}]$. A sampling plan is optimal, if it balances these two countermoving expenses in such a way that their product becomes minimal.

**Example**

Back to the example of a $np$-control chart. From (3.80) we immediately obtain the loss function:

$$L(n,c) = \left( \frac{1}{\sum_{m=c+1}^{n} \binom{n}{m} p_{II}^m (1-p_{II})^{n-m}} - \frac{1}{2} \right) \cdot \left( an + \sum_{m=c+1}^{n} \binom{n}{m} p_I^m (1-p_I)^{n-m} \right) \quad (3.81)$$

As generally the optimal sample size $n^*$ is not so large, minimization could be performed by means of a straightforward search algorithm leading to an approximate $\Pi^*$-optimal sampling plan $(\hat{n}^*, \hat{c}^*)$.

Once $(\hat{n}^*, \hat{c}^*)$ is available, an approximate $\Pi^*$-optimal sampling interval is obtained by means of (3.32).

# 7 EXTENSIONS

The results obtained in the previous section are based on the restrictions of

- a fixed sample size $n$
- an independent decision function $\gamma$

During the last decade discussion in SPC has been centered on dependent decision functions, e.g., CUSUM (cumulative sum) or EWMA (exponentially weighted moving average) procedures, rather than on independent decisions. Moreover, in case of expensive sampling a fixed sample size $n$ may not be appropriate, but a sequential sampling procedure may be. Thus, it would be desirable to generalize our results.

In the general case there is no fixed sample size but the sample size is a random variable depending on the decision function $\gamma$. This is expressed by denoting the sample size by $\mathcal{N}_\gamma$ with expectations
$E[\mathcal{N}_\gamma|State\ I] = n_I$ and $E[\mathcal{N}_\gamma|State\ II] = n_{II}$.

For the case of a fixed sample size, $\mathcal{N}_\gamma$ is degenerate and we have:

$$n = n_I = n_{II} = E[\mathcal{N}_\gamma] \tag{3.82}$$

For independent decision functions we have according to (3.27) and (3.28):

$$\alpha = \frac{E[A_F]}{E[A_I]} \tag{3.83}$$

$$\frac{1}{1-\beta} = E[A_{II}] \tag{3.84}$$

Inserting (3.82), (3.83) and (3.84) into (3.80) yields:

$$L(n_I, \ell_1, \cdots, \ell_m) = \left(E[A_{II}] - \frac{1}{2}\right)\left(an_I + \frac{E[A_F]}{E[A_I]}\right) \tag{3.85}$$

With (3.85) we have derived an objective function for determining the economic design for general sampling plans. Although it is based on a full economic model, there is only one economic input parameter, namely the relative sampling cost, entering explicitly the objective function. This single economic parameter includes the cost of sampling and the cost of a false alarm and intuitively it is clear that knowledge about these two quantities is a minimum requirement for adjusting the sampling plan to the economic environment.

Next, the proceeding in the general case is illustrated by means of two examples: the first refers to the case of dependent decisions and the second to the case where the sample size is a random variable.

## 7.1 CUSUM Procedures

Let $\Gamma_C$ denote the set of admissible CUSUM decision functions, where each $\gamma_C \in \Gamma$ is determined by the $m$ design parameters $(\ell_1, \cdots, \ell_m)$.

CUSUM procedures lead to dependent decision functions and are based on a fixed sample size, $n$. According to Taylor (1968), the required expectations in (3.85) may be approximated by:

$$E[A_F] \approx \frac{E[A_I]}{ARL(I)} \tag{3.86}$$

$$E[A_{II}] \approx ARL(II) \tag{3.87}$$

Inserting (3.86) and (3.87) in (3.85) yields:

$$L(n, \ell_1, \cdots, \ell_m) = \left(ARL(II) - \frac{1}{2}\right)\left(an + \frac{1}{ARL(I)}\right) \tag{3.88}$$

with $ARL(I) = ARL(I)_{(n,\ell_1,\cdots,\ell_m)}$ and $ARL(II)_{(n,\ell_1,\cdots,\ell_m)}$

Let $(\hat{n}_C^*, \hat{\gamma}_C^*)$ be an approximate $\Pi^*$-optimal sampling plan, obtained by minimizing (3.88), and $ARL^*(I)$, $ARL^*(II)$ and $L(\hat{n}_C^*, \hat{\gamma}_C^*)$, the corresponding values of the average run lengths and the loss function, respectively.

Then an approximate $\Pi^*$-optimal sampling interval is obtained by (3.32), (3.86) and (3.87):

$$\hat{h}_C^* = \frac{1}{ARL^*(II) - \frac{1}{2}} \frac{E[\tau_0]}{\sqrt{\frac{b + \frac{ARL^*(II)}{ARL^*(I)}}{L(\hat{n}_C^*, \hat{\gamma}_C^*)} - 1}} \tag{3.89}$$

## 7.2 Sequential Procedures

Let $\Gamma_S$ denote the set of admissible sequential decision functions, and $(n_I, n_{II}, \gamma_S)$ denote a corresponding sequential sampling plan, with

- $n_I = E[\mathcal{N}_{\gamma_S} | State\ I]$
- $n_{II} = E[\mathcal{N}_{\gamma_S} | State\ II]$

For $\gamma_S \in \Gamma_S$ the error probabilities $\alpha$ and $\beta$ exist, and therefore the only change in (3.85) refers to the sample size when operating in State I:

$$L(n_I, \gamma_S) = \left(\frac{1}{1-\beta} - \frac{1}{2}\right)\left(a \cdot n_I + \alpha\right) \tag{3.90}$$

Again, by minimizing (3.90) an approximate $\Pi^*$-optimal sequential sampling plan $(\hat{n}_I^*, \hat{n}_{II}^*, \hat{\gamma}_S^*)$ is obtained. For a special case it is shown in Hryniewicz (1989), that the approximation is very good.

Unfortunately, formula (3.32) for determining an approximately optimal sampling interval derived earlier cannot be used right away in the case where the sample size is a random variable with expectation depending on the actual state of the process. For deriving an analogous formula, the objective function $\Pi^*$ has to be modified taking into account the sequential procedure.

By means of Wald's identity we obtain the expected sample size per renewal cycle in State I and State II, respectively:

$$E[\sum_{i=1}^{A_I} \mathcal{N}_{\gamma s} | State\ I] = E[\mathcal{N}_{\gamma s} | State\ I] \cdot E[A_I] = n_I E[A_I] \tag{3.91}$$

and

$$E[\sum_{i=1}^{A_{II}} \mathcal{N}_{\gamma s} | State\ II] = E[\mathcal{N}_{\gamma s} | State\ II] \cdot E[A_{II}] = n_{II} \cdot E[A_{II}] \tag{3.92}$$

With (3.91) and (3.92) the long run profit per item $\Pi^*$ is obtained:

$$\Pi^*(h, n_I, n_{II}, \gamma_S) = \frac{b^* - e^* E[A_F] - a^* \Big( n_I E[A_I] + n_{II} E[A_{II}] \Big)}{E[A_I + A_{II}] h v} + g_{II} \tag{3.93}$$

or in its transformed version:

$$\begin{aligned} \Pi(h, n_I, n_{II}, \gamma_S) &= \frac{\Big(\Pi^*(h, n_I, n_{II} \gamma_S) - g_{II}\Big) v E[\tau_0]}{e^*} \qquad (3.94) \\ &= \frac{E[\tau_0]}{h} \left\{ \frac{b - E[A_F]}{E[A_I + A_{II}]} - a n_I - \frac{a(n_{II} - n_I) E[A_{II}]}{E[A_I + A_{II}]} \right\} \end{aligned}$$

and finally the first simplification:

$$\Pi_1(h, n_I, n_{II}, \gamma_S) =$$

$$\frac{E[\tau_0]}{h} \left\{ \frac{\left(b + \frac{\alpha}{2}\right) - \alpha \cdot \frac{E[\tau_0]}{h}}{\frac{E[\tau_0]}{h} + \left(\frac{1}{1-\beta} - \frac{1}{2}\right)} - a n_I - \frac{a(n_{II} - n_I) \frac{1}{1-\beta}}{\frac{E[\tau_0]}{h} + \left(\frac{1}{1-\beta} - \frac{1}{2}\right)} \right\} \tag{3.95}$$

An approximate $\Pi^*$-optimal sampling interval denoted by $\hat{h}_S^*$ in the case where a sequential sampling plan $(\hat{n}_I^*, \hat{n}_{II}^*, \hat{\gamma}_S^*)$ is used is obtained from (3.95) by straightforward calculation:

$$\hat{h}_S^* = 2\,\frac{1-\hat{\beta}^*}{1+\hat{\beta}^*}\,\frac{E[\tau_0]}{\sqrt{\frac{b+\frac{\hat{\alpha}^*+a(n_I-n_{II})}{1-\beta}}{L(\hat{n}_I^*,\hat{n}_{II}^*,\hat{\gamma}_S^*)}-1}} \tag{3.96}$$

where $\hat{\alpha}^*$, $\hat{\beta}^*$ and $L(\hat{n}_I^*, \hat{n}_{II}^*, \hat{\gamma}_S^*)$ are the error probabilities and the value of the loss function, respectively, for the approximately $\Pi^*$-optimal sequential sampling plan.

## 7.3 Perspective

The methodology developed here could form a basis for a new evaluation and a new evolution of the *economic approach* in Statistical Process Control.

A comparison with the *statistical approach* reveals the following features:

- **Sampling Plan**

  1. The assumptions about the underlying process distributions are identical for either of the approaches.
  2. The *statistical approach* requires additionally specification of certain values with respect to error probabilities or average run lengths without giving sufficiently justified hints how to do it. Moreover, error probabilities and average run lengths are statistical concepts and, therefore, decisions about them should be reserved to experts.
  3. The *economic approach* requires specification of two economic parameters, namely the cost for sampling and the cost of a false alarm. Determination of either of them is an inherent part of the job of practitioners operating a production process.

- **Sampling Interval**

  1. Generally, the *statistical approach* does not include the possibility of determining in a rational way the sampling interval.

2. Determination of the most appropriate sampling interval is part of the economic methodology. For adjusting the sampling interval to a given situation, the average frequency of disturbances and the average benefit of an intervention must be known, at least approximately.

In summary, the preconditions to be met for the *economic approach* seem to be easier and more appropriate from a practical point of view than those in the case of the *statistical approach*. Additionally, the *economic approach* is more adapted to the situation at hand and yields more results.

However, for a successful introduction of the *economic approach* to industrial practice it must be presented in a user-friendly and industry-conforming way. Thus, standard-like economic procedures for Statistical Process Control should be developed and this goal may be achieved based on the models and methodology introduced here.

## Acknowledgements

This research was supported by Grant Be 1338/2-1 from the Deutsche Forschungsgemeinschaft (DFG) and by Grant CP93: 12074 from the European Communities. Moreover I am very much indebted to my colleague Dr. Vladimir Dragalin for many valuable discussions which led to substantial improvements of the original manuscript.

## REFERENCES

[1] Arnold, B. F., "Optimal Control Charts and Discrimination Between 'Acceptable' and 'Unacceptable' States," *Sankhya 51*, Series B, pp 375-389, 1989.

[2] Arnold, B. F., "An Economic $\bar{x}$-Chart Approach to the Joint Control of the Means of Independent Quality Characteristics," *ZOR - Methods and Models of Operations Research* 34, pp 59-74, 1990.

[3] Arnold, B. F. and v. E. Collani, "Economic Process Control," *Statistica Neerlandica* 41, pp 89-97, 1987.

[4] Arnold, B. F. and v. E. Collani, "On the Robustness of $\bar{x}$-Charts," *Statistics* 20, pp 149-159, 1989.

[5] Banerjee, P. K. and M. A. Rahim, "The Economic Design of Control Charts: A Renewal Theory Approach," *Engineering Optimization* 12, pp 63-73, 1987.

[6] Banerjee, P. K. and M. A. Rahim, "Economic Design of $\bar{x}$ Control Charts Under Weibull Shock Models," *Technometrics* 30, pp 407-414, 1988.

[7] Chung, K. J., "A Simplified Procedure for the Economic Design of $\bar{x}$ Charts," *International Journal of Production Research* 28, pp 1239-1246, 1990.

[8] Chung, K. J., "An Efficient Procedure for the Economic Design of np-Charts," *International Journal of Quality and Reliability Management* 9, pp 58-68, 1992.

[9] Chung, K. J., "An Economic Study of $\bar{x}$ Charts with Warning Limits," *Computers in Industrial Engineering* 24, pp 1-7, 1993.

[10] Chung, K. J., "An Algorithm for Computing the Economically Optimal $\bar{x}$-Control Chart for a Process with Multiple Assignable Causes," *European Journal of Operational Research* 72, pp 350-363, 1994.

[11] Chung, K. J. and C.-N. Lin, "The Economic Design of Dynamix $\bar{x}$ - Control Charts Under Weibull Shock-Model," *International Journal of Quality an Reliability Management* 10, pp 41-56, 1993.

[12] v. Collani, E., "Kostenoptimale Prüfpläne für die laufende Kontrolle eines normalverteilten Merkmals," *Metrika* 28, pp 211-236, 1981.

[13] v. Collani, E., "A Simple Procedure to Determine the Economic Design of an $\bar{x}$ Control Chart," *Journal of Quality Technology* 18, pp 145 - 151, 1986.

[14] v. Collani, E., "Economic Control of Continuously Monitored Production Processes," *Rep. Stat. Appl. Res., JUSE*, 34, pp 1-18, 1987.

[15] v. Collani, E., "A Unified Approach to Optimal Process Control," *Metrika* 35, pp 145-159, 1988.

[16] v. Collani, E., "The Economic Design of Control Charts", *Stuttgart: Teubner*, 171 pages, 1989.

[17] v. Collani, E., "Economically Optimal *c*- and *np*-Control Charts," *Metrika*36, pp 215-232, 1989.

[18] v. Collani, E. and V. Dragalin, "Determination of the Economic Design of CUSUM Charts for Controlling a Normally Distributed Process Mean," *Technical Reports of the Würzburg Research Group on Quality Control*, No. 69, 1966.

[19] v. Collani, E. and K. Dräger, "A Simplified Economic Design of Control Charts for Monitoring the Nonconforming Probability," *Economic Quality Control* 10, pp 231-292, 1995.

[20] v. Collani, E., E.M. Saniga, Ch. and Weigand, "Economic Adjustment Designs for $\bar{x}$ Control Charts," *IIE Transactions* 26, pp 37-43, 1994.

[21] v. Collani, E. and J. Sheil, "An Approach to Controlling Process Variability," *Journal of Quality Technology* 21, pp 87-96, 1989.

[22] v. Collani, E. and J. Treml, "Control of a Two-Dimensional Process-Quality-Indicator by Means of a Screening Procedure," *Economic Quality Control* 8, pp 167-194, 1993.

[23] v. Collani, E. and Ch. Weigand, "Economic Machine Adjustment in the Case of Product Screening," *Statistical Papers* 33, pp 171-184, 1992.

[24] Costa, A. F. B., "Joint Economic Design of $\bar{x}$ and $R$ Control Charts for Processes Subject to Two Independent Assignable Causes," *IIE Transactions* 25, pp 27-33, 1993.

[25] Del Castillo, E. and D. C. Montgomery, "Optimal Design of Control Charts for Monitoring Short Production Runs," *Economic Quality Control* 8, pp 225-240, 1993.

[26] Duncan, A. J., "The Economic Design of $\bar{x}$ Control Charts Used to Maintain Current Control of a Process," *Journal of the American Statistical Association* 51, pp 228-242, 1956.

[27] Frahm, P., "$\bar{x}$ - Control Charts and Age Replacement Policies," *Economic Quality Control* 7, pp 85-96, 1992.

[28] Frahm, P., *Alterserneuerung und Blockerneuerung unter Einbeziehung von Stichprobenkontrollen,* Dr.-Thesis, Würzburg, 152 pages, 1994.

[29] Ho, C. and K.E. Case, "Economic Design of Control Charts: A Literature Review for 1981 - 1991," *Journal of Quality Technology* 26, pp 39-53, 1994.

[30] Hryniewicz, O., "The Economic Design of a Certain Class of Control Charts: A General Approach," *Technical Reports of the Würzburg Research Group on Quality Control*, No. 11, 1988.

[31] Hryniewicz, O., "A Simple and Generally Applicable Approximation Technique for the Determination of the Economic Design of Control Charts," *Technical Reports of the Würzburg Research Group on Quality Control*, No. 15, 1988.

[32] Hryniewicz, O., "Economic Design of Attribute Control Charts Based on Double Sampling Plans," *Technical Reports of the Würzburg Research Group on Quality Control*, No. 17, 1989.

[33] Hryniewicz, O., "The Performance of Differently Designed $p$-Control Charts in the Presence of Shifts of Unexpected Size," *Economic Quality Control* 4, pp 7-18, 1989.

[34] Keats, J. B. and J. R. Simpson, "Comparison of $\bar{x}$ and the CUSUM Control Charts in an Economic Model," *Economic Quality Control* 9, pp 203-220, 1994.

[35] Kurc, K., "The Performance of Differently Designed $\bar{x}$ Control Charts in the Presence of a Shift of Unexpected Size," *Economic Quality Control* 6, pp 3-15, 1991.

[36] Lorenzen, T. J. and L.C. Vance, "The Economic Design of Control Charts: A Unified Approach," *Technometrics* 28, pp 3-10, 1986.

[37] McWilliams, T. P., "Economic Control Chart Designs and the In-control Time Distribution: A Sensitivity Analysis," *Journal of Quality Technology* 21, pp 103 - 110, 1989.

[38] McWilliams, T. P., "Economic, Statistical, and Economic-Statistical $\bar{x}$ Chart Designs," *Journal of Quality Technology* 26, pp 227-238, 1994.

[39] Montgomery, D. C., "The Economic Design of Control Charts: A Review and Literature Survey," *Journal of Quality Technology* 12, pp 75-87, 1980.

[40] Montgomery, D. C., "The Economic Design of an $\bar{x}$ Control Chart," *Journal of Quality Technology* 14, pp 40-43, 1982.

[41] Montgomery, D.C., *Introduction to Statistical Quality Control,* 2nd ed, New York: Wiley, 1991.

[42] Montgomery, D.C., "The Use of Statistical Process Control and Design of Experiments in Product and Process Improvement," *IIE Transactions* 24, pp 4-17, 1992.

[43] Montgomery, D.C., J.C.C. Torng, J.K. Cochran, and F.P. Lawrence, "Statistically Constrained Economic Design of the EMWA Control Chart," *Journal of Quality Technology* 27, pp 250-256, 1995.

[44] Moskowitz, H. and R. Plante, "Effect of Quality Loss Function on the Economic Design of $\bar{x}$ Process Control Charts," *European Journal of Operational Research* 72, pp 333 - 349, 1994.

[45] Mukherjee, S. P. and A. Mandal, "Economic Design of $\bar{x}$-Chart Revisited", *Calcutta Statist, Bulletin* 43, pp 109 - 118, 1993.

[46] Nandi, S. N., "Economic Design of np-Control Charts When There is a Multiplicity of Assignable Causes," *IAPQR Transactions* 17, pp 1 - 9, 1992.

[47] Rahim, M. A., "Economically Optimal Determination of the Parameters of $\bar{x}$-Charts with Warning Limits When Quality Characteristics Are Non-Normally Distributed," *Engineering Optimization* 7, pp 289 - 301, 1984.

[48] Rahim, M. A., "Economic Model of $\bar{x}$ Chart Under Non-Normality and Measurement Errors," *Computers and Operations Research* 12, pp 291 - 299, 1985.

[49] Rahim, M. A., "Determination of Optimal Design Parameters of Joint $\bar{x}$ and $R$ Charts," *Journal of Quality Technology* 21, pp 65 - 70, 1989.

[50] Rahim, M. A., "Economic Design of $\bar{x}$ Control Charts Assuming Weibull In-Control Times," *Journal of Quality Technology* 25, pp 296-305, 1993.

[51] Rahim, M. A. and P. K. Banerjee, "A Generalized Model for the Economic Design of $\bar{x}$-Control Charts for Production Systems with Increasing Failure Rate and Early Replacement," *Naval Research Logistics* 40, pp 787-809, 1993.

[52] Rahim, M. A., R. S. Lashkari, and P. K. Banerjee, "Joint Economic Design of Mean and Variance Control Charts," *Engineering Optimization* 14, pp 65-78, 1988.

[53] Roy, S. D. and S. Kakoty, "An Economic Design of CUSUM Control Chart for Controlling the Mean of a Normal Distribution," *IAPQR Transactions* 18, pp 29-42, 1993.

[54] Saniga, E. M., "Joint Economically Optimal Design of $\bar{x}$ and R Control Charts," *Management Science* 24, pp 420-431, 1977.

[55] Saniga, E. M., "Joint Economic Design of $\bar{x}$ and R Control Charts with Alternative Process Models," *AIIE Transactions* 11, pp 254-260, 1979.

[56] Saniga, E. M., "Heuristic Economical Design for $\bar{x}$ and $R$ Control Charts," *Frontiers in Statistical Quality Control* 3, pp 220-228, 1987.

[57] Saniga, E. M., "Economic Statistical Control-Chart Designs with an Application to $\bar{x}$ and $R$ Charts," *Technometrics* 31, pp 313-320, 1989.

[58] Saniga, E.M., D.J. Davis, and T.P. McWilliams, "Economic, Statistical, and Economic-Statistical Design of Attribute Charts," *Journal of Quality Technology* 27, pp 56-73, 1995.

[59] Saniga, E. M. and D. C. Montgomery, "Economically Quality Control Policies for a Single Cause System," *AIIE Transactions* 13, pp 258-264, 1981.

[60] Saniga, E. M. and T. P. McWilliams, "Economic, Statistical, and Economic-Statistical Design of Attribute Charts," *Journal of Quality Technology* 27, pp 56 - 73, 1995.

[61] Svoboda, L., "Economic Design of Control Charts: A Review and Literature Survey (1979-1989)," *In: Statistical Process Control in Manufacturing.* Eds. J.B. Keats and D.C. Montgomery. New York: Marcel Dekker, 1991.

[62] Tagaras, G., "Economic $\bar{x}$ Charts with Asymmetric Control Limits," *Journal of Quality Technology* 21, pp 147-154, 1989.

[63] Tagaras, G., "Power Approximation in the Economic Design of Control Charts," *Naval Research Logistic Quarterly* 36, pp 639-654, 1989.

[64] Tagaras, G. and H. L. Lee, "Economic Design of Control Charts with Different Control Limits for Different Assignable Causes," *Management Science* 34, pp 1347-1366, 1988.

[65] Taylor, H. M., "The Economic Design of Cumulative Sum Control Charts," *Technometrics* 10, pp 479-488, 1968.

[66] Vaughan, T. S. and M. H. Peters, "Economic Design of Fraction Nonconforming Control Charts with Multiple State Changes," *Journal of Quality Technology* 23, pp 32-43, 1991.

[67] Vance, L. C., "A Bibliography of Statistical Quality Control Chart Techniques, 1970-1980," *Journal of Quality Technology* 15, pp 59-62, 1983.

[68] Weigand, Ch., "A New Approach for Optimal Control of a Production Process," *Economic Quality Control* 7, pp 225-251, 1992.

[69] Woodall, W. H., "The Statistical Design of Quality Control Charts," *The Statistician* 34, pp 155-160, 1985.

[70] Woodall, W. H., "The Design of CUSUM Quality Control Charts," *Journal of Quality Technology* 18, pp 99-102, 1986.

[71] Woodall, W. H., "Weakness of the Economic Design of Control Charts," *Technometrics* 28, pp 408-409, 1986.

[72] Woodall, W. H., "Conflicts Between Deming's Philosophy and the Economic Design of Control Charts," *Frontiers in Statistical Quality Control* 3, pp 242-248, 1987.

[73] Woodall, W. H., and F. W. Faltin, "An Overview and Perspective on Control Charting," in: *Statistical Applications in Process Control and Experimental Design*. Eds. J.B. Keats and D.C. Montgomery. New York: Marcel Dekker, 1995.

# 4

# ECONOMIC DESIGN OF TIME-VARYING AND ADAPTIVE CONTROL CHARTS

**G. Tagaras**

*Department of Mechanical Engineering,*
*Aristoteles University of Thessaloniki,*
*54006 Thessaloniki,*
*Greece.*

## ABSTRACT

The continuously increasing computational power of modern computers and the tremendous advances in automated inspection systems have led to the development of more elaborate and flexible control charts, which can be much more effective than their ancestors, the traditional Shewhart charts. Thus, a new direction in the design of control charts has appeared in recent years, based on the premise that the control of production processes can be improved if the chart parameters, namely the sampling interval, sample size and control limit spread, are not kept fixed and constant but are allowed to change during production. There exists already a considerable number of publications dealing with the design of such charts. Some of them examine their statistical properties, while others adopt an economic approach.

The purpose of this paper is to present a survey of publications on the **economic** design of control charts with variable parameters. A distinction is made between "charts with time-varying parameters" and "adaptive charts". The first category includes charts with parameters that change in a predetermined fashion as the production process evolves. On the other hand, adaptive control charts allow some of their parameters to change during production, taking into account new sample information as it becomes available. The different formulations and results are described and compared. The paper concludes by summarizing the findings so far and proposing fruitful areas for further research.

**Key words:** control charts, economic design, variable chart parameters, adaptive charts

# 1 INTRODUCTION

The design and operation of control charts require the determination of three parameters: the sampling interval $h$, the sample size $n$ and the control limit coefficient $k$, which is the number of standard deviations of the sample statistic separating each control limit and the center line. The typical approach to both the statistical and the economic design of control charts assumes that $h$, $n$ and $k$ are kept constant for the duration of the production run, which is often considered to be effectively infinite. In the traditional statistical design case, the choice of chart parameters is based on statistical considerations, such as acceptable Type I and Type II errors. In economic design, an appropriate cost function is formulated and optimized with respect to $h, n, k$.

A significant part of recent research on the design of control charts has followed a new direction, based on the premise that statistical process control can be improved if the chart parameters are allowed to change during production. This has been motivated by the dramatic increase in available computational power and the significant advances in automated inspection systems, which render practical implementation of the proposed models feasible. As is the case with fixed-parameter charts, one part of the research on variable-parameter charts deals with their statistical properties while the other studies their design from an economic perspective. This paper focuses on the presentation of developments in the area of economic design of control charts with variable parameters. A brief summary of models and results in the area of statistical design of control charts with variable parameters is provided in the introductory section of Prabhu, Montgomery and Runger (1994), as well as in other related publications.

The opportunity for increasing the effectiveness of control charts by relaxing the constraint of having fixed parameters has been apparent for a long time. Bather (1963), Taylor (1965, 1967) and Carter (1972) have studied the problem of optimally monitoring certain classes of production processes and derived theoretical results adopting a Bayesian approach. Crowder (1992) elaborated on the work of Bather (1963) placing emphasis on short production runs. Using the process control framework and terminology of the previous paragraph, it can be said that the models in Bather (1963), Taylor (1965, 1967) and Crowder (1992) correspond to control charts with fixed $h$ and $n$ and variable $k$ and the model in Carter (1972) allows for variable $n$ and $k$ but keeps $h$ constant. The practical value of these earlier models has been limited by the complexity of the resulting process control rules.

Although the scope of more recent publications, which will be presented in the remainder of this paper, is not as broad, their orientation is more practical. Their ultimate objective is to provide the means for designing control charts with clear rules that can be implemented in practice and to prove their superiority relative to the traditional control charts with fixed parameters. In the same spirit, to clarify the differences among the several models from an implementation point of view, the variable-parameter control charts are classified here into two categories: "charts with time-varying parameters" and "adaptive charts". The latter are also called dynamic and can be further classified as either Bayesian or non-Bayesian charts for reasons that will become apparent in the next section.

This paper has three main objectives. The first is to expose and explain the basic differences between the different types of control charts with variable parameters. The second objective is to provide a comprehensive presentation of the models for optimal economic design of these control charts, including a comparison with conventional charts with constant parameters. The third objective is to promote interest in this new area and propose topics for further investigation.

The first objective of the paper is pursued in Section 2. Section 3 presents models for the optimization of control charts with time-varying parameters. The existing literature on the economic design of adaptive charts is surveyed in Section 4. In the last section, the main contributions and results are summarized and opportunities for future research are identified.

# 2 TYPES OF CHARTS WITH VARIABLE PARAMETERS

In the category of charts with time-varying parameters belong control charts with parameters that change in a predetermined fashion as the production process evolves. Let $h_i$ be the time interval between the $(i-1)^{th}$ and $i^{th}$ samples, $n_i$ be the $i^{th}$ sample size and $k_i$ be the control limit coefficient at the $i^{th}$ sample. Minimization of the relevant cost model results in the determination of values of $h_i, n_i, k_i$ for $i = 1, 2, ...$, which are optimal in the context of each model's specific assumptions. These values are obtained before the beginning of the process to be controlled and are not updated during operation, as new sample information is collected. In terms of implementation, this means that the process control scheme is static, albeit with unequal parameters. At the time that sample $i$ of size $n_i$ is taken, there are only two alternative courses of action:

- if the sample statistic is plotted outside the control limits (determined by $k_i$), a signal is issued and an investigation begins to locate an assignable cause and then repair it, if the alarm proves to be true, in which case a new cycle begins;
- if the sample statistic is plotted inside the control limits, no signal is issued and the next sample is to be taken after $h_{i+1}$ time units of operation with chart parameters $n_{i+1}$ and $k_{i+1}$.

A control chart is considered to belong to the category of adaptive or dynamic charts if one, two or even all three chart parameters are allowed to change in real time, based on current sample information. Furthermore, if the adaptive control scheme is based on Bayesian updating of the knowledge about the state of the process after each sample statistic is recorded, the control chart is called Bayesian. Minimization of the relevant cost model for a Bayesian process control scheme results in an optimal policy, which is a function of the updated probability that the process is out of control. If evaluation of sample $i$ indicates that this posterior probability is "high", a signal is issued and search for an assignable cause is initiated; otherwise, no signal is issued but the parameters $h_{i+1}, n_{i+1}$ and $k_{i+1}$ depend on the value of that posterior probability. Given that the posterior probability can take multiple (theoretically infinite) values, there are multiple (practically finite) alternative courses of action, i.e., combinations of $h, n, k$ values. The in-control area of the chart at sample i is essentially divided into multiple regions, each corresponding to a different set of $h_{i+1}, n_{i+1}$ and $k_{i+1}$. The operation of an adaptive non-Bayesian chart is similar in that there are also multiple possible combinations of $h_{i+1}, n_{i+1}$ and $k_{i+1}$ values, but the non-Bayesian chart differs from the Bayesian in that the choice of $h, n, k$ for the following sample in the non-Bayesian scheme depends on the current value of the chart statistic.

Figures 1 and 2 show examples of non-Bayesian and Bayesian adaptive charts respectively, in the simple case where the sample size is the only adaptive parameter and is allowed to take one of two values: a large sample size $n_l$ is to be used when there is strong evidence that the process may be in an out-of-control state, while a smaller sample size $n_s$ is used otherwise. The adaptive non-Bayesian chart of Figure 1 resembles an ordinary control chart with control limits LCL, UCL, and warning limits LWL, UWL, defining two regions between the control limits. If the current chart statistic, e.g., the sample mean, is plotted in Region 1 of the chart, i.e., close to the center line CL, $n_s$ will be used next; if the chart statistic falls in Region 2, consisting of the two disjoint areas far from the center line, it is likely that the process is under the influence of an assignable cause and the next sample size will be $n_l$. The Bayesian chart of Figure 2 differs from conventional control charts in the quantity plotted, which is the posterior probability that the process is out of control. The critical value of the probability plays the role of the non-Bayesian chart's control limit coefficient $k$ and defines the in-control area of the Bayesian chart. The warning value, also expressed as probability, divides the in-control area into two regions, corresponding to $n_l$ and $n_s$; the large sample size will be used when the out-of-control probability is relatively high (Region 2). Practical implementation of the Bayesian chart necessitates the availability of an on-line computer, in order to perform the necessary Bayesian updating and determine the parameters for the following sample in real time.

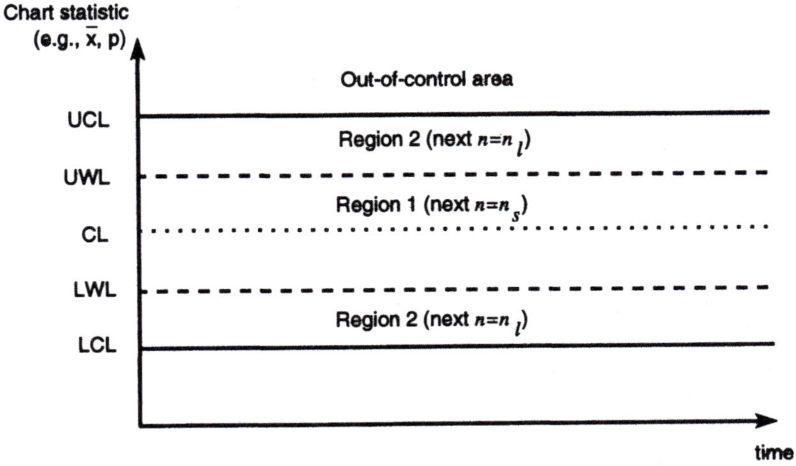


**Figure 1** Non-Bayesian control chart with adaptive sample size, $n$

To further understand the connection between traditional and Bayesian charts and the inherent adaptiveness of the latter, consider the Bayesian chart of Fig-

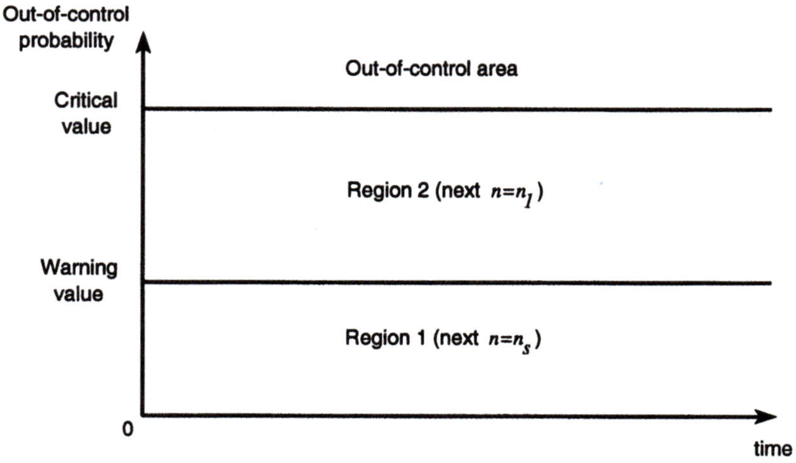


**Figure 2** Bayesian control chart with adaptive sample size, $n$

ure 3 that has no warning limit and assume it is used with fixed sample sizes $n$ taken at fixed sampling intervals $h$. At any sampling instance $i$ the critical probability can be translated into an equivalent coefficient $k_i$ of a traditional chart, but the value of $k_i$ depends on the prior out-of-control probability. Consequently, even though the critical probability is constant, the control limit coefficient is not. Moreover, the equivalent $k_{i+1}$ will depend on the out-of-control probability at sample $i$. Thus, from a traditional control chart point of view this process monitoring scheme is adaptive because the parameter $k$ is. Going back to the chart of Figure 2, it can be said now that, in addition to the sample size, the control limit coefficient is adaptive as well.

In light of the above, one final clarifying remark is in order before proceeding with the detailed presentation of the models in the next section. As one of the referees pointed out, statistical process control by means of a control chart can be viewed as a combination of inspection and control policies. The inspection policy is defined by the sampling interval $h$ and the sample size $n$ and determines how evidence on the process is accumulated. The control policy dictates how the accumulated evidence is used, e.g., to declare the process out of control. In the case of traditional non-Bayesian charts the control policy is defined by the control limit coefficient, while in Bayesian charts it is expressed by the critical probability value. The presentation in this paper could indeed have been based on such a distinction and would have been different then. In-

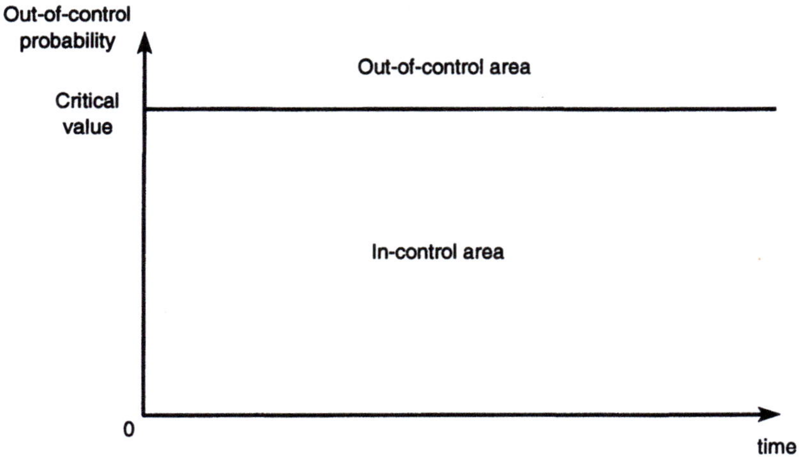


**Figure 3** Bayesian control chart with fixed sample size and sampling interval

stead, it was eventually decided to present the material from a conventional control chart design point of view, as this perspective is more familiar to and acceptable by the majority of academicians and practitioners of process control. Having resolved to adopt the conventional perspective, a Bayesian chart like the one depicted in Figure 3 is considered dynamic for the purposes of this survey, although both the inspection policy and the control policy are static in a Bayesian sense.

# 3 CHARTS WITH TIME-VARYING PARAMETERS

The major motivation for studying control charts with time-varying parameters comes from concerns regarding the failure mechanism of production processes. The vast majority of models for the economic design of control charts assume that assignable causes occur during an interval of time according to a Poisson process. In other words, the occurrence time of the assignable cause is an exponentially distributed random variable. Although the memoryless exponential distribution, characterized by constant failure rate (CFR), may adequately represent the life distribution of electronic components or systems composed of large numbers of components, there are many mechanical processes for which an increasing failure rate distribution (*IFR*) is more appropriate, due to phenomena of fatigue, wear, buildup, etc. For these processes, the probability of shifting to an out-of-control state is an increasing function of operation time. It is therefore reasonable to expect that progressively more frequent sampling or tighter control limits, for example, might lead to a more economical process control scheme. This idea may be valuable even under a constant failure rate. For example, if the process is known to start in control after a restoration, it may be advantageous to use a relatively long first sampling interval and thus avoid incurring unnecessary inspection costs.

Models for the economic design of control charts with time-varying limits have been proposed in Banerjee and Rahim (1988), Parkhideh and Case (1989), Rahim and Banerjee (1993) and Rahim (1994), all dealing with the design of Shewhart-type $\overline{x}$-charts. The models of Banerjee and Rahim treat only the sampling interval as a time-varying parameter, assuming that the sample size and control limit coefficient remain constant for the duration of the process, while Parkhideh and Case allow all three chart parameters to vary over time. In the remainder of this section, the specific model assumptions, formulations and results are presented in more detail.

In the series of papers by Banerjee and Rahim (1988), Rahim and Banerjee (1993) and Rahim (1994), the production process under consideration is subject to the occurrence of an assignable cause, which shifts the process mean $\mu$ by a known amount, either upwards or downwards, but does not affect the process standard deviation $\sigma$. The time that the process remains in the in-control state follows a Weibull distribution in Banerjee and Rahim (1988), but the distribution is generalized to any *IFR* distribution in Rahim and Banerjee (1993) and Rahim (1994). Let $f(t)$ denote the density function and $F(t)$ the cumulative distribution function of the failure time. The process is not self-correcting. The time to sample and chart one item is assumed to be negligible. In Banerjee and Rahim (1988) and Rahim and Banerjee (1993) production is assumed to cease during the searches and repair, while Rahim (1994) assumes that production ceases only during repair and that the search time for the assignable cause, whether true or false, is negligible.

The process is monitored by an $\overline{x}$-chart with control limit coefficient $k$. Random samples of constant size $n$ are drawn at times $\omega_1 = h_1, \omega_2 = (h_1 + h_2), \omega_3 = (h_1 + h_2 + h_3)$, and so on. To facilitate optimization, a restriction is imposed on the lengths of sampling intervals, motivated by the observation that uniform sampling intervals for Markovian shock models signify constant integrated hazard over each integral. Therefore, the $h_i$'s are chosen in such a way that the integrated hazard over each interval is kept constant, i.e.,

$$\int_{w_i}^{w_{i+1}} r(t)dt = \int_0^{w_1} r(t)dt, \quad \text{for} \quad i = 1, 2, ..., \tag{4.1}$$

where the hazard $r(t)$ is defined by

$$r(t) = \frac{f(t)}{1 - F(t)} \tag{4.2}$$

Maintaining a constant integrated hazard implies that the probability of shift in an interval, given no shift until its start, is constant for all intervals. It is clear from (4.1) that the intervals $h_i, i > 1$, can be computed in terms of $h_1$. Furthermore, $h_1 \geq h_2 \geq h_3 \geq ...$, because of the *IFR* property of the failure time distribution.

In the first paper of the series, Banerjee and Rahim (1988) examine the typical infinite-horizon problem of economic design of $\overline{x}$-charts. The failure time distribution is assumed to be Weibull with density function

$$f(t) = \lambda \eta t^{\eta-1} exp(-\lambda t^{\eta}), \qquad t > 0 \tag{4.3}$$

where $\lambda > 0$ is the scale parameter and $\eta \geq 1$ is the shape parameter of the distribution. It follows from (4.1) that the sampling intervals $h_i, i > 1$, are expressed in terms of $h_1$ through the relationship:

$$h_i = \left[ i^{1/\eta} - (i-1)^{1/\eta} \right] h_1 \tag{4.4}$$

A production cycle starts with the process in control, immediately after a true alarm and correction of the problem; it ends after the control chart correctly detects the occurrence of an assignable cause and the process is brought back to the in-control state. Since this sequence of monitoring and adjustment is a renewal reward stochastic process, the expected cost per time unit $ECT$ is the ratio of the expected cost per production cycle, $E(C)$, to the expected duration of the cycle, $E(T)$. Banerjee and Rahim (1988) develop the expressions for $E(C), E(T)$ and $ECT = E(C)/E(T)$, taking the following types of cost into account:

- cost of sampling and testing,

- cost of false alarms,
- cost of locating an assignable cause and repairing the process,
- relative cost of operating in the out-of-control state, due to the production of lower quality output compared to production in control.

The resulting *ECT* is a complicated function of the decision variables $h_1$, $n$ and $k$ and no analytical solution to the optimization problem could be found. Therefore, it is proposed to determine the optimal chart parameters by means of a pattern search technique.

The authors discuss numerical results and sensitivity of the design to the Weibull parameters, based on a single combination of time, cost and shift parameters. For that particular example, the reported cost advantage of the optimal time-varying (nonuniform) sampling scheme over the corresponding optimal uniform sampling scheme is between 5% and 17%, depending on the values of the Weibull parameters. It is also demonstrated that, keeping the mean time to failure constant, the optimal sampling intervals are decreasing in $\lambda$ and increasing in $\eta$, the optimal sample size decreases as $\eta$ decreases, while the control limit coefficient $k$ remains practically unaffected by changes in $\lambda$ and $\eta$. With regard to the time-varying sampling intervals, it is worth noting that their length is rapidly decreasing and stabilizing, especially when $\eta = 3$ or larger: applying expression (4.4) with $\eta = 3$ yields $h_2 = 0.26h_1, h_3 = 0.18h_1, h_4 = 0.145h_1, h_5 = 0.123h_1$ and so forth.

Although these findings offer some evidence that the time-varying scheme may lead to substantial cost improvements in the case of an *IFR* distribution of failure time, it would be dangerous to generalize from results coming from analysis of a single numerical example. It may be argued, for instance, that part of the relative improvement over the uniform sampling scheme is due to the small assumed magnitude of the out-of-control shift in $\mu$ ($0.5\sigma$), which leads to unusually large sample sizes (between 20 and 30 in this case) and, consequently, high sampling costs. Whether or not such an argument is valid, it might be worthwhile to study systematically the conditions under which use of the proposed charts is expected to result in the highest benefits.

Rahim and Banerjee (1993) generalize the contribution of Banerjee and Rahim (1988) along two directions. First, a general *IFR* distribution is assumed for the duration of the in-control period. Second, the model allows the possibility of terminating a production cycle at a certain time $w_m = h_1 + h_2 + ... + h_m$, even if no failure has been detected yet. Thus, the time horizon is essentially finite, but not prespecified. The objective function ECT expresses again expected cost per time unit, but in this case a salvage value for the working equipment is considered and subtracted from the total cost.

In addition to the chart parameters $h_i (i = 1, 2, ..., m)$, $n$ and $k$, parameter $m$ is also a decision variable, which indirectly determines the maximal duration of the production cycle. However, the decision variables $h_i (i = 1, 2, ..., m)$ are reduced to $h_1$, since for a general *IFR* distribution function $F(t)$ expression (4.1) is written

$$1 - F(\omega_i) = [1 - F(\omega_1)]^i, \quad i = 1, 2, 3, ..., m \tag{4.5}$$

where

$$\omega_i = \sum_{j=1}^{i} h_j, \quad i = 1, 2, ..., m \tag{4.6}$$

Thus $\omega_i$ and $h_i$ can be computed numerically in terms of $h_1$. Optimization of ECT over $h_1, n, k, m$ is achieved using a pattern search technique.

Limited numerical results are given for the cases of Weibull and Gamma distributions of the time the process remains in control. In Gamma models, it is observed that the value of $h_i$ stabilizes very quickly. To facilitate implementation, the authors suggest a special time-varying scheme, where all sampling intervals subsequent to the initial interval $h_1$ are the same, $h_2 = h_3 = ... = h$, and they show that the cost of this special scheme is very close to the optimal cost of the scheme with sampling intervals derived according to (4.5). The other conclusions from the experimentation are analogous to those of the earlier paper. Truncating the production run at a certain time seems to be economically justified. However, the observed savings over the respective uniform sampling schemes, with or without considering early termination of the production cycle, are rather modest, namely between 1.5% and 7.5%.

Rahim (1994) enriches the previous model by integrating it with the Economic Production Quantity (EPQ) model and simultaneously considering production set-up cost and inventory holding cost, in addition to the costs associated with process control. The decision variables and the optimization method are exactly as in Rahim and Banerjee (1993). An optimal production lot size can be computed directly from the optimal maximal production run time $\omega_m$, but if a true alarm from the control chart interrupts the process, the lot size will remain incomplete.

The numerical results confirm the earlier findings and the reported savings over uniform sampling schemes vary between 3% and 6%. However, the time, cost and shift parameters of the numerical examples are almost identical to those in the first two papers in this series, hence the word of caution against drawing general conclusions is reiterated.

The work of Parkhideh and Case (1989) constitutes a different approach to the economic design of time-varying $\overline{x}$-charts from that of Banerjee and Rahim, in that all three chart parameters are allowed to change over time. The problem setting and assumptions, types of costs considered, model development and optimization method in Parkhideh and Case (1989) are very similar to those in Banerjee and Rahim (1988), with only a few minor differences, e.g., the process is not shut down during the search for an assignable cause. After deriving the relevant cost per time unit function ECT, Parkhideh and Case (1989) argue that whether any or all of the sampling intervals $h_i$, sample sizes $n_i$ and control limit coefficients $k_i (i = 1, 2, ...)$ should be increasing, constant or decreasing must be determined by minimization of ECT. However, to reduce the prohibitively large (theoretically infinite) number of decision variables, they also impose restrictions on the values of $h_i, n_i, k_i$, through the following relationships:

$$h_i = h_1(f_h)^{i-1}, \quad i = 2, 3, ...$$
$$n_i = n_1(f_n)^{i-1}, \quad i = 2, 3, ...$$
$$k_i = k_1(f_k)^{i-1}, \quad i = 2, 3, ...$$

where $f_h, f_n, f_k$ are factors for the sampling interval, sample size and control limit coefficient respectively, that determine the values of the chart parameters throughout the control chart operation. Thus, the problem is reduced to minimizing ECT with respect to only six decision variables, namely $h_1, n_1, k_1, f_h, f_n, f_k$.

Fifteen numerical examples are considered and solved to permit economic comparisons between the proposed time-varying $\overline{x}$-chart and the conventional $\overline{x}$-chart with constant parameters. The cost reduction provided by the proposed chart varies between 1% and 15%, depending mostly on the sampling costs and the parameters of the Weibull distribution for the failure time. The cost improvement is more pronounced when sampling costs are high, when the mean of the time until failure increases and when the distribution differs more significantly from the exponential.

The above findings, which are discussed in that paper, are quite interesting and intuitively appealing. What is also worth commenting upon, though, is the pattern of the proposed design, as it results from the optimization of the numerical examples. As expected, the optimal sampling intervals and control limit coefficients are decreasing in time ($f_h < 1, f_k < 1$), in order to detect the anticipated shift earlier. However, contrary to the authors' conjecture when justifying the imposed relationships between the chart parameters, the optimal sample sizes also decrease as time of process operation increases ($f_n < 1$). In any case, all these changes in chart parameters are extremely slow, since the optimal values of the factors $f_h, f_n, f_k$ are very close to 1 (larger than 0.99) in almost all 15 examples. Only $f_h$ takes values as low as 0.90 under certain conditions (Weibull parameter $\eta = 6$), which implies $h_i$'s that are still very far from the respective values suggested in Banerjee and Rahim (1988) (for $\eta = 6$, the constant hazard rate restriction implies $h_2 = 0.12h_1$). The magnitude of this difference cannot be attributed to simultaneously changing the other chart parameters, since the rate of change is very low ($f_n, f_k > 0.99$). Note that the $n_i$'s, in particular, remain constant for the first many samples, since they have to be integers; in example 1, the optimal size of the first sample is $n_1 = 5$ and $f_n = 0.9989854$, hence the first 103 samples are of size 5 and the next 248 samples are of size 4.

To recapitulate, the two approaches to the economic design of Shewhart-type $\overline{x}$-charts with time-varying parameters differ mainly in the chart parameters that are considered as decision variables and the relationships they impose on these parameters. Optimization of the respective models results in substantially differing policies. The models in Banerjee and Rahim (1988), Rahim and Banerjee (1993) and Rahim (1994) recommend constant sample sizes and control limit coefficients and rapidly decreasing sampling intervals, while Parkhideh and Case (1989) recommend slowly decreasing sampling intervals, sample sizes and control limit coefficients, at least for Weibull models. Despite that difference, both approaches result in designs that are consistently more economical than the respective conventional charts with fixed parameters, with reported cost advantages mostly about 5%-10%, which can go as high as 16%. At first observation and comparing examples with similar parameters of the underlying Weibull distribution ($\eta = 3$), it can be said that the designs proposed by

Rahim and Banerjee are probably superior to those suggested by the model of Parkhideh and Case. Intuition reinforces this statement, since the former depart more seriously from the traditional fixed-parameter designs, due to the abrupt decrease in the length of the sampling intervals. However, the time, cost and shift parameters of the numerical examples considered are so different, that a systematic and extensive investigation is needed before meaningful comparative results can be documented.

# 4 ADAPTIVE CHARTS

The motivation and philosophy behind the development of models for the economic design of adaptive charts are quite different from the case of charts with predetermined time-varying parameters. The basic principles here are: a) all available information should be used for effective monitoring of the production process and b) the process control scheme should be flexible enough to respond to that information by adapting in real time. The first principle is shared by several types of charts, most notably CUSUM and EWMA charts. However, none of these charts takes the time dimension into account, except for the sequence of observations. The second principle is adopted when some chart parameters for the following sample depend on the values of the previous (or simply the latest) sample statistics. As has already been mentioned in the introductory section, obeying that principle signifies that there are more than two alternative courses of action available to the decision maker at every sampling instance.

Non-Bayesian adaptive charts are based on the second principle, but not necessarily on the first. Their design has been typically statistical, at least so far. The only two exceptions are the papers by Flaig (1991) and Park and Reynolds (1994), both examining control charts with adaptive sample size and fixed sampling interval and control limit coefficient. Flaig (1991) proposes an adaptive $\overline{x}$-chart divided in four regions. Depending on the region where the sample statistic is plotted, there are four possibilities: issue a signal, continue operation with next sample size $n'$, continue operation with next sample size $n''$, or continue operation with next sample size $n'''$. After deriving expressions for some statistical measures of performance, Flaig (1991) outlines a very simple and rather limited economic analysis. Park and Reynolds (1994) study in more depth the economic design of an $\overline{x}$-chart with two possible values for the sample size and find that the cost savings over a static $\overline{x}$-chart can be as high as over 25%. Since the chart analyzed by Park and Reynolds is the traditional Shewhart-type $\bar{x}$-chart with the addition of warning limits as in Figure 1, the inspection and control decisions depend only on the most current sample mean, not on all the available information from previous samples.

The two basic principles of adaptive charts are harmoniously brought together in a Bayesian process control context, where the knowledge about the state of the process is continuously updated through computation of the posterior probability that the process has shifted to an out-of-control state. The optimality of Bayesian techniques has been shown formally in the early theoretical papers, which are referenced in the introduction. The remainder of this section presents the recent contributions of Tagaras (1994, 1996), Calabrese (1995) and Porteus and Angelus (1996), which attempt to provide insights and workable solutions to the problem of economically optimizing Bayesian, dynamic control charts. Tagaras (1994, 1996) examines the design of $x$-charts in the first paper ($n = 1$) and $\bar{x}$-charts in the second one and allows all applicable chart parameters to change dynamically as new information becomes available. Calabrese (1995) and Porteus and Angelus (1996) study the design of p-charts for attributes. In the paper by Calabrese (1995), the sample size and sampling interval are kept fixed and the control limit coefficient is the only adaptive parameter. Porteus and Angelus (1996) allow for dynamically changing sampling intervals and control limit coefficients, but inspection is performed on one unit at a time, hence the sample size is essentially fixed at $n = 1$. The specific model assumptions, formulations and results are presented in more detail in the following paragraphs.

Tagaras (1994) was the first to use the Bayesian framework explicitly for the modeling and economic optimization of a typical control chart. He considers a production process characterized by a single out-of-control state, where the mean is shifted from the nominal value $\mu_0$ to a new value $\mu_1 = \mu_0 + \delta\sigma(\delta > 0)$ and $\sigma$ remains unchanged. The time of occurrence of the assignable cause is assumed to be an exponentially distributed random variable. The process is set up for a finite production run to produce a prespecified lot size and production ceases during searches and restoration. The assumptions of constant $\sigma$ and exponential distribution are not critical to the model development and are maintained to facilitate comparisons with analogous fixed-parameter models.

Since only single measurements may be taken (constant sample size $n = 1$) and only a positive shift in the process mean is possible, the proposed $x$-chart is one-sided with its single (upper) control limit, UCL, at $\mu_0 + k_i\sigma$. The sampling intervals and control limit coefficients constitute the decision variables of the optimization problem, which considers the usual process-control related types of costs (sampling, false alarms, locating and repairing an assignable cause, operating out of control), that are incurred during a production run of finite length $H$.

To determine the dynamically optimal sampling intervals and control limit coefficients, Tagaras (1994) first shows that the decision rule

"issue a signal if $x >$ UCL",

which leads to process interruption and investigation, is equivalent to the rule

"issue a signal if $\tilde{p} > \lambda$",

where $\tilde{p}$ is the probability that the process is in the out-of-control state after measurement $x$ is taken into account and $\lambda$ is some critical probability value. The probability $\tilde{p}$ is computed from Bayes' rule using the prior probabilities that the process is in control or out of control, while $\lambda$ is a function of these same prior probabilities and the control limit coefficient. Based on that observation, he then formulates a dynamic programming (DP) model with state variable $p_i$, namely the posterior probability that the process is out of control after measurement at stage $i$ and, possibly, restoration (in case $x >$ UCL, i.e., $\tilde{p} > \lambda$). State transitions are determined through Bayesian updating of the probability that the process is out of control at the instances before and after a measurement is taken. The peculiarity of the DP formulation is that the stage variable $i$ is defined as the potential $i^{\text{th}}$ inspection. Thus, while for computational purposes the total production run interval $H$ is divided into $m$ equal subintervals of length $h_{min}$, equal to the minimum possible interval between inspections, the actual sampling interval can be any multiple of $h_{min}$ and it may vary within the same production run. Consequently, only a subset of the possible $m$ stages may be visited starting from the initial stage in the beginning of the production run.

After a measurement at stage $i$, the optimal next sampling interval and control limit coefficient are determined from the iterative functional equation of the dynamic programming formulation. Detailed expressions can be found in Tagaras (1994), and are omitted here for the sake of brevity. Since the state space and decision space are theoretically infinite, they need to be appropriately quantized for practical optimization purposes. Several suggestions and guidelines are provided in the paper, based on numerical experience.

Results from 24 examples with a variety of time, cost and shift parameters show that the optimal dynamic one-sided $x$-chart is considerably more economical than the optimal conventional one-sided $x$-chart with fixed parameters. The cost improvement ranged from 3% to 26% in the 24 examples, with an average of 14.5%. This improvement cannot be attributed to the finite horizon of the problem, since larger savings were observed for longer production runs. It was also observed that the superiority of the dynamic chart is more pronounced when the failure rate decreases, the shift in the process mean is larger and/or the cost of out-of-control operation increases. Finally, some intuitive and useful properties of the optimal solution were identified from the numerical results, but were not proven formally: a) the optimal next sampling interval is a non-increasing function of the posterior probability $p_i$; *b*) for given sampling interval, the optimal next control limit coefficient is also non-increasing in $p_i$. Integrating these conjectures in the computational procedure leads to a great reduction in the computational requirements, through elimination of dominated parts of the decision space at every stage and state.

The substantial economic superiority of the one-sided dynamic $x$-chart over its static counterpart provided the motivation for an examination of the economic design and characteristics of more general dynamic control charts in the sequel paper by Tagaras (1996). The first extension concerns the sample size, which is also treated as a dynamic chart parameter. In other words, Tagaras (1996) studies the economic design of fully adaptive one-sided $\overline{x}$-charts for "short" (finite) production runs, which are appropriate when only positive (or only negative) shifts of the process mean to a known value are possible. The development of the dynamic programming formulation parallels that of Tagaras (1994) with some minor modifications. Optimization is performed in a similar manner, but finer quantization of the state space for low values of $p_i$ is recommended and more detailed guidelines are provided for the quantization of the decision space, which now includes sample sizes as well. Results from the optimization of 40 numerical examples confirm and generalize the earlier findings of Tagaras (1994) for $x$-charts. However, the magnitude of the cost advantage of adaptive one-sided $\overline{x}$-charts with respect to the respective Shewhart-type $\overline{x}$-charts is smaller in this case, with an average percentage improvement below 10% in the 40 cases that were examined.

The second part of the paper by Tagaras (1996) attempts to extend the DP approach to the economic design of adaptive two-sided $\overline{x}$-charts for the detection of both positive and negative shifts in the process mean. Two assignable causes are considered, one resulting in a process mean $\mu_1 > \mu_0$ (out-of-control state 1) and another resulting in a process mean $\mu_2 < \mu_0$ (out-of-control state 2). Then, the state of the process is defined by a pair of posterior probabilities $(p_{i1}, p_{i2})$ that at sample $i$ the process is in state 1 and state 2 respectively. The problem is further complicated by the need to maintain asymmetric control limits at every inspection, since at any given time the probabilities that the process mean is $\mu_1$ or $\mu_2$ will be different in general. Therefore, there are now four groups of adaptive chart parameters: the sampling intervals, the sample sizes, the control limit coefficients for the upper control limit and the control limit coefficients for the lower control limit. Although the procedure for computing state transition probabilities and the iterative DP equation are presented in the paper, it is clear that the computational requirements for the optimization of the adaptive two-sided $\bar{x}$-chart are so large, due to the expansion of the state and decision spaces, that no numerical investigation is undertaken. Thus, the question of how advantageous the use of adaptive two-sided $\bar{x}$-charts can be for monitoring finite production runs has been left unanswered.

In concurrent but independent work, Calabrese (1995) develops and analyzes a Bayesian model for the economic design of partially adaptive p-charts. The sampling intervals $h$ and sample sizes $n$ are given and fixed, so that control limits are the only decision variables. His model assumes exponential distribution of the time between occurrences of assignable causes and a single out-of-control state, characterized by increased fraction defective with respect to in-control operation. After the number of nonconformances in a sample has been determined the probability that the process is out of control is updated using Bayes' formula. The process is assumed to stop during search and restoration and the length of the production run is finite. The usual types of costs are considered, except for sampling costs, which are constant for fixed $h$ and $n$ and are not included in the main model. They are introduced and discussed later, though, in the context of a numerical illustration.

Calabrese (1995) uses the partially observable Markov decision process methodology to develop the N-period cost function and to prove structural properties of the optimal solution. He shows that a control limit policy is optimal at every stage, but the optimal value of the control limit depends on the number of periods (sampling intervals) remaining in the production horizon. He expresses the control limit not in terms of a control limit coefficient but directly in terms of the out-of-control probability. Thus, Calabrese (1995) essentially recommends the use of a Bayesian chart like the one in Figure 3 over a conventional p-chart.

Numerical results are presented for two sampling intervals and five different sample sizes, but with only one set of time, cost and shift parameters. It is observed that the optimal value of the control limit stabilizes as the number $N$ of periods in the production horizon and respective DP formulation grows as large as about 10. Similarly, the incremental hourly costs of adding a period to the horizon also stabilize and can be taken as an estimate of the long run average cost per time unit. An analogous behavior in relatively long production runs is reported by Tagaras (1994). Comparisons with conventional static Shewhart-type p-charts are based on these long run estimates of average cost per time unit. When sampling costs are included in the computations, the percentage cost advantage of the Bayesian procedure is about 10% - 13%. However, optimization with respect to $h$ and $n$ is restricted to a limited choice of 24 candidate pairs (4 sampling intervals times 6 sample sizes). It is not clear whether a more comprehensive optimization method would widen or narrow the distance between the optimal Bayesian policy and the optimal conventional p-chart in this particular numerical illustration, nor can one accurately predict the magnitude of the potential savings from use of the Bayesian approach in a larger set of combinations of time, cost and shift parameters.

Porteus and Angelus (1996) study Bayesian process control for attributes in a setting similar to that of Calabrese (1995) in terms of operating assumptions, but very different in terms of decision parameters. Specifically, their model also assumes exponential distribution of the time between occurrences of assignable causes and a single out-of-control state, characterized by increased fraction defective with respect to in-control operation. The length of the production run is finite but not given; it is affected by the choice of production lot size, which is a decision variable in the model. Whether the process stops or continues production during inspection and/or restoration depends also on a decision, to be made based on available information about the state of the process. All this information is encapsulated by the probability that the process is out of control, which is updated after production of every single unit using Bayes' rule. The process is monitored by means of this probability $p_i$. The effective sampling interval is adaptive and may take many values, depending on the trajectory of $p_i$. On the other hand, only one unit may be inspected at a time, i.e., the sample size is fixed at $n = 1$. This is partly a consequence of the assumption that there are no economies of scale in inspection. Given that $n = 1$ and inspection is by attributes, the sample "fractions" nonconforming will be either 0 (conforming unit) or 1 (nonconforming unit) and any attempt to correspond a critical probability value to a control limit of an equivalent p-chart becomes meaningless. In this case, it is better to monitor the evolution of the process by plotting the successive values of $p_i$ on a Bayesian chart with precomputed, time-varying limits.

Porteus and Angelus (1996) formulate the optimization problem in a finite-horizon dynamic programming framework. In addition to the standard quality costs associated with process control, they consider setup and holding costs. In that respect, their model reminds the time-varying model of Rahim (1994), which also deals with lot sizing issues. The optimal lot size is determined after obtaining the optimal dynamic statistical process control policy for successively larger values of the lot size, until the minimum value of the total cost objective function is found. Since the total cost function is not necessarily a convex (or even quasi-convex) function of the lot size, the authors caution against the use of algorithms identifying only a local optimum.

Six numerical examples are solved and discussed in detail. The findings are expressed in the form of nine "opportunities to improve statistical process control". Some of these opportunities are peculiar to the assumed production context, as they pertain to the decisions to stop or continue production during inspection and/or restoration. Most of the proposed opportunities, though, are more generally applicable and they corroborate the suggestions and conclusions of earlier work of other researchers, that has already been presented in this paper. For example, opportunity/suggestion 1 in Porteus and Angelus (1996) implies a relatively long first sampling interval, exactly like Banerjee and Rahim (1988). Opportunity/suggestion 3, "utilize evidence from more than one inspection to justify a restoration", is obviously in accordance with the Bayesian reasoning adopted by Tagaras (1994, 1996) and Calabrese (1995). Opportunity/suggestion 5, "hesitate to restore the process at the end of a production run," is consistent with an analogous remark of Calabrese (1995) and so forth.

Direct cost comparisons with static process control policies are not provided in Porteus and Angelus (1996). Rather, the authors reference the results of Tagaras (1994, 1996), to argue that significant savings can be achieved by using dynamic process control. Then, they state explicitly that they focus on describing ways to achieve these savings, rather than on computing the amount of potential savings in their particular setting.

# 5 DISCUSSION, CONCLUSIONS, AND FUTURE RESEARCH

Upon cursory examination, the various newly proposed control charts with variable chart parameters may look similar in concept and implementation. One of the objectives of this paper was to clarify the basic differences between **static** charts with parameters that change over time according to a predetermined pattern and dynamic charts with parameters adapting in real time to the available information from some or all previous sample statistics. In addition, a distinction was made between non-Bayesian and Bayesian approaches to the analysis and design of dynamic charts. A second objective was to present in some detail the models for optimal economic design of these classes of control charts, trying to explain their setting and assumptions, to uncover their similarities and differences and to compare them with conventional control charts with constant parameters.

The exposition of the previous chapters is summarized in Tables 1, 2 and 3, which contain the major characteristics and cost comparisons for the static charts with time-varying parameters, dynamic non-Bayesian charts and Bayesian charts respectively. The following general conclusions may be drawn from these tables, as well as from the extensive presentation in the main text:

1. When an *IFR* distribution describes more accurately the statistical behavior of the occurrence time of assignable causes than the memoryless exponential distribution, using time-varying chart parameters results in more cost-effective monitoring of the production process. In such cases, it seems that the most important parameter is the sampling interval, which should be progressively shorter in anticipation of the shift to an out-of-control state.

2. The probability that the process is out of control, as computed from Bayes' formula, encapsulates all relevant information in statistical process monitoring. Therefore, the Bayesian approach to process control leads to the optimal economic design of control charts. The cost advantage with respect to conventional Shewhart-type charts seems to be usually in the order of 10% to 15%, but it greatly depends on the combination of problem parameters.

3. Bayesian control charts, plotting the estimated probability that the process is out of statistical control, may have the potential to become a powerful

addition to the quality assurance toolkit, that will facilitate and improve on-line monitoring of production processes.

Overall, it is deemed that rapid progress has been made in the development of economic models for the design of control charts with variable parameters. The foundations have been laid and a new research area has been created in statistical process control. However, the topic is very far from being exhausted. Many issues have been raised, which have not received adequate treatment and consequently they constitute open research questions. In addition, there exist interesting cases, as well as variations of the published models, that have not been investigated at all. The third objective of this paper was to identify and present these research opportunities. The following paragraphs in this section describe a number of possible extensions of the existing literature on control charts with variable parameters.

In the area of control charts with time-varying parameters, all four publications that have been reviewed deal with the economic design of $\overline{x}$-charts for monitoring processes with a single out-of-control state, under specific restrictions on the relationships between successive values of the chart parameters. Numerical results and comparisons with Shewhart-type charts are based on a very limited set of problem parameters. Therefore, a list of candidate topics for further study may include:

- Modeling and optimization of the economic design of other types of control charts, like p-charts, CUSUM and EWMA charts with time-varying parameters.
- Modeling and optimization allowing the existence of multiple assignable causes, which result in different out-of-control states.
- Examination and evaluation of alternative patterns for changing the chart parameters over time. An important criterion here would be the practicality of the monitoring mechanism. For example, Rahim and Banerjee (1993) have found that using a long initial interval and equal short intervals thereafter reaps most of the benefits of a variable scheme, while the numerical results of Parkhideh and Case (1989) suggest that the sample size and control limit coefficient should change at a very slow rate. Then, it is reasonable to study the performance of a chart with only two or three carefully chosen values for each chart parameter. It is noteworthy that the

| | Banerjee and Rahim (1988) | Parkhideh and Case ( 1989) | Rahim and Banerjee (1993) | Rahim (1994) |
|---|---|---|---|---|
| Control chart | $\overline{x}$-chart | $\overline{x}$-chart | $\overline{x}$-chart | $\overline{x}$-chart |
| Failure time | Weibull | Weibull | General *IFR* | General *IFR* |
| OC* states | 1 | 1 | 1 | 1 |
| Production run | Inifinite | Infinite | Finite | Finite |
| Varying parameters | $h$ | $h-n-k$ | $h$ | $h$ |
| Fixed parameters | $n-k$ | - | $n-k$ | $n-k$ |
| Lot sizing issues | No | No | Yes | Yes |
| Cost savings** | 5%-17% | 1% - 15% | 1.5%-7.5 % | 3%-6% |

*out-of-control

**with respect to corresponding chart with fixed parameters

**Table 1** Models for economic design of control charts with time-varying parameters

| | Flaig (1994) | Park and Reynolds (1996) |
|---|---|---|
| Control chart | $\overline{x}$-chart | $\overline{x}$-chart |
| Failure time | Not considered | Exponential |
| Out-of-control states | 1 | Multiple |
| Production run | Infinite | Infinite |
| Adaptive parameters | $n$ (3 values) | $n$ (2 values) |
| Fixed parameters | $h-k$ | $h-k$ |
| Lot sizing issues | No | No |
| Cost savings* | No details given | 3%-26% |

***with respect to corresponding chart with fixed parameters**

**Table 2** **Models for economic design of adaptive non-Bayesian control charts**

| | Tagaras (1994) | Tagaras (1996) | Calabrese (1995) | Porteus and Angelus (1996) |
|---|---|---|---|---|
| Control chart | x-chart | $\overline{x}$-chart | p-chart | p-chart |
| Failure time | Exponential | Exponential | Exponential | Exponential |
| Out-of-control states | 1 | 1 and 2 | 1 | 1 |
| Production run | Finite/Infinite* | Finite | Finite/Infinite* | Finite/Infinite* |
| Adaptive parameters | $h-k$ | $h-n-k$ | $k$ | $h-k$** |
| Fixed parameters | $n=1$ | - | $h-n$ | $n=1$ |
| Lot sizing issues | No | No | No | Yes |
| Cost savings *** | 3%-26% | 1%-15% | 10%-13% | Not reported |

*infinite horizon treated as long, finite horizon

**trivial because of $n = 1$, refer to text

***with respect to corresponding chart with fixed parameters

**Table 3** Models for economic design of adaptive Bayesian control charts

majority of the literature on statistical design of control charts with variable parameters has adopted such a simple approach (Prabhu et al.(1994)).

- Systematic numerical experimentation and comparisons with corresponding conventional charts, under a broad set of time, cost and shift parameters. The purpose of such a study would be to identify cases where investment in charts with time-varying parameters is expected to yield the highest dividends.
- Extension of the concept of time varying sampling intervals to processes characterized by constant failure rate. If the process is known with certainty to start in control after a restoration, it may be reasonable to delay the first inspection, especially if sampling and false alarm costs are high. An economic model is needed to evaluate the performance of such a process control scheme and compare it with fixed-parameter charts.

The area of economic design of non-Bayesian adaptive charts is under-developed. Although there are many publications on the statistical design of these charts, only two papers on economic design could be found and surveyed. An obvious need and opportunity for future research exists in the area of modeling and optimization of the economic design of other types of adaptive charts, notably charts with adaptive sampling intervals, which have been studied extensively from a purely statistical perspective. Looking at the titles and abstracts in the programs of recent and upcoming conferences, we can infer that investigation of that area is already in progress and more results will soon be published.

In the area of Bayesian process control, the challenges and opportunities are even greater. The mathematical difficulties in modeling certain situations and the computational complexities call for both theoretical and algorithmic contributions. Moreover, the reported comparisons with static, fixed-parameter charts may not be completely fair, as they have not been made against static charts utilizing previous sample information, like the CUSUM chart. In order to judge whether the additional complexity of Bayesian process control is justified, it is necessary to have a better feeling about the expected economic advantage with respect to the best conventional alternative. With these remarks in mind, we have compiled the following list of open research questions, associated with the economic design of dynamic charts in a Bayesian process control context:

- Modeling and optimization of the economic design of fully adaptive charts for attributes under different assumptions. Recall that Calabrese (1995) keeps $h$ and $n$ constant and Porteus and Angelus (1996) assume that $n = 1$.

- Efficient algorithms for optimization of adaptive $\bar{x}$-charts with two control limits, when the process may shift to two different out-of-control states. Tagaras (1996) formulated this problem and outlined an algorithm, but admitted that its computational requirements were excessive.

- Formulation and optimization of the economic design of adaptive charts under non-Markovian models for the occurrences of assignable causes. Tagaras (1994) claims that the Markovian assumption is not critical for the development of the finite horizon model, but the algorithmic and computational implications of such an extension are not obvious.

- Direct consideration of Bayesian process control with infinite horizon. Tagaras (1994), Calabrese (1995) and Porteus and Angelus (1996) treat the infinite-horizon case indirectly, allowing the number of stages in the respective dynamic programming formulations to grow as large as needed for the model characteristics to stabilize. In addition to lacking theoretical elegance, this approach may also prove to be practically ineffective in many cases, due to the associated computational burden. Therefore, it would be worthwhile to develop a model for an infinite stage production process subject to Bayesian process control and propose an efficient optimization algorithm.

- Systematic numerical experimentation and comparisons with static charts, under a broad set of time, cost and shift parameters. This investigation should have two inter-related objectives: first, to provide more reliable estimates of the magnitude of the cost savings that can be expected from adoption of dynamic charts, in order to evaluate their attractiveness; second, to identify problem settings where dynamic charts have the greatest potential for improvement of the monitoring scheme. The comparison should extend to static charts like CUSUM and EWMA, which are known to be more effective than simple Shewhart charts in many cases, as well as to adaptive non-Bayesian charts. Calabrese (1995) argues that the more complicated conventional methods cannot produce equally large savings as Bayesian charts, but his argument needs additional support and generalization.

In closing, we believe that the tremendous advances in automated inspection, measurement and data analysis along with the significant savings that can be realized through the utilization of control charts with time-varying and adaptive parameters dictate the continuation of systematic research efforts towards better understanding of their properties and more efficient algorithms for optimization of their economic design.

# REFERENCES


[1] Banjeree, P.K. and M.A. Rahim, "Economic Design of $\bar{x}$-Control Charts Under Weibull Shock Models," *Technometrics* , 30, pp 407-414, 1988.

[2] Bather, J.A., "Control Charts and Minimization of Costs," *Journal of the Royal Statistical Society*, Series B, 25, pp 49-80, 1963.

[3] Calabrese, J.M., "Bayesian Process Control for Attributes." *Management Science* , 41, pp 637-645, 1995.

[4] Carter, P.L., "A Bayesian Approach to Quality Control," *Management Science* , 18, pp 647-655, 1972.

[5] Crowder, S.V., "An SPC Model for Short Production Runs: Minimizing Expected Cost," *Technometrics* , 34, pp 64-73, 1992.

[6] Flaig, J.J., "Adaptive Control Charts," in *Statistical Process Control in Manufacturing* , ed. by J.B. Keats and D.C. Montgomery, Marcel Dekker, New York, NY, pp 111-122, 1991.

[7] Park, C., and M.R. Jr. Reynolds, "Economic Design of a Variable Sample Size $\bar{x}$-Chart," *Communications in Statistics - Simulation and Computation*, 23, pp 467-483, 1994.

[8] Parkhideh, B., and K.E. Case, "The Economic Design of a Dynamic $\bar{x}$-Control Chart," *IIE Transactions*, 21, pp 313-323, 1989.

[9] Porteus, E.L., and A. Angelus, "Opportunities for Improved Statistical Process Control," accepted *Management Science*, 1996.

[10] Prabhu, S.S., D.C. Montgomery, and G.C. Runger, "A Combined Adaptive Sample Size and Sampling Interval $\bar{x}$ Control Scheme," *Journal of Quality Technology*, 26, pp 164-176, 1994.

[11] Rahim, M.A., "Joint Determination of Production Quantity, Inspection Schedule, and Control Chart Design", *IIE Transactions*, 26(6), pp 2-11, 1994.

[12] Rahim, M.A. and P.K. Banerjee, "A Generalized Model for the Economic Design of $\bar{x}$-Control Charts for Production Systems with Increasing Failure Rate and Early Replacement," *Naval Research Logistics*, 40, pp 787-809, 1993.

[13] Tagaras, G., "A Dynamic Programming Approach to the Economic Design of $x$-Charts," *IIE Transactions*, 26(3), pp 48-56, 1994.

[14] Tagaras, G., "Dynamic Control Charts for Finite Production Runs," *European Journal of Operational Research*, 91, pp 38-55, 1996.

[15] Taylor, H.M., "Markovian Sequential Replacement Processes," *Annals of Mathematical Statistics*, 36, pp 1677-1694, 1965.

[16] Taylor, H.M., "Statistical Control of a Gaussian Process," *Technometrics*, 9, pp 29-41, 1967.

# 5

# ECONOMICALLY OPTIMAL DESIGN OF $\overline{X}$-CONTROL CHARTS ASSUMING GAMMA DISTRIBUTED IN-CONTROL TIMES

**M. A. Rahim**

*University of New Brunswick,*
*Fredericton, New Brunswick,*
*Canada E3B 5A3.*

## ABSTRACT

This paper is motivated by the idea of perfect switching of a repairable equipment adherent to statistical process control. The problem can be viewed as a combination of the inspection policy and the control policy. The state of the randomly failing equipment can only be determined by sampling inspection. The output of the product quality is assumed to be normally distributed and monitored by an $\bar{x}$-control chart. The paper determines economically optimum design parameters of $\bar{x}$-control charts. A gamma distribution of the in-control periods having an increasing hazard rate is assumed, an age-dependent salvage value of the equipment is introduced. The possibility of an early replacement of the equipment before its failure is considered. The hazard rate is defined to be the probability density of failure at time $t$ to given survival up to that time. Results of using both truncated and non-truncated production cycles are shown. A truncated production cycle begins when a new component of the equipment is installed. It ends with a repair or after a specified number of sampling intervals, whichever occurs first. A non-truncated production cycle is defined in the usual way. It begins when a new component is installed and ends after a shift due to component failure is detected. The process is brought back to the in-control state only by replacement. A single assignable cause model is assumed. Minimizing the expected cost per hour, the optimal values of the design parameters (i.e., sample size, sampling intervals, control limit coefficient and number of inspection intervals) are determined under five different inspection schemes. The sensitivity of the model is examined. Economic benefits of truncated/non-truncated non-uniform schemes are shown.

**Key words:** economic design of control charts, gamma shock models, increasing hazard rates, variable inspection scheme, truncated production cycle

# 1 INTRODUCTION

The pioneering work of Duncan (1956) for the economic design of $\overline{x}$-control charts and its numerous extensions (including the unified model of Lorenzen and Vance, (1986), assumed an exponentially distributed (Markovian) shock model (i.e., the amount of time the process remains in control has an exponential distribution) and a uniform inspection scheme (i.e., one where the lengths of the inspection intervals are constant). The Markovian shock model has a constant hazard rate and since there is no advantage in preventive replacements failing under constant hazard rate, a uniform inspection scheme has always been recommended. There are many devices for which a constant hazard rate is appropriate. Considerable attention has also recently been devoted to the optimal economic design of $\overline{x}$-control charts under non-Markovian shock models (for example, Heikes et al.(1974), Montgomery and Heikes (1976), Hu (1984), Banerjee and Rahim (1987, 1988), McWilliams (1989, 1992), Parkhideh and Case (1989), Montgomery (1991, 1992), Collani et al.(1992), and Rahim and Banerjee (1993)). Banerjee and Rahim (1988) assumed a Weibull distributed shock model with an increasing hazard rate and provided a non-uniform inspection scheme where the lengths of the sample intervals are chosen to maintain a constant integrated hazard rate over each sampling interval. The concept of failure rate, hazard rate and constant integrated hazard rate may need some clarification. This is as follows. The failure rate is defined to be the rate at which failures occur within a certain interval $(t_1, t_2)$. It is defined as the probability that a failure occurs in the specified interval $(t_1, t_2)$ per unit time, given that it has not occurred prior to $t_1$, the beginning of the interval. The hazard rate is the instantaneous failure rate. It is a conditional function of the failure probability density function, the conditional relationship being the reliability function. Maintaining a constant integrated hazard rate implies that the probability of failure in an interval given no failure until it starts, is constant for all intervals. They showed that a non-uniform sampling scheme and a decreasing process inspection interval scheme resulted in a lower cost than that of a uniform inspection scheme. Based on this work, Rahim (1993) provided a FORTRAN computer program for the optimal economic design of $\overline{x}$-control charts. A production cycle is defined in the usual way. It begins when a new component is installed and ends after a shift due to component failure is detected and the process is brought back to the in-control state by replacement. Weibull distribution has been widely applied to study many non-Markovian process failure mechanisms. However, there are many other probability distributions that are useful in the fields of reliability and quality control engineering. One such distribution is gamma that allows a non-constant hazard rate and has a number of important applications. For example, consider a standby redundant system having two components with a perfect switch. While component 1 is on, component 2 is off, and when component 1 fails, the switch automatically turns component 2 on. If each component has a life described by an exponential distribution with parameter $\lambda$ then the system life is gamma distributed with parameter $\nu = 2$ (Montgomery, 1991). Tadikamalla (1979) provided an inspection policy for the gamma failure distributions. In his policy the inspection cost and the failure cost were considered. However, the quality control costs were not considered. In developing an economic model for determining the design parameters of an $\bar{x}$-chart, Gibra (1971) assumed that the length of time between the occurrence of the assignable cause and its elimination was gamma distributed. As in the Weibull case, gamma distribution becomes exponential when $\nu = 1$ . For a discussion of renewal theory models, Soland (1968) pointed out the renewal functions for Weibull cases do not converge to their asymptotic expression as fast as gamma cases. Furthermore, he studied the number of battery replacements required for a fleet of 1,000 jeeps. The lifetime distribution of the batteries was found to be adequately represented by a gamma distribution rather than a Weibull distribution. The performance quality measure of those jeeps during the next year period was the major objective. One can imagine many real life distributions of mechanical components which follow gamma distribution. As did Duncan (1956), Gibra (1971) considered the length of the production cycle to be infinite. Rahim and Banerjee (1993) developed a generalized model for the economic design of $\bar{x}$-control charts for a production system, where an age-dependent salvage value of the equipment was introduced and the possibility of an early replacement of the equipment before its failure was considered. As a result the length of the production cycle is considered to be finite or truncated. A truncated production cycle was defined as follows. It begins when a new component is installed and ends with a repair or after a specified number of $m$ sampling intervals, whichever occurs first. That is, a notion of optimal truncated time was introduced. The idea that it could be more economically beneficial to scrap/replace a machine *prior* to its failure was found to be very interesting. In this paper, an economic model of an $\bar{x}$-control chart under gamma shock models is presented. A FORTRAN program is developed. This program is compatible for PCs. Both truncated/non-truncated uniform and non-uniform sampling schemes are shown. This option could have been incorporated into the author's original program (Rahim, 1993) if it had been ready on time.

# 2 DEFINITIONS AND ASSUMPTIONS

The output quality characteristic of the product variable is assumed to be normally distributed with mean $\mu_0$ and variance $\sigma_0^2$. The production process begins in a state of statistical control and is subject to a single assignable cause which may shift the process mean $\mu_0$ to $\mu_1 = \mu_0 + \delta\sigma_0$, where $\delta$ is the shift parameter. However, the process variance remains constant. An $\bar{x}$-control chart is used to maintain current control of the process. But, the process failure mechanism is considered to be a standby redundant system having two or more identical components. The system life is thus distributed a gamma with scale parameter $\lambda$ and shape parameter $\nu$. The density function for the in-control period is given by

$$f(t) = \frac{\lambda(\lambda t)^{\nu-1}}{\nu} e^{-\lambda t} \quad t \geq 0, \lambda > 0 \tag{5.1}$$

The cumulative distribution function is given by

$$F(t) = 1 - \Sigma_{j=0}^{\nu-1} e^{-\lambda t} (\lambda t)^j / j! \tag{5.2}$$

and the hazard rate is defined by

$$r(t) = f(t)/\overline{F}(t) \tag{5.3}$$

where $\overline{F}(t) = 1 - F(t)$. An exponential failure distribution has a constant hazard rate, $\lambda$. The process is monitored by drawing a random sample of size $n$ at times $h_1, h_1 + h_2$, and so on. The production cycle ends either with a repair after detecting a true alarm or at the $m$th sampling (at time $\omega_m = \Sigma_{j=1}^{m} h_j$) whichever occurs first. If no true alarm is found by the time $\omega_{m-1}$ then the cycle is allowed to continue for an additional time $h_m$. At time $\omega_m$ the old component is replaced by a new one. It is assumed there is no cost of sampling and charting during the $m$th sampling. The expected length of the production cycle $E(T)$ consists of the following periods: in-control period (it includes the period during which production stops for false alarms); the time between the shift to out of control and when the first sample point falls outside the control limit; and the time to search for an assignable cause and repair the process. The expected cost per cycle $E(C)$ consists of the following costs: the cost for producing nonconforming items while the process is in-control as well as out of control; the cost of false alarms which includes the cost of searching and the cost of down time if production ceases during the search; the cost of locating an assignable cause and repairing the process, which includes the cost of an appropriate down time; and the cost of sampling and testing less the salvage value for the working machine.

**Values to be Specified**

| | | |
|---|---|---|
| $Z_0$ | = | the expected search time associated with a false alarm |
| $Z_1$ | = | the expected search and repair time if a failure is detected |
| $D_0$ | = | the expected cost per hour due to producing a nonconforming item when the process is in control |
| $D_1$ | = | the expected cost per hour due to producing a nonconforming item when the process is out of control |
| $W$ | = | the expected cost of locating an assignable cause and repairing the process |
| $Y$ | = | the expected cost of false alarms |
| $a$ | = | the fixed cost per sample |
| $b$ | = | the variable cost per unit sample |
| $S_0$ | = | salvage value of the machine at age zero |
| $\delta$ | = | the shift parameter |

Expressions for the expected cycle time $E(T)$ and the expected cost per cycle $E(C)$ are given in the Appendix.

# 3 PROGRAM DESCRIPTION

The objective of the program is to derive the optimal decision variables. These are the sample size $n$, the control limit coefficient $k$, the number of inspection intervals $m$, and the length of $j$th sampling interval $h_j$ $(j = 1, 2, ..m)$. The program achieves this by minimizing the expected cost per hour $ECT(m) = \frac{E(C)}{E(T)}$.

The search algorithm is similar to the one used by Rahim (1989, 1993). For a non-uniform scheme, the lengths of the sampling intervals are chosen to maintain a constant integrated hazard rate over each sampling interval. The assumption of constant integrated hazard rate may be explained more clearly by the following expression.

$$\int_{\omega_j}^{\omega_{j+1}} r(t)dt = \int_0^{\omega_1} r(t)dt \text{ for } j = 1, 2, ..., m-1 \tag{5.4}$$

That is, maintaining a constant integrated hazard over each sampling interval is equivalent to stating that the probability of shift in an interval, given no shift until its start, is a constant for all intervals. The following Lemma will provide us with a similar and equivalent expression for (5.4).

**Lemma**

As

$$\int_{\omega_j}^{\omega_{j+1}} r(t)dt = \ln \frac{1 - F(\omega_j)}{1 - F(\omega_{j+1})} \tag{5.5}$$

holds, (5.4) can be shown to be equivalent to

$$\overline{F}(\omega_j) = [\overline{F}(\omega_1)]^j \quad j = 1, 2, ..., m \tag{5.6}$$

The proof of the lemma is shown in the Appendix.

Thus, $\omega_j$ can be obtained numerically, and can be used to calculate the values of $h_2, h_3, ....$ in terms of $h_1$. For a gamma shock model, when $\nu = 1$, we will have

$$e^{(-\lambda \omega_j)} = [e^{(-\lambda \omega_1)}]^j \quad j = 1, 2, .., m \tag{5.7}$$

That is,

$$h_j = h_1 \, \forall j, \; j = 1, 2, .., m \tag{5.8}$$

When $\nu = 2$, equation (5.6) provides

$$(1 + \lambda \omega_j)e^{-\lambda \omega_j} = [(1 + \lambda \omega_1)e^{-\lambda \omega_1})]^j, \; j = 1, 2, .., m \tag{5.9}$$

Similarly when $\nu = 3$, we have

$$[1 + \lambda\omega_j + \frac{\lambda^2\omega_j^2}{2}]e^{-\lambda\omega_j} = \{(1 + \lambda\omega_1 + \frac{\lambda^2\omega_1^2}{2}e^{-\lambda\omega_1}\}^j, \; j = 1, 2, ..., m \quad (5.10)$$

and so on.

For details the readers are referred to Rahim and Banerjee (1993). In the case of a Weibull shock model, we provided an explicit expression for $h_j$ in terms of $h_1$. In gamma shock models, however, explicit solutions of Equation (5.9) or (5.10) for $h_j$ are tedious. In this paper, Inspection Scheme A is referred to as a general non-uniform scheme for determining the values of $h_j$. It is observed that the value of $h_i$ stabilizes very quickly. However, in order to demonstrate that the non-uniform inspection scheme provides a lower cost than the uniform inspection scheme, even though all intervals subsequent to the initial interval are the same, we assume that $h_2 = h_3 = ... = h_m$. This inspection scheme is referred to as a special non-uniform scheme (Inspection Scheme B). In total, five different schemes are presented in this program.

| | | |
|---|---|---|
| 1) | Inspection Scheme A: | Truncated, General Non-Uniform (i.e., $h_1 \geq h_2 \geq h_3 \geq ... > h_m$) |
| 2) | Inspection Scheme B: | Truncated, Special Non-Uniform (i.e., $h_1 \geq h_2 = h_3 = ... = h_m$) |
| 3) | Inspection Scheme C: | Truncated Uniform (i.e., $h_1 = h_2 = h_3 = ... = h_m$) |
| 4) | Inspection Scheme D: | Non-Truncated, Special Non-Uniform (i.e., $m = \infty, h1 \geq h2 = h3 = ...$) |
| 5) | Inspection Scheme E: | Non-Truncated Uniform (i.e., $m = \infty, h1 = h2 = h3 = ...$) |

# 4 PROGRAM OPERATION

The program prompts the user for

- The expected search and repair times previously defined.
- The six kinds of costs previously defined.
- The shift parameter $\delta$ at which the power of the test is to be evaluated.

- The salvage value of the machine at time zero.
- The type of search scheme (uniform or non-uniform) desired.
- The initial values of $k$ and $h_1$.

The output consists of the optimal control limit coefficient $k$, the first sampling interval $h_1(h)$, the probability of a point falling outside the control limits when the process is in control $\alpha$, the probability a point falling outside the control limits after a shift of $\delta$ (the power), and the value of the expected cost per hour. These quantities are computed for the range of sample sizes $n$. The program determines the overall optimal design parameters $(n, k, h_1)$ corresponding to the minimum cost value for a given $m$ for Inspection Schemes A, B and C. The program then searches for the optimal value of $m$, determined by the inequalities $ECT(m-1) \geq ECT(m) \leq ECT(m+1)$. Inspection Schemes D and E assume $m$ to be infinite. The program is limited to the gamma type shock model with parameters $\lambda$ and $\nu = 1, 2, 3$ having a non-decreasing hazard rate.

# 5 EXAMPLES

**Example 1.** *Inspection Scheme A: Truncated, General Non-Uniform*

Assume that the values of the time parameters, cost parameters, and shift parameter are as follows: $Z_0 = 0.25$ hours, $Z_1 = 1.00$ hours, $D_0 = \$50.00$, $D_1 = \$950.00$, $W = \$1100.00$, $Y = \$500.00$, $a = \$20.00$, $b = \$4.22$, $S_0 = \$1100$, $\delta = 0.50$ and $\lambda = 0.05$. Suppose that the process-failure mechanism is governed by a gamma distribution with parameters $\lambda$ and 2. A non-uniform sampling scheme is desired. The resulting optimal plan is $n = 33$, $h_1 = 10.05$ hours, and $k = 1.57$. The characteristics of this plan are $\alpha = 0.1156$, power $= 0.9030$, the expected cost per hour is \$164.18, and the optimal number of sampling intervals $m$ is 4.

**Example 2.** *Inspection Scheme B: Truncated, Special Non-Uniform*

Using the same parameter values as in Example 1, the resulting scheme for a special non- uniform sampling yields the resulting optimal plan as $n = 24$, $h_1 = 7.98$ hours, $h_2 = 4.44$ hours, and $k = 1.56$. The characteristics of this plan are

$\alpha = 0.1198$, power $= 0.8144$, the expected cost per hour is \$165.58, and the optimal number of sampling intervals is 5. The resulting plan under this special scheme may provide a good approximation to the optimal one.

**Example 3.** *Inspection Scheme C: Truncated Uniform*

Using the same parameter values as in Example 1, the resulting scheme for a uniform sampling yields $n = 24, h = 5.00$ hours, and $k = 1.57$. The characteristics of this plan are $\alpha = 0.1165$, power $= 0.8106$, the expected cost per hour is \$169.07, and the optimal number of sampling intervals is 5. Thus, the non-uniform sampling scheme results in a 3% lower cost than a uniform sampling scheme.

**Example 4.** *Inspection Scheme D: Non-Truncated, Special Non-Uniform*

The resulting plan is $n = 28, h1 = 6.73$ hours, and $k = 1.60$ with the expected cost per hour \$171.22. Thus, the economic benefit of a truncated scheme over a non-truncated scheme is \$5.64 per hour (i.e., the cost difference between the Schemes B and D).

**Example 5.** *Inspection Scheme E: Non-Truncated Uniform*

The outcome of the plan is $n = 28, h = 3.76$ hours, and $k = 1.62$. The expected minimum cost is \$174.02 per hour, and is higher than the expected minimum cost of \$171.22 from Scheme D and the expected minimum cost of \$169.07 from Scheme C.

# 6 DISCUSSION ABOUT THE EXAMPLES' RESULTS

Based on the examples' results, the economic benefit of a truncated scheme is better than that of a non-truncated scheme. And a non-uniform inspection scheme is better than a uniform scheme for both truncated and non-truncated production processes. The optimal expected costs for schemes B, C, D and E, show that there is a trade-off between the truncated production cycle and

| | Inspection Scheme | | |
|---|---|---|---|
| Production Cycle | Non-Uniform | Uniform | HEBNOU* |
| Truncated | $165.58 (B) | $169.07 (C) | $3.49 |
| Non-Truncated | $171.22 (D) | $174.02 (E) | $2.80 |
| HEBTSONTS** | $5.64 | $4.95 | |

***Hourly Economic Benifit of Non-Uniform Over Uniform**

****Hourly Economic Benefit of Truncated Scheme Over Non-Truncated Scheme**

**Table 1** Comparison of expected costs of economic design of $\overline{x}$-control charts under truncated and non-truncated gamma shock models for non-uniform and uniform inspection schemes

non-uniform inspection. Table 1 makes more clear the relative significance of the optimization results of the above examples.

The worst scenario is the non-truncated uniform scheme (Scheme E). If one moves from Scheme E to Scheme D, the saving is $2.80 per hour of operation, whereas if one moves from Scheme D to Scheme B, the saving is $5.64 per hour. However, if one implements Scheme B instead of Scheme E, the economic benefit is $8.44 per hour. The above results provide insight into the economic benefit of a pure truncation scheme ($4.95 per hour of operation), and a pure non- uniform scheme ($2.80 per hour of operation). The combined effect of a truncated and non- uniform scheme would be $0.69 ($8.44 - $2.80 - $4.95, or $5.64 - $4.95).

On the basis of expected costs, scheme A is the best. If power instead of cost is considered, scheme A is still the best. The question may arise whether this result is purely by chance or whether it can be proven. Intuitively, the higher the power, the lower the probability of Type II error, the lower the chance of failing to detect the true alarm when it does exist and hence reducing the duration and operation cost of an unnoticed out-of-control period. Due to the complex nature of the cost function, the mathematical proof that the resulting optimal solution of Scheme A provides higher power or vice-versa, is an area which requires further investigation and is left out. However, for further numerical analysis, Table 2 is prepared. Table 2 shows that for every set considered, Scheme A provides the minimum expected cost per unit time as well as the highest power among various schemes.

# 7 SENSITIVITY OF THE MODEL

In this section we conducted a comparative analysis of the various inspection schemes on the design parameters and the expected cost. Thereafter, a sensitivity analysis of the input factors on the number of inspection intervals is performed. Finally, the effect of the model specification in the choice of failure distribution parameters is investigated.

## 7.1 Comparison of Various Inspection Schemes

Comparison of the expected cost per unit time ($ECT$) for various inspection schemes, as shown in Table 2 provides an interesting relationship. As seen in Table 2, in all example sets (except case C of set 3) it is found that $ECT(A) \leq ECT(B) \leq ECT(C) \leq ECT(D) \leq ECT(E)$. It has been also observed that the resulting optimal first sampling interval $h_1$ for scheme A $\geq$ scheme B $\geq$ scheme C $\geq$ scheme D $\geq$ scheme E. Furthermore, it has been found that the second and the subsequent sampling interval thereafter, $h_2$ for the scheme B $\geq$ scheme D. The sample size $n$ for the schemes B and C, and for the schemes D and E are found to be same in all cases. However, scheme A yields larger sample size in all cases considered in this study. The effects of the inspection schemes on the control limit coefficient are found to be in the following orders: scheme B $\leq$ scheme C $\leq$ scheme D $\leq$ scheme E. The results show that the optimal $\alpha$ and $\beta$ are not necessarily smaller from an economic point of view. It may be need further explanation and justification. The $\alpha$ is called the producer's risk and the second of risks, $\beta$, called the consumer's risk. The consumer's risk increases as the control limits are widened, and decreases as they are narrowed. Ultimately, in choosing control limits a manager must consider these risks and the cost associated with them (Adam and Ebert, 1992). If the costs of undetected shifts are high relative to the costs of correcting the process, narrow limits (lower consumer risks, $\beta$) are appropriate. If the costs of restoring the process to the desired state are high compared with the costs of producing defective output, wider limits (lower than producer's risk, $\alpha$) are more appropriate.

Now, the question may arise, under certain conditions, will all of these above relationships exist? To answer this question, much more numerical experimentation will be necessary to justify their validity. Nevertheless, the present study provided the basis for further investigation in this area of research.

## 7.2 Sensitivity Analysis of Input Factors on the Number of Inspection Intervals $m$

A lot of research has been done on the sensitivity of all the input factors to the economic design parameters (for example, Duncan (1956), Goel et al.(1968), Chiu (1975), Saniga (1977), Koo and Case (1990), and others). However, little attempt so far has been made to determine the relationship between the number of inspection intervals $m$ and the input factors. To examine the effects of the input factors on $m$, Table 3 is prepared. It is easy to observe the following three features in Table 3:

1. The number of inspection intervals, $m$, decreases as the expected search and repair time, $Z_1$; the fixed cost per sample, $a$; the variable cost per unit sample, $b$; or the per hour operating cost while the process is in control, $D_0$ increases.
2. The number of inspection intervals, $m$, remains constant as the expected search time for a false alarm, $Z_0$, the per hour operating cost while the process is out of control, $D_1$; the expected cost of false alarms, $Y$; or the salvage value, $S_0$ increases.
3. The number of inspection intervals, $m$, increases as the shift parameter, $\delta$, or the expected cost of repairing the process, $W$, increases.

## 7.3 Sensitivity of the Optimal Design to Gamma Distribution Parameters

In this section we conducted a sensitivity analysis of the model to the gamma distribution parameters on the economic design and the expected cost. Keeping the mean time to failure constant and varying the distribution parameter values, several sets of economic designs are obtained as shown in Table 4. Three values of mean time to failure are chosen: 3.96, 8.86 and 19.81. A non-truncated production cycle is assumed; however, a non-uniform sampling scheme is used.

In order to keep the same mean time to failure, both the scale parameter $\lambda$ and the shape parameter $\nu$ have to increase or decrease simultaneously. This would cause the hazard rate to increase rapidly. In most cases, the increase in expected cost may be caused by the increase in the hazard rate. Keeping the value of $\nu$ constant and decreasing the value of $\lambda$ (for example in sets of 1, 3

| Distribution Parameters | | | Inspection Scheme | Economic Design of $\overline{x}$-Control Charts with Input Factors $\delta = 0.50, Z_0 = 0.25, Z_1 = 1.0, D_0 = 50.0, D_1 = 950.00$ $a = 20, b = 4.22, Y = 500.0, W = 1100, S_0 = 1100.00$ | | | | | | | |
|---|---|---|---|---|---|---|---|---|---|---|---|
| Set | $\lambda$ | $\nu$ | | $n$ | $h_1$ | $h_2$ | $k$ | $m$ | $\alpha$ | $1-\beta$ | $E(C)/E(T)$ |
| 1 | .05 | 2 | A | 33 | 10.05 | - | 1.57 | 4 | 0.1156 | 0.9030 | \$164.18 |
| | | | B | 24 | 7.98 | 4.44 | 1.56 | 5 | 0.1198 | 0.8144 | \$165.58 |
| | | | C | 24 | 5.00 | - | 1.57 | 5 | 0.1165 | 0.8106 | \$169.07 |
| | | | D | 28 | 6.73 | 3.69 | 1.60 | - | 0.1091 | 0.8516 | \$171.22 |
| | | | E | 28 | 3.76 | - | 1.62 | - | 0.1058 | 0.8482 | \$174.02 |
| 2 | .08 | 2 | A | 32 | 7.45 | - | 1.55 | 4 | 0.1221 | 0.9001 | \$200.65 |
| | | | B | 24 | 6.03 | 3.41 | 1.54 | 5 | 0.1242 | 0.8191 | \$202.14 |
| | | | C | 24 | 3.84 | - | 1.55 | 5 | 0.1213 | 0.8160 | \$206.52 |
| | | | D | 27 | 5.25 | 2.93 | 1.57 | - | 0.1163 | 0.8480 | \$206.54 |
| | | | E | 27 | 3.02 | - | 1.58 | - | 0.1135 | 0.8450 | \$210.45 |
| 3 | .10 | 2 | A | 33 | 6.66 | - | 1.51 | 3 | 0.1316 | 0.9138 | \$221.39 |
| | | | B | 24 | 5.29 | 3.01 | 1.53 | 5 | 0.1269 | 0.8220 | \$222.90 |
| | | | C | 23 | 3.36 | - | 1.53 | 5 | 0.1265 | 0.8078 | \$227.75 |
| | | | D | 27 | 4.69 | 2.65 | 1.56 | - | 0.1189 | 0.8505 | \$226.72 |
| | | | E | 27 | 2.75 | - | 1.57 | - | 0.1160 | 0.8076 | \$231.25 |

**Table 2** Comparison of optimal design for various inspection schemes

| Input Parameters | | | | | | | | | | Number of Inspection Intervals |
|---|---|---|---|---|---|---|---|---|---|---|
| Case | $\delta$ | $Z_0$ | $Z_1$ | $a$ | $b$ | $D_0$ | $D_1$ | $Y$ | $W$ | $S_0$ | $m$ |
| 1 | 0.50 | 0.25 | 1.00 | 20.00 | 4.22 | 50.00 | 950.00 | 500.00 | 1100.00 | 1100.00 | 4 |
| 2 | 0.75 | 0.25 | 1.00 | 20.00 | 4.22 | 50.00 | 950.00 | 500.00 | 1100.00 | 1100.00 | 13 |
| 3 | | 0.50 | 1.00 | 20.00 | 4.22 | 50.00 | 950.00 | 500.00 | 1100.00 | 1100.00 | 13 |
| 4 | | | 2.00 | 20.00 | 4.22 | 50.00 | 950.00 | 500.00 | 1100.00 | 1100.00 | 9 |
| 5 | | | | 30.00 | 4.22 | 50.00 | 950.00 | 500.00 | 1100.00 | 1100.00 | 7 |
| 6 | | | | | 6.22 | 50.00 | 950.00 | 500.00 | 1100.00 | 1100.00 | 4 |
| 7 | | | | | | 100.00 | 950.00 | 500.00 | 1100.00 | 1100.00 | 3 |
| 8 | | | | | | | 1200.00 | 500.00 | 1100.00 | 1100.00 | 3 |
| 9 | | | | | | | | 800.00 | 1100.00 | 1100.00 | 3 |
| 10 | | | | | | | | | 1500.00 | 1100.00 | 7 |
| 11 | | | | | | | | | | 1200.00 | 7 |

**Table 3** Sensitivity analysis of the input factors on the number of sampling intervals $m$ assuming $\lambda = 0.05$ and $\nu = 2$

| Parameters | | | | Economic Design Using Non-truncated Production Cycle and Non-Uniform Sample Scheme | | | | | |
|---|---|---|---|---|---|---|---|---|---|
| Set | $\lambda$ | $\nu$ | Mean | $n$ | $h_1$ | $k$ | $\alpha$ | $1-\beta$ | $E(C)/E(T)$ |
| 1 | 0.5050 | 2 | 3.96 | 22 | 2.07 | 1.34 | 0.1790 | 0.8418 | \$458.10 |
| 2 | 0.7575 | 3 | 3.96 | 24 | 2.34 | 1.25 | 0.2099 | 0.8842 | \$457.29 |
| 3 | 0.2257 | 2 | 8.86 | 25 | 3.08 | 1.48 | 0.1385 | 0.8459 | \$323.16 |
| 4 | 0.3380 | 3 | 8.86 | 30 | 4.16 | 1.34 | 0.1813 | 0.9196 | \$334.23 |
| 5 | 0.1009 | 2 | 19.81 | 27 | 4.67 | 1.56 | 0.1191 | 0.8507 | \$227.58 |
| 6 | 0.1514 | 3 | 19.81 | 37 | 8.19 | 1.35 | 0.1758 | 0.9543 | \$257.12 |

**Table 4** Sensitivity of the optimal design to the gamma distribution parameters

and 5 of Table 4), it is found that the optimal design parameters sample size $n$, first sampling intervals $h_1$ and the control limit coefficients $k$ increased while the Type I error and the expected cost per unit time decreased. In other words, as the mean time to failure increases, the expected per hour cost decreases as expected.

# 8 CONCLUSIONS

This paper presents the economic design of $\bar{x}$-control charts under gamma shock models. The salvage value and the replacement policy of the equipment are both assumed to be age-dependent. Five different inspection schemes are presented. Based on the examples' results, the economic benefits of a non-truncated production cycle are reported. Numerical studies show that Scheme A provides the lower expected cost as well as the higher power of the five schemes. Although it is unlikely that this result is purely coincidental, a rigorous mathematical proof is beyond the scope of this paper and remains the subject matter for further investigation. A sensitivity analysis of the optimal design to the distribution parameters is performed. Finally, the relationship between the number of inspection intervals, $m$, with other input factors is discussed.

## ACKNOWLEDGEMENT

Financial support for this research was provided by the Natural Science and Engineering Research Council of Canada, whose assistance is gratefully acknowledged. The author is very grateful to the referees for their valuable comments and suggestions which greatly improved this article.

# REFERENCES

[1] Adam, E.E., Jr. and R. J. Ebert, *Production Operations Management*, Prentice Hall, Fifth Edition, 1992.

[2] Banerjee, P. K. and M. A. Rahim, "The Economic Design of Control Charts: A Renewal Theory Approach", *Engineering Optimization*, 12, pp 63-73, 1987.

[3] Banerjee, P. K. and M. A. Rahim, "Economic Design of $\bar{x}$-Control Charts Under Weibull Shock Models." *Technometrics*, 30, pp 407-414, 1988.

[4] Chiu, W.K., "Minimum Cost Control Schemes", *International Journal of Production Research*, 13, pp 341-349, 1975.

[5] Collani, E. V., P. Frahm, and P. Garbriel, "Economic Inspection and Renewal Policies in the Case of Unperfect Renewals", *Economic Quality Control*, 7, pp 195-212, 1992.

[6] Duncan, A. J., "The Economic Design of X Charts Used to Maintain Current Control of a Process", *Journal of the American Statistical Association*, 51, pp 228-242, 1956.

[7] Gibra, I.N., "Economically Optimal Determination of the Parameters of $\bar{x}$-Control Charts", *Management Science*, 17, pp 635-646, 1971.

[8] Goel, A.L., S.C. Jain and S.M. Wu, "An Algorithm for the Determination of the Economic Design of $\bar{x}$-Charts Based on Duncan's Model", *Journal of American Statistical Association*, 62, pp 304-320, 1968.

[9] Heikes, R. G., D. C. Montgomery, and J. Y. H. Yeung, "Alternative Process Models in Economic Design of T2 Control Charts", *AIIE Transactions*, 6, pp 55-61, 1974.

[10] Hu, P. W., "Economic Design of an $\bar{x}$-Control Chart Under Non-Poisson Process Shift." Abstract, TIMS/ORSA Joint National Meeting, San Francisco, May 14-16, pp 87, 1984.

[11] Koo, T.Y. and K.E. Case, "Economic Design of $\bar{x}$-bar Control Charts for use in Monitoring Continuous Flow Process", *International Journal of Production Research*, 28, pp 2001-2011, 1990.

[12] Lorenzen, T. J. and L. C. Vance, "The Economic Design of Control Charts: A Unified Approach",*Technometrics*, 28, pp 3-10, 1986.

[13] McWilliams, T. P., "Economic Control Chart Designs and the In-Control Time Distribution: A Sensitivity Study", *Journal of Quality Technology*, 21, pp 103-110, 1989.

[14] McWilliams, T. P., "Economic Control Chart Models with Cycle Durations Constraints", *Economic Quality Control*, 7, pp 164-194, 1992.

[15] Montgomery, D. C., *Introduction to Statistical Quality Control*, Second Edition, John Wiley & Sons, pp 428-429, 1991.

[16] Montgomery, D. C., "The Use of Statistical Process Control and Design of Experiments in Product and Process Improvement", *IIE Transactions*, 24, pp 4-17, 1992.

[17] Montgomery, D. C. and R. G. Heikes, "Process Failure Mechanisms and Optimal Design of Fraction Defective Control Charts", *AII Transactions*, 8, pp 467-474, 1976.

[18] Parkhideh, B. and K. E. Case, "The Economic Design of a Dynamic $\bar{x}$-Control Chart", *IIE Transactions*, 21(4), pp 313-323, 1989.

[19] Rahim, M. A., "Determination of Optimal Design Parameters of Joint X and R Charts", *Journal of Quality Technology*, 21, pp 65-70, 1989.

[20] Rahim, M. A., "Economic Design of x Control Charts Assuming Weibull In-Control Times", *Journal of Quality Technology*, 25, pp 296-305, 1993.

[21] Rahim, M. A. and P. K. Banerjee, "A Generalized Model for the Economic Design of $\bar{x}$-Control Charts for Production Systems with Increasing Failure Rate and Early Replacement",*Naval Research Logistics*, 40, pp 787-809, 1993.

[22] Saniga, E.M., "Joint Economically Optimal Design of $\bar{x}$ and $R$ Control Charts", *Management Science*, 24, pp 420-431, 1977.

[23] Soland, R.M., "Availability of Renewal Functions for Gamma and Weibull Distributions with Increasing Hazard Rate", *Operations Research*, 16, pp 536-543, 1968.

[24] Tadikamalla, P.R., "An Inspection on Policy for the Gamma Failure Distributions", *Journal of Operational Society*, 30, pp 77-80, 1979.

# APPENDIX A

# DERIVATIONS

## A.1 DERIVATION OF $E(T)$ AND $E(C)$ FOR INSPECTION SCHEMES A, B AND C

The power for detecting a shift of $\delta$ is denoted by $1 - \beta$, the probability of false alarm is denoted by $\alpha$ and we define $\overline{F}(\omega_j) = 1 - F(\omega_j)$ and $\Delta F(\omega_j) = F(\omega_j) - F(\omega_{j-1})$. The general expressions for $E(T)$ and $E(C)$ for a truncated Gamma shock model are

$$E(T) = \Sigma_{j=1}^{m} h_j F(\omega_{j-1}) + \alpha Z_0 \Sigma_{j=1}^{m-1} + \beta \Sigma_{j=1}^{m-1} \Delta F(\omega_j) \Sigma_{i=j+1}^{m} h_i \beta^{i-j-1} + Z_1 \tag{A.1}$$

and

$$E(C) = D_0 \Sigma_{j=1}^{m} h_j \overline{F}(\omega_{j-1}) + \alpha Y \Sigma_{j=1}^{m-1} \overline{F}(\omega_j) + (D_0 - D_1) \int_0^{\omega_m} x f(x)\, dx$$

$$+ D_1 \beta [\Sigma_{j=1}^{m-1} \Delta F(\omega_j) \Sigma_{i=j+1}^{m} h_i \beta^{i-j-1}] + (a + bn)[1 + \Sigma_{j=1}^{m-2} \overline{F}(\omega_j)$$

$$+ \beta \Sigma_{j=1}^{m-2} \nabla F(\omega_j) \{(1-\beta) \Sigma_{i=1}^{m-1-j} i \beta^{i-1} + (m-1-j) \beta^{m-1-j}\} + W - \overline{F}(\omega_m) S(\omega_m). \tag{A.2}$$

We assume $S(\omega_m) = S_0 e^{(-\omega_m)}$. For details the readers are referred to Rahim and Banerjee (1993).

## A.2 DERIVATION OF $E(T)$ AND $E(C)$ FOR INSPECTION SCHEMES D AND E

When $\nu = 2$

$$E(T) = h_1 + (\alpha Z_0 + h_2)\frac{e^{\lambda h_1}}{1-e^{-\lambda h_2}}[1 + \lambda h_1 + \frac{\lambda h_2 e^{\lambda h_2}}{1-e^{-\lambda h_2}}] + h_2\frac{\beta}{1-\beta} + Z_1 \quad \text{(A.3)}$$

$$E(C) = (a + bn + \alpha Y + D_1 h_2)\frac{e^{-\lambda h_1}}{1-e^{-\lambda h_2}}[1 + \lambda h_1 + \frac{\lambda h_2 e^{-\lambda h_2}}{1-e^{-\lambda h_2}}]$$

$$+\frac{a+bn}{1-\beta} + \frac{2}{\lambda}D_0 + D_1(h_1 + h_2\frac{\beta}{1-\beta} - \frac{2}{\lambda}) + W. \quad \text{(A.4)}$$

The corresponding expressions for $E(C)$ and $E(T)$ under a uniform sampling scheme (i.e., $h_1 = h_2 = ... = h$) are given by

$$E(T) = h + (\alpha Z_0 + h)\frac{e^{-\lambda h}}{1-e^{-\lambda h}}[1 + \frac{\lambda h}{1-e^{-\lambda h}}] + \frac{h\beta}{1-\beta} + Z_1 \quad \text{(A.5)}$$

and

$$E(C) = (a + bn + \alpha Y + D_1 h)\frac{e^{-\lambda h}}{1-e^{-\lambda h}}[1 + \frac{\lambda h}{1-e^{-\lambda h}}] + \frac{a+bn}{1-\beta} + \frac{2}{\lambda}D_0$$
$$+D_1[\frac{h}{1-\beta} - \frac{2}{\lambda}] + W. \quad \text{(A.6)}$$

When $\nu = 3$,

$$E(T) = h_1 + Z_1 + (\alpha Z_0 + h_2)[\frac{e^{-\lambda h_1}}{1-e^{-\lambda h_2}}(1 + \lambda h_1 + \lambda^2\frac{h_1^2}{2}] + \frac{\lambda h_2 e^{-\lambda(h_1+h_2)}}{(1-e^{-\lambda h_2})^2}$$
$$[(1 + \lambda h_1 + \frac{\lambda h_2}{1(1-e^{-\lambda h_2})^2}] + h_2\frac{\beta}{1-\beta} \quad \text{(A.7)}$$

and

$$E(C) = (a + bn + \alpha Y + D_1 h_2)[\frac{e^{-\lambda h_1}}{1 - e^{-\lambda h_2}}(1 + \lambda h_2 + \lambda^2 \frac{h_1^2}{2}] + \frac{\lambda h_2 e^{-\lambda(h_1+h_2)}}{(1 - e^{-\lambda h_2})^2}$$
$$[(1 + \lambda h_1 + \frac{\lambda h_2}{2(1 - e^{-\lambda h_2})^2}] + \frac{a + bn}{1 - \beta} + \frac{3}{\lambda} D_0 + D_1(h_1 + h_2 \frac{\beta}{1 - \beta}$$
$$- \frac{3}{\lambda}) + W \qquad (A.8)$$

Substituting $h_1 = h_2$ in (A.7) and (A.8), $E(T)$ and $E(C)$ under uniform sampling can easily be obtained.

# A.3 DERIVATION OF $E(T)$ AND $E(C)$ FOR THE EXPONENTIAL SHOCK MODEL

If $\nu = 1$, $f(t)$ becomes the exponential density function of the time to failure distribution used in Duncan's model. Then the expressions for $E(T)$ and $E(C)$ for the exponential shock model under a uniform sampling scheme are given by

$$E(T) = he^{(-\lambda h)}/(1 - e^{(-\lambda h)}) + \alpha Z_0 e^{(-\lambda h)}/(1 - e^{(-\lambda h)}) + h1/(1 - \beta) + Z_1 \quad (A.9)$$

and

$$E(C) = D_0(h + h exp(-\lambda h)/(1 - exp(-\lambda h)) + \alpha Y exp(-\lambda h)/(1 - exp(-\lambda h)$$
$$+(D_0 - D_1)/\lambda + (D_1 - D_0)h/(1 - exp(-\lambda h)) + D_1 h\beta/(1 - \beta)$$
$$+(a + bn)(1/(1 - \beta) + exp(-\lambda h)/(1 - exp(-\lambda h))) + W. \qquad (A.10)$$

Proof of Lemma:

Consider $I = \int_a^b \frac{f(t)}{1-F(t)}\, dt$

Put $1 - F(t) = u$. Therefore $-f(t)dt = du$.

Consequently $I = \int_{1-F(b)}^{1-F(a)} \frac{1}{u} = \ln[\frac{1-F(a)}{1-F(b)}]$

Therefore from (4), we get for $j = 2$

$$\ln[\frac{1}{1-F(h_1)}] = \ln[\frac{1-F(h_1)}{1-F(h_1+h_2)}]$$

or

$$\overline{F}(h_1+h_2) = [\overline{F}(h_1)]^2$$

This method can be used to prove the lemma for larger values of $j$.

# 6

# CONSTRAINED OPTIMIZATION MODELS FOR DETERMINING ECONOMIC CONTROL CHART PARAMETERS

**T. P. McWilliams**

*School of Management,*
*Arizona State University West,*
*USA.*

## ABSTRACT

This survey paper presents recent research on the design of constrained economic control charts, where control chart parameters are chosen to minimize expected hourly quality-related costs. Constraints may be placed on average run lengths while the process is in and out-of-control, on the average time to signal a shift in the process parameter, or on a percentile of the distribution of the time which the process spends in an out-of-control state. The concept is applied to control charts based on sampling by attributes and by variables. A variety of numerical examples are presented which illustrate applications of constrained designs.

**Key words:** statistical process control, control chart design, economic design, constrained optimization

## 1 INTRODUCTION

Control charts are widely used to maintain statistical control of a manufacturing process which is subject to assignable causes which induce shifts in process parameters such as the mean or the standard deviation of a quality characteristic of interest. To establish a control process, an appropriate chart ($\overline{x}, R, p$, etc.) or set of charts is selected and the specific chart parameters are chosen. For example, setting up an $\overline{x}$ chart requires selection of the sample size $n$, the time between samples $h$, and the control limit $L$, expressed in standard deviation units.

One traditional approach to parameter selection is to use "rule of thumb" values such as $n = 5, h = 1$ hour, and $L = 3\sigma_{\bar{x}}$ for $\bar{x}$ charts. Another approach is to use purely statistical criteria to determine the choice of $n, h$ and $L$, choosing values which result in an acceptably large average run length (ARL) when in control and acceptably small ARLs calculated at one or more out-of-control states, as represented by various shifts in the process parameter being monitored. This can also be viewed as controlling the "false alarm" or Type I error probability $\alpha$ when in control and the power $p$ or $1 - \beta$ at various out-of-control states. Duncan (1956) gives general guidelines for $\bar{x}$-chart design.

A more sophisticated technique for selecting control chart parameters is the use of an *economic control chart* model, where a total cost function is developed which reflects all costs which are dependent on the control chart design parameters. These parameters are then chosen to minimize the cost function. This approach was originally proposed by Duncan (1956), and there is now a considerable amount of material in the quality literature regarding economic design. Literature surveys can be found in Ho and Case (1994), Svoboda (1991), and Montgomery (1980).

While the concept of cost minimization is appealing, it has been pointed out by Woodall (1985, 1986) that control charts designed in this manner may have undesirable or unacceptable statistical properties such as a low ARL when in control or a high ARL when out-of-control. As an example, consider Example 1 from Gibra's (1978) article on economic $np$-charts. The economic plan reported by Gibra has an in control ARL of $1/\alpha = 1/.3187 = 3.138$, so a false out-of-control signal will occur, on average, at about every third sample. Woodall points out that this large number of false alarms could lead to an increase in process variability through overadjustment, and that regular false alarms tend to destroy confidence in the control process. Woodall (1987) also mentions that the economic control chart's focus on short-term profit maximization does not support the modern long-term goal of continuous improvement, which could be achieved in this case via reduction of costs and reduction in the frequency and size of shifts in the process parameter of interest.

The statistical properties of an economic control chart could be improved by artificially inflating selected costs. For example, inflating the cost of a false alarm should drive the cost minimizing plan's in control ARL upwards, while inflating the cost of continuing production when the process is out-of-control should drive the out-of-control ARL downwards. However, if this is done then the cost function becomes an artificial measure. A better approach is to determine the parameter values which minimize a *constrained* cost function, where constraints are placed on statistical performance parameters such as ARLs, false alarm probabilities, or assignable cause detection probabilities. Models which use this approach are discussed in Section 3.

# 2 AN ECONOMIC CONTROL CHART MODEL

To provide a framework for discussing constrained economic control chart models, we begin with the "unified" model proposed by Lorenzen and Vance (1986). This model was chosen from many which appear in the quality literature because of its generality. The model can be used to determine design parameters for different types of control charts, such as $\overline{x}$-charts, $np$-charts, or EWMA charts, and it allows for various assumptions regarding process shutdowns during the search for or correction of an assignable cause. McWilliams (1996) shows that a variety of earlier $\overline{x}$-chart models (Chiu and Wetherill (1974), Duncan (1956, 1971), Montgomery (1982)) and $np$-chart models (Chiu (1975, 1976), Duncan (1978), Gibra (1978, 1981)) are special cases of the Lorenzen-Vance (LV) model. He presents a table showing parameter values and parameter equivalencies which can be used to express any of these models as a version of the LV model.

In the LV model, a process is initially in control and is subject to the occurrence of a single assignable cause. The in control period is assumed to have a random length which follows an exponential distribution with mean $1/\lambda$. The control charting process involves taking a sample of $n$ observations from the process output every $h$ hours. It takes $E$ hours to sample and chart one item. A search for an assignable cause, which is assumed to cause the process parameter to shift by an expected $\Delta$ standard deviations, is undertaken if the charted variable ($\overline{x}$, $np$, etc.) exceeds control limits. Note that in the case of $np$-chart examples, we deviate from the original LV notation in that we do not use the parameter $\Delta$. Instead, the impact of the assignable cause is to shift the process from nonconforming proportion $p_0$ to nonconforming proportion $p_1$. Control limits are specified in terms of $L$, the number of standard deviations above or below the process center line, or by the accept value $c$ in the case of $np$-charts. Relevant time parameters are $T_0$, the expected search time when a false alarm occurs, $T_1$, the expected time to discover the assignable cause, and $T_2$, the expected time to repair the process.

Regarding costs, let $C_0$ and $C_1$ represent, respectively, the hourly quality cost incurred when the process is in and out-of-control. These costs are due to production of nonconforming items. Let $Y$ represent the cost of a false alarm, $W$ the cost to locate and repair the assignable cause, $a$ the fixed cost associated with taking a sample, and $b$ the cost per unit sampled. Then the expected hourly cost for this model has three components: the cost due to production of nonconforming items, the cost of false alarms and for locating and repairing a true assignable cause, and finally the cost of sampling. Lorenzen and Vance show that each of these costs can be expressed as a ratio of an expected cost per cycle to the expected cycle length (discussed below), so the total cost is the sum of these three ratios:

$$\begin{aligned} C \; = \; & \frac{C_0/\lambda + C_1(-\tau + nE + h(ARL_2) + \delta_1 T_1 + \delta_2 T_2)}{ECL} \\ & + \frac{sY/ARL_1 + W}{ECL} \\ & + \frac{[(a+bn)/h]\,[1/\lambda - \tau + nE + h(ARL_2) + \delta_1 T_1 + \delta_2 T_2]}{ECL}. \end{aligned} \tag{6.1}$$

The term in the denominator is the expected cycle length ($ECL$), representing the average time between successive "renewals" when the process is brought back into a state of control:

$$ECL = \frac{1}{\lambda} + (1 - \delta_1)\frac{sT_0}{ARL_1} - \tau + nE + h(ARL_2) + T_1 + T_2. \tag{6.2}$$

The terms $ARL_1$ and $ARL_2$ represent, respectively, average run lengths when in and out-of-control, while $\tau$ represents the expected time, within the sampling interval which contains the assignable cause, to the occurrence of that assignable cause. The term $s$ is the expected number of samples taken while in control. Based on the assumption of an exponential time to occurrence, Lorenzen and Vance show that

$$\tau = \frac{1 - (1 + \lambda h)e^{-\lambda h}}{\lambda(1 - e^{-\lambda h})} \text{ and } s = \frac{e^{-\lambda h}}{1 - e^{-\lambda h}}.$$

Finally, $\delta_1$ and $\delta_2$ are indicator variables used to show the status of the process during search or repair. Set $\delta_1 = 1$ if production continues during searches for an assignable cause, 0 otherwise; and set $\delta_2 = 1$ if production continues during correction of the assignable cause, 0 otherwise. The use of these variables is in part responsible for the general applicability of the LV model. Note that the LV cost function can also be derived using a renewal theory approach, as shown by Banerjee and Rahim (1987).

In cases where successive charted variables are statistically independent, average run lengths are easily calculated according to

$$ARL_1 = 1/\alpha \text{ and } ARL_2 = 1/p$$

where, for any sample, $\alpha$ is the false alarm probability when in control and $p$ is the assignable cause detection probability, or power, when out-of-control. If, as is the case for the exponentially weighted moving average (EWMA) chart, successive values are not independent, then the calculation of $ARL_1$ and $ARL_2$ is more complex.

In the economic control chart approach, it remains to identify appropriate values for the input parameters and then to find the values of $n, h$, and $L$ which minimize expression (6.1). Due to the complexity of the function being minimized, this is generally done using a computer search routine.

# 3 CONSTRAINED OR ECONOMIC-STATISTICAL CONTROL CHART MODELS

Several approaches have been suggested for controlling the statistical behavior of the economic control chart. For example, Saniga (1989) proposes imposing the following set of constraints on the cost minimization problem:

1. An upper bound on $\alpha$: $\alpha \leq \alpha_u$,
2. A lower bound on $p$: $p \geq p_l$,
3. An upper bound on the average time to signal an expected shift $(ATS)$ : $ATS \leq ATS_u$. In the case of independent chart values, note that $ATS = h/p$.
4. A series of lower bounds on the power $ps_i$ to detect a shift at $m$ other shift levels of interest: $ps_i \geq ps_{li}, i = 1, 2, \cdots, m$.

Note that in the case of independent charted variables, constraints on $\alpha$ or on power values can be easily re-expressed as constraints on average run lengths.

Saniga's work used an attribute control chart model and corresponding cost function developed by Chiu (1975). This model is a special case of the generalized Lorenzen-Vance model, and Saniga's suggested constraints apply equally well to the LV model. Note that if constraints (4) are used, then the model is being extended to control powers or average run lengths at shift values other than the single shift considered by LV. Saniga points out that other types of constraints can be imposed in addition to or in place of those listed above. For example, Woodall (1985) suggested imposing a lower bound on the ARL, which would correspond to an upper bound on power, if a shift in the process parameter occurs which is so small that we would prefer it go undetected. Also, in other works such as Saniga, Davis, and McWilliams (1995), the ATS constraint is replaced with separate constraints on $1/p$ and $h$ and power bounds are re-expressed as ARL bounds.

As an alternative but less popular approach, Gibra (1971), in an earlier $\overline{x}$-chart article dealing with constrained design, suggested constraining the distribution of the number of nonconforming items produced within a "quality cycle." Once again, Gibra's model is a special case of the LV model. Imposing Gibra's constraint was shown to be equivalent to constraining a percentile of the distribution of $T_{out}$, where $T_{out}$ represents the random length of time during which the process is allowed to remain in an out-of-control state. For example, the 95th percentile of the distribution of $T_{out}$ might be constrained to be less than or equal to one hour, which implies that $T_{out}$ can only exceed one hour in, at most, 5% of all cycles. Implementation of this approach requires knowledge of the distribution, rather than just the expected value, of the time required to locate and correct an assignable cause. It was assumed that this random time followed an Erlang distribution.

McWilliams (1992) applied the concept of constraining $T_{out}$ to the more general LV model, presenting a variety of numerical examples which demonstrated cost/benefit tradeoffs for the constrained approach. He pointed out that the $T_{out}$ constraint is similar in concept to Saniga's $ATS$ constraint, but that the $ATS$ constraint only controls process behavior *on average* while, by focusing on an upper percentile of the distribution of $T_{out}$, the $T_{out}$ constraint does a better job of controlling variability and making sure that very high out-of-control times have low probabilities. On the other hand, the $ATS$ constraint only requires knowledge of the *mean* time to locate and correct an assignable cause, while the $T_{out}$ approach requires knowledge of the distribution. This makes the $ATS$ constraint easier to implement.

# 4 IMPLEMENTATION OF THE ECONOMIC-STATISTICAL CONTROL CHART

As in the case of the design of a pure or unconstrained economic control chart, the first implementation step is to identify appropriate values for all input parameters. Then, a set of reasonable constraints must be chosen. Finally, the problem of determining the control chart parameters which minimize expression (6.1), subject to the constraints, must be solved. Due to the complexity of the cost function, closed form solutions are not readily available so computer search techniques are typically used. Not all authors agree on the best approach to this multivariate search scheme, and some comments regarding optimization schemes can be found in McWilliams (1992a, 1996).

Several economic-statistical control chart design programs have recently been published in the *Journal of Quality Technology* and are readily available to quality practitioners. All are based on the LV cost function. McWilliams (1994) presented a program which finds economic, statistical, and economic-statistical designs for $\overline{x}$-charts, while a program by Saniga, Davis, and McWilliams (1995) finds these designs for *np*– and *c*-charts. Finally, a Montgomery, Torng, Cochran, and Lawrence (1995) article on an economic-statistical EWMA chart model is accompanied by a program (Torng, Cochran, Montgomery, and Lawrence, 1995) which facilitates the use of this model.

# 5 EXAMPLES

## 5.1 Attributes Control Charts: *np*-chart design

Consider Example 1 from Gibra (1978), the example used by Woodall in his discussion of the statistical performance of economic designs. An *np*-chart is used to control a process, and production stops during the search for or repair of an assignable cause. Input parameters, expressed in terms of the LV model's notation, are: $E = .005$ hours; $T_0 = T_1 = 0.2$ hours; $T_2 = 2.0$ hours; $C_0 = \$0$; $C_1 = \$600; Y = \$5; W = \$75; a = \$2; b = \$.10$; $\lambda = .0125; p_0 = .02; p_1 = .10$; and $\delta_1 = \delta_2 = 0$. Gibra reports an economic design of $n = 19$, $c = 0$, and $h = 1.02$ hours for an hourly cost of \$ 10.99. Note that program ATT from

Saniga, Davis, and McWilliams (1995) found a design having a slightly lower cost: $n = 16, c = 0, h = .90$ hours for an hourly cost of \$ 10.95. This design has an $\alpha$-value of .276 for an in control ARL of $ARL_1 = 1/.276 = 3.623$. As Woodall points out, this ARL may not be acceptable. This would naturally lead to consideration of a constrained design.

Table 1 shows the result of imposing a variety of constraints on the in control ARL, the out-of-control ARL, and, in one case, the ATS. In case 2, the constraint $ARL_1 \geq 20$ is imposed to improve performance when in control, and the constraint $ARL_2 \leq 1.25$ is imposed to assure that the original out-of-control ARL is not compromised. The economic-statistical design of $n = 59, c = 3$, and $h = 1.27$ hours has an hourly cost of \$ 15.27, which is considerably more expensive than the economic design's \$ 10.95 cost.

Cases 3 through 6 involve other combinations of constraints on $ARL_1$ and $ARL_2$. We consider imposing stricter constraints on $ARL_1$ while, in some cases, relaxing the constraint on $ARL_2$. Finally, in Case 7 a constraint on the intersample interval $h$ is added. This, together with the constraints on $ARL_2$, serves to constrain the ATS to $1.00(1.25) = 1.25$ hours.

Table 1 does not provide a "best" design, but rather illustrates cost/benefit tradeoffs obtained from imposing various constraints. The final decision regarding which design to use is of course up to the user.

In the examples of Table 1, constraints are only imposed at the in control $p$—value of .02 and at the expected out-of-control $p$-value of .10. For example, in Case 3 the constraints are $ARL_1 \geq 50$ at $p = .02$ and $ARL_2 \leq 2$ at $p = .10$. Since parameter shifts to other values may also be possible, consider imposing the following additional constraints in Case 3:

1. The constraint $ARL \geq 35$ at $p = .03$. As suggested by Woodall (1985), this provides assurance that a small, possibly unimportant shift in the process parameter will *not* be detected quickly.

2. The constraints $ARL \leq 2.5$ at $p = .08$ and $ARL \leq 1.5$ at $p = .20$. These constraints provide assurance of quick detection of assignable causes which lead to shifts in the process parameter to values other than the expected shifted value of .10.

Table 2 shows the control chart designs, costs, and average run lengths for the economic (unconstrained) design and for the economic-statistical designs

| Case # | Constraints | $n$ | $c$ | $h$ (rounded) | Cost | $ARL_1$ (in control) | $ARL_2$ (out-of-control) |
|---|---|---|---|---|---|---|---|
| 1 | None | 16 | 0 | 0.90 hrs.<br>1.0 | \$ 10.95<br>10.99 | 3.62 | 1.23 |
| 2 | $ARL_1 \geq 20$;<br>$ARL_2 \leq 1.25$ | 59 | 3 | 1.27<br>1.0 | 15.27<br>15.63 | 32.8 | 1.17 |
| 3 | $ARL_1 \geq 50$;<br>$ARL_2 \leq 2$ | 29 | 2 | 0.73<br>0.50 | 15.21<br>16.15 | 50.4 | 1.77 |
| 4 | $ARL_1 \geq 100$;<br>$ARL_2 \leq 2$ | 42 | 3 | 0.87<br>1.0 | 16.40<br>16.53 | 102 | 1.62 |
| 5 | $ARL_1 \geq 100$;<br>$ARL_2 \leq 1.5$ | 65 | 4 | 1.24<br>1.0 | 16.74<br>17.05 | 102 | 1.25 |
| 6 | $ARL_1 \geq 100$;<br>$ARL_2 \leq 1.25$ | 85 | 5 | 1.48<br>1.50 | 17.86<br>17.86 | 137 | 1.16 |
| 7 | $ARL_1 \geq 100$;<br>$ARL_2 \leq 1.25$<br>$h \leq 1.00$ | 78 | 5 | 1.00 | 18.62 | 208 | 1.24 |

**Table 1** Examples of economic-statistical control chart designs

of Case 3 along with Case 3 supplemented by the three constraints described above. Imposed constraints appear in parentheses below ARL values. $h$-values were rounded to the nearest half hour. Note that supplementing the original Case 3 constraints is not particularly costly, raising the hourly cost from \$ 16.15 to \$ 17.35. The ARL functions for the three designs can be found in Figure 1. As the figure indicates, the main effect of supplementing the constraints is

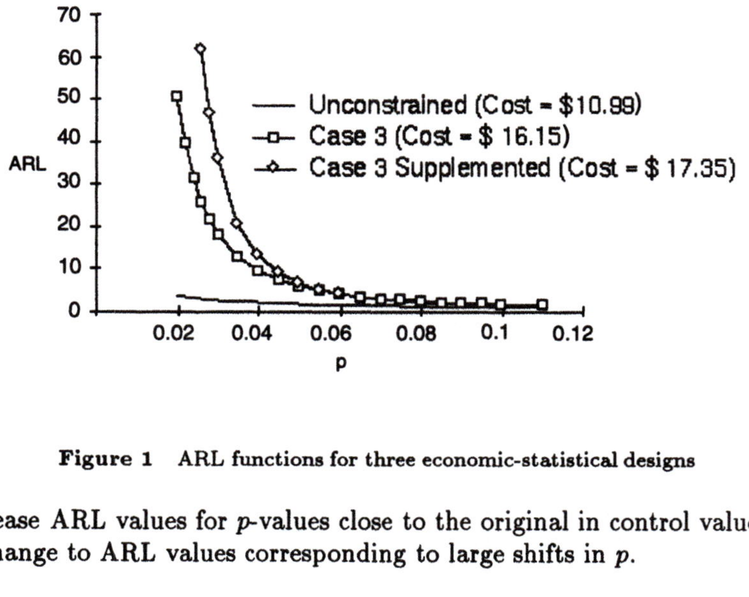


**Figure 1** ARL functions for three economic-statistical designs

to increase ARL values for $p$-values close to the original in control value, with little change to ARL values corresponding to large shifts in $p$.

## 5.2 Variables Control Chart: $\bar{x}$-chart design

Consider Example 1 from Duncan (1956). Input parameters are $\lambda = 0.01, \Delta = 2, E = 0.05, T_0 = T_1 = 0, T_2 = 2, \delta_1 = \delta_2 = 1, C_0 = 0, C_1 = \$100, Y = \$50, W = \$25, a = \$0.50$, and $b = \$0.10$. The program presented by McWilliams (1994) is used to determine economic and economic-statistical $\bar{x}$-chart designs, and to compare the performance of these designs with the usual heuristic design $n = 5, L = 3.0$, and $h = 1$ hour. For the economic-statistical design, specified constraints are $ARL \geq 200$ when in control (zero shift), $ARL \geq 150$ for a minor shift of 0.2 standard deviations, $ARL \leq 3$ for a shift of 1 standard deviation, and $ARL \leq 1.05$ for a shift of 2 standard deviations.

Table 3 shows design parameters, costs, and selected ARL values for the three designs. For the economic-statistical design, constraints appear in parenthe-

| | $n$ | $c$ | $h$ | Cost | ARL at $p = .02$ | ARL at $p = .03$ | ARL at $p = .08$ | ARL at $p = .10$ | ARL at $p = .20$ |
|---|---|---|---|---|---|---|---|---|---|
| Case 1 (Unconstrained) | 16 | 0 | 1.0 | \$ 10.99 | 3.62 | 2.59 | 1.36 | 1.23 | 1.03 |
| Case 3 | 29 | 2 | 0.50 | 16.15 | 50.4 ($\geq 50$) | 18.1 | 2.42 | 1.77 ($\geq 2$) | 1.05 |
| Case 3 with supplements | 57 | 4 | 1.0 | 17.35 | 177 ($\geq 50$) | 35.7 ($\geq 35$) | 2.07 ($\leq 2.5$) | 1.46 ($\leq 2.0$) | 1.01 ($\leq 1.5$) |

**Table 2** *np*-chart design parameters, costs, and selected *ARL* values

| | $n$ | $L$ | $h$ | Cost | In control ARL | ARL at $.2\sigma$ shift | ARL at $1\sigma$ shift | ARL at $2\sigma$ shift |
|---|---|---|---|---|---|---|---|---|
| Heuristic Design | 5 | 3 | 1.0 | \$ 4.12 | 370.4 | 177.7 | 4.50 | 1.08 |
| Economic Design | 5 | 3.1 | 1.4 | 4.01 | 516.7 | 238.9 | 5.16 | 1.09 |
| Economic-Statistical Design | 7 | 3.05 | 1.6 | 4.14 | 437.0 ($\geq 200$) | 166.0 ($\geq 150$) | 2.92 ($\leq 3.0$) | 1.01 ($\leq 1.05$) |

**Table 3** $\bar{x}$-chart design parameters, costs, and selected ARL values

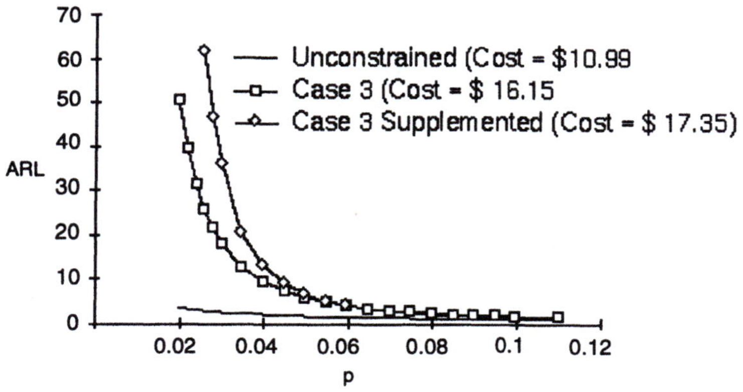


**Figure 2** ARL functions for three $\bar{x}$-chart designs

ses below the ARL values. Note that with these cost and time parameters, the heuristic design happens to be nearly economically optimal, with a cost of $4.12 vs $4.01 for the economic design. Also, both the heuristic and economic designs already meet the specified lower bounds on ARL when in control and for a $0.2\sigma$ shift, and come close to meeting the upper ARL bounds at larger shifts. As a result, there is only a small cost penalty over the economic design ($4.14 vs. $4.01) when using the economic-statistical design which of course meets these constraints.

# REFERENCES

[1] Banerjee, P.K. and, M.A. Rahim, "The Economic Design of Control Charts: A Renewal Theory Approach", *Engineering Optimization,* 12, pp 63-73, 1987.

[2] Chiu, W.K., "Economic Design of Attribute Control Charts", *Technometrics,* 17, pp 81-87, 1975.

[3] Chiu, W.K., "Economic Design of *np* Charts for Processes Subject to a Multiplicity of Assignable Causes", *Management Science,* 23, pp 404-411, 1976.

[4] Chiu, W.K. and, G.B. Wetherill, "A Simplified Scheme for the Economic Design of $\bar{x}$-Charts", *Journal of Quality Technology,* 6, pp 63-69, 1974.

[5] Duncan, A.J., "The Economic Design of $\bar{x}$-Charts Used to Maintain Current Control of a Process", *Journal of the American Statistical Association,* 51, pp 228-242, 1956.

[6] Duncan, A.J., "The Economic Design of $\bar{x}$-Charts When There is a Multiplicity of Assignable Causes", *Journal of the American Statistical Association,* 66, pp 107-121, 1971.

[7] Duncan, A.J., "The Economic Design of $p$-Charts to Maintain Current Control of a Process: Some Numerical Results", *Technometrics,* 20, pp 235-243, 1978.

[8] Gibra, I.N., "Economically Optimal Determination of the Parameters of $\bar{x}$-Control Chart", *Management Science,* 17, pp 635-646, 1971.

[9] Gibra, I.N., "Economically Optimal Determination of the Parameters of $np$-Control Charts", *Journal of Quality Technology,* 10, pp 12-19, 1978.

[10] Gibra, I.N., "Economic Design of Attribute Control Charts for Multiple Assignable Causes", *Journal of Quality Technology,* 13, pp 93-99, 1981.

[11] Ho, C. and, K.E. Case, "Economic Design of Control Charts: A Literature Review for 1981-1991", *Journal of Quality Technology,* 26, pp 39-53, 1994.

[12] Lorenzen, T.J., and L.C. Vance, "The Economic Design of Control Charts: A Unified Approach", *Technometrics,* 28, pp 3-10, 1986.

[13] McWilliams, T.P., "Economic, Statistical, and Economic-Statistical Chart Designs", *Journal of Quality Technology,* 26, pp 227-238, 1994.

[14] McWilliams, T.P., "Economic Control Charts: Relating the Model and Finding the Optimal Design", *Statistical Applications in Process Control,* edited by J.B. Keats and D.C. Montgomery. Marcel Dekker, New York, NY, 1996.

[15] McWilliams, T.P., "Comments on Lorenzen and Vance (1986)", *Technometrics,* 34, pp 248-249, 1992a.

[16] McWilliams, T.P., "Economic Control Chart Models with Cyclic Duration Constraints", *Economic Quality Control,* 7, 164-194, 1992b.

[17] Montgomery, D.C., "The Economic Design of Control Charts: A Review and Literature Survey", *Journal of Quality Technology,* 12, pp 75-87, 1980.

[18] Montgomery, D.C., "Economic Design of an $\bar{x}$–Control Chart", *Journal of Quality Technology,* 14, pp 40-43, 1982.

[19] Montgomery, D.C., J.C.-C. Torng, J.K. Cochran, and, F.P. Lawrence, "Statistically Constrained Economic Design of the EWMA Control Chart", *Journal of Quality Technology,* 27, pp 250-256, 1995.

[20] Saniga, E.M., "Economic Statistical Control-Chart Designs With an Application to $\bar{x}$ and $R$ Charts", *Technometrics,* 31, pp 313-320, 1989.

[21] Saniga, E.M., D.J. Davis, and, T.P. McWilliams, "Economic, Statistical, and Economic-Statistical Design of Attribute Charts", *Journal of Quality Technology,* 27, pp 56-73, 1995.

[22] Svoboda, L., "Economic Design of Control Charts: A Review and Literature Survey (1979 - 1989)", *Statistical Process Control in Manufacturing,* edited by J.B. Keats and D.C. Montgomery. Marcel Dekker, New York, NY, 1991.

[23] Torng, J.C.-C., J.K.Cochran, D.C. Montgomery, and, F.P. Lawrence, "Implementing Statistically Constrained Economic EWMA Control Charts", *Journal of Quality Technology,* 27, pp 257-264, 1995.

[24] Woodall, W. H., "The Statistical Design of Quality Control Charts", *The Statistician,* 34, pp 155-160, 1985.

[25] Woodall, W. H., "Weaknesses of the Economic Design of Control Charts (letter to the editor)", *Technometrics,* 28, pp 408-410, 1986.

[26] Woodall, W. H., "Conflicts Between Deming's Philosophy and the Economic Design of Control Charts", *Frontiers in Statistical Quality Control,* 3, edited by H.-J. Lenz, G.B. Wetherill, and P.-Th. Wilrich. Physica-Verlag, Heidelberg, 1987.

# PART III

## ECONOMIC SELECTION

7

# ECONOMIC SELECTION OF THE MEAN AND UPPER LIMIT FOR A CONTAINER-FILLING PROCESS UNDER CAPACITY CONSTRAINTS

**J. Liu, K. Tang and Y. H. Chun**

*Department of Information Systems and Decision Sciences,*
*E.J. Ourso College of Business Administration,*
*Louisiana State University,*
*Baton Rouge, LA 70803-6316,*
*USA.*

## ABSTRACT

Consider a container-filling process with a lower product specification limit. It is assumed that the items with contents below the lower specification limit cannot be shipped to customers. To screen out nonconforming items, every filling result is examined by an automatic weighing machine through a conveyor belt. We consider the two-level problem, in which a lower specification limit is used to screen out nonconforming items and an artificial upper limit is used to screen out overfilled items. Both nonconforming and overfilled items are re-processed until they become accepted items. In addition, we also consider a capacity constraint that requires that the total number of conforming items produced by the production process meet a specified demand. We illustrate the effectiveness of our model with an example problem and compare the result with that of other models applied to the capacitated case.

**Key words:** optimal target value, process optimization, process capacity

## 1 INTRODUCTION

Consider a container-filling process with a lower product specification limit. It is assumed that the items with contents below the lower specification limit cannot be shipped to customers. To screen out nonconforming items, every filling result is examined by an automatic weighing machine through a conveyor belt. The producer may set the process mean (target value) higher to reduce the chance of producing nonconforming items. This action, however, results in a higher material cost because the average amount filled in the containers is larger. Considering this tradeoff, the producer must determine the process mean either to minimize the production cost or to maximize the profit of operating the process. This problem has been studied very extensively, under various revenue functions and rework schemes. Literature reviews on this problem were given by Tang and Tang (1994) and Al-Sultan and Rahim (1994).

The studies in the container-filling problem can be generally classified into two categories, depending on whether an artificial upper limit is used to screen out overfilled items for re-processing. The first category is the single-level problem, in which only the lower product specification limit is used to screen out nonconforming items. It is generally assumed that conforming items can be sold in a regular market at a fixed price and that nonconforming items are scrapped, reworked, or sold in a secondary market at a reduced price. Relevant papers include Hunter and Kartha (1977), Nelson (1978), Carlsson (1984), Bisgaard, Hunter and Pallesen (1984) and Golhar (1987).

The second category is the two-level problem, in which both a lower product specification limit and an artificial upper limit are used. The latter is used to screen out "overfilled" items for reprocessing in order to reduce the material cost. Consider, for example, the liquor, perfume, pharmacy, food, and chemical industries, where the value of material filled into a container is more expensive than rework disbursements. This approach was first studied by Bettes (1962). In his model, both nonconforming items and overfilled items are reprocessed at a fixed cost, and the optimal process mean and the upper limit are determined simultaneously. Golhar and Pollock (1988) studied a situation in which it is assumed that, in the reprocessing process, after the materials are recovered, the empty container is cleaned for use in another filling attempt. They illustrated numerically the saving in material cost by using the two-level approach and also provided a computer program (Golhar 1988) for calculating the optimal process mean and the artificial upper limit.

In addition to the studies based on screening, Carlsson (1989) discussed a situation in which the lots produced by a production process are subjected to lot-by-lot acceptance sampling by variables. Boucher and Jafari (1991) studied the same problem except that an attributes sampling plan is used to decide whether a lot is accepted. Pulak and Al-Sultan (1996a) extended this problem by considering rectifying inspection and Al-Sultan and Pulak (1996) considered the additional issue of process improvement by process variance reduction. Al-Sultan (1994) developed an algorithm for finding the optimal process means for two machines in a series configuration when outgoing items are subject to sampling inspection. Al-Sultan and Al-Fawzan (1996) studied the benefit of variance reduction for a process with random linear drift. A FORTRAN based computer package was developed by Pulak and Al-Sultan (1996b) for most of the important models in the literature.

As pointed out by Schmidt and Pfeifer (1991), an implicit but critical assumption in the Bettes and the Golhar and Pollock models is that the capacity of the process is unlimited. Schmidt and Pfeifer stressed the importance of the capacity constraint and considered the opportunity cost of the bottleneck capacity for every filling attempt. They used the expected profit per fill attempt as the objective to derive a closed-form solution for the upper control limit as an economic judgment that reworking is beneficial only if the material value of a container is higher than the unit rework and capacity opportunity costs.

In general, the Golhar and Pollock (GP) model can be applied to the case where the production facility has a much larger capacity than the demand. On the other hand, the Schmidt and Pfeifer (SP) model is used to optimally utilize a bottleneck capacity. Neither of these models addressed a commonly seen situation in which a producer has to use the production capacity to satisfy a given demand. In other words, the solution obtained from these models may overload the production line and/or delay scheduled orders.

The objective of this paper is to develop a model for optimal selection of the process mean and the upper limit to control the production cost and to satisfy the demand when the production capacity is fixed and limited. We show that incorporation of the capacity constraint into the model is important, particularly when the demand is very close to the capacity. In addition, we illustrate the effectiveness of our model with an example and a sensitivity analysis, by comparing the solution with those of the GP and SP models.

## 2 MODEL DEVELOPMENT

Consider a container-filling process with a maximum capacity of producing $r$ items of a specific product per unit time. Suppose the process can only be used to produce the product and the demand for the product is $D$ per unit time. Let $X$ denote the content that an item receives and let $L$ denote its lower specification limit: an item is nonconforming if its $X$ value is smaller than $L$. In addition to $L$, an artificial limit $U$ is used to screen out overfilled items for reprocessing. That is, an item is accepted only if its $X$ is between $L$ and $U$. Otherwise, the item is rejected and subject to reprocessing. The material cost of producing an accepted item is $cX$, where $c$ is the unit material cost. It is also assumed that the contents of nonconforming and overfilled items are recovered because of the expensive material cost. Let $R$ denote the per-item reprocessing cost, which is the monetary disbursements incurred during the reprocessing process, including retrieving the materials, cleaning the container, re-arranging a filling attempt, etc. Assume that $X$ follows a normal distribution with a variance of $\sigma^2$ and a mean to be determined by the producer. Let $f(x;\mu,\sigma^2)$ denote the probability density function of $X$. The probability of producing an accepted item in each filling attempt is given by

$$p = \int_L^U f(x;\mu,\sigma^2)dx, \tag{7.1}$$

and the yield rate of the process is $\lambda = r \cdot p$. It is also assumed that all the demand will be satisfied in such a way that the expected total number of accepted items produced is equal to the total demand and no backlog is allowed. Note that use of the expected accepted items is reasonable, especially in high-speed production, because the production output can be treated as approximately constant. Therefore, in order to satisfy the demand, $\lambda$ has to be equal to or larger than $D$.

Furthermore, let $\overline{UC}(\mu, U)$ denote the expected total cost of producing an accepted item. If an item is accepted in the first filling attempt, the production cost is $cX$; otherwise, the reprocessing process on this item starts and continues until it is accepted. The net cost of the latter case is the sum of $\overline{UC}(\mu, U)$ and $R$. Let $UC(X;\mu,U)$ denote the cost of producing an accepted item with content $X$, which is determined as follows:

$$UC(X;\mu,U) = \begin{cases} cX & \text{if } L \leq X \leq U \\ \overline{UC}(\mu,U) + R & \text{otherwise} \end{cases} \tag{7.2}$$

Thus, the expected total cost of producing an accepted item is:

$$\begin{aligned}\overline{UC}(\mu, U) &= \int_L^U cxf(x;\mu,\sigma^2)dx + \int_{-\infty}^{L} \left[\overline{UC}(\mu,U), +R\right] f(x;\mu,\sigma^2)dx \\ &+ \int_U^{\infty} \left[\overline{UC}(\mu,U) + R\right] f(x;\mu,\sigma^2)dx \qquad (7.3)\end{aligned}$$

Define $t_1 = (U-\mu)/\sigma$ and $t_2 = (L-\mu)/\sigma$. It can be verified that $\overline{UC}(\mu, U)$ is given by

$$\overline{UC}(\mu, U) = c\mu - R + \{R - c\sigma\left[\phi(t_1) - \phi(t_2)\right]\} / \left[\Phi(t_1) - \Phi(t_2)\right] \qquad (7.4)$$

where $\phi(.)$ and $\Phi(.)$ are the standard normal density and distribution functions, respectively.

Considering the capacity constraint, the model is

$$\min_{\mu,U} \overline{UC}(\mu, U) = c\mu - R + \{R - c\sigma\left[\phi(t_1) - \phi(t_2)\right]\} / \left[\Phi(t_1) - \Phi)(t_2)\right] \qquad (7.5)$$

Subject to: $r.p \geq D$.

Our objective is to find the optimal process mean $\mu^*$ and upper limit $U^*$ which minimize $\overline{UC}(\mu, U)$ under the capacity constraint.

Equation (6.4) can be written as

$$\overline{UC}(\mu, U) = cE(X|L \leq X \leq U) + R(1-p)/p \qquad (7.6)$$

Note that, since $E(X|L \leq X \leq U)$ is the expected content per accepted item, the first term on the right side of the last equation is the expected material cost per accepted item. The total number of times that a rejected item is reprocessed before it is accepted follows a geometric distribution, and the expected number of times it is reprocessed is $(1-p)/p$. Therefore, the per-accepted-item expected reprocessing cost is $R(1-p)/p$, and the total per-item expected cost is the sum of the expected material cost and the expected reprocessing cost.

The first-order partial derivatives of (6.6) with respect to $t_1$ and $t_2$ are, respectively,

$$\frac{\partial \overline{UC}(\mu, U)}{\partial t_1} = t_1\left[\Phi(t_1) - \Phi(t_2)\right] - R/(c\sigma) + \Phi(t_1) - \Phi(t_2), \qquad (7.7)$$

and

$$\frac{\partial \overline{UC}(\mu, U)}{\partial t_2} = [\Phi(t_1) - \Phi(t_2)]^2 + [t_2\Phi(t_2) - t_1\Phi(t_1)]\,[\Phi(t_1) - \Phi(t_2)] + R/(c\sigma) - [\Phi(t_1) - \Phi(t_2)]^2 \quad (7.8)$$

Setting these partial derivatives equal to zero, we can obtain the optimal solution to the model when the capacity constraint is ignored. If the yield rate of this solution is larger than the demand rate, the solution is still optimal under the capacity constraint. If the resulting yield rate is smaller than the demand rate, however, a larger process mean and/or a larger upper control limit must be used. Because of the complexity of this problem, the solution must be obtained by a direct search procedure. To ensure obtaining optimal solutions, complete enumeration is used in this paper to develop a computer program for finding $\mu^*$ and $U^*$. Using the program, Table 1 is developed for $t_1^*$ and $t_2^*$ as functions of $M = R/(c\sigma)$ for selected values of $K = D/r$. Note that because $r.p \geq D, p \geq K$.

In the table, the solutions associated with $K = 70\%$ and $M \leq 0.7$ are identical and the capacity constraint is binding (i.e., $r.p = D$). In other words, in these cases, the process yield rate has to be set larger than the level associated with the minimum per-accepted-item production cost in order to satisfy the demand. When $K = 70\%$ and $M > 0.7$, excess capacity exists. As discussed in the next section, the solutions are identical to those of the GP model. In the table, the missing $t_1^*$ and $t_2^*$ are identical to those corresponding to $K = 70\%$ and the same $M$ value. For example, for $K = 75\%$ and $M = 1.4$, $t_1^* = 2.0487$ and $t_2^* = -0.884$, which are the same values corresponding to $K = 70\%$ and $M = 1.4$. This phenomenon is also caused by excess capacity when K/M is small. On the other hand, the capacity constraints associated with the cases in the unshaded area are binding. The production cost per accepted item in these cases is higher than those in the cases in which excess capacity is available.

# 3 RELATIONSHIP TO OTHER MODELS

In order to show the effect of the capacity constraint, we compare the solution resulting from the model developed in the previous section with those of the GP and SP models. In this section, we briefly discuss the relationship between

| M | K 70% $t_1^*$ | 70% $t_2^*$ | 75% $t_1^*$ | 75% $t_2^*$ | 80% $t_1^*$ | 80% $t_2^*$ | 85% $t_1^*$ | 85% $t_2^*$ | 90% $t_1^*$ | 90% $t_2^*$ | 95% $t_1^*$ | 95% $t_2^*$ |
|---|---|---|---|---|---|---|---|---|---|---|---|---|
| 0.1 | 1.5483 | -0.7088 | 1.7861 | -0.7962 | 2.1144 | -0.9049 | 2.6588 | -1.0534 | 3.8415 | -1.2819 | 4.5001 | -1.6449 |
| 0.2 | 1.5483 | -0.7088 | 1.7861 | -0.7962 | 2.1144 | -0.9049 | 2.6588 | -1.0534 | 3.8415 | -1.2819 | 4.5001 | -1.6449 |
| 0.3 | 1.5483 | -0.7088 | 1.7861 | -0.7962 | 2.1144 | -0.9049 | 2.6588 | -1.0534 | 3.8415 | -1.2819 | 4.5001 | -1.6449 |
| 0.4 | 1.5483 | -0.7088 | 1.7861 | -0.7962 | 2.1144 | -0.9049 | 2.6588 | -1.0534 | 3.8415 | -1.2819 | 4.5001 | -1.6449 |
| 0.5 | 1.5483 | -0.7088 | 1.7861 | -0.7962 | 2.1144 | -0.9049 | 2.6588 | -1.0534 | 3.8415 | -1.2819 | 4.5001 | -1.6449 |
| 0.6 | 1.5483 | -0.7088 | 1.7861 | -0.7962 | 2.1144 | -0.9049 | 2.6588 | -1.0534 | 3.8415 | -1.2819 | 4.5001 | -1.6449 |
| 0.7 | 1.5483 | -0.7088 | 1.7861 | -0.7962 | 2.1144 | -0.9049 | 2.6588 | -1.0534 | 3.8415 | -1.2819 | 4.5001 | -1.6449 |
| 0.8 | 1.5465 | -0.7095 | 1.7861 | -0.7962 | 2.1144 | -0.9049 | 2.6588 | -1.0534 | 3.8415 | -1.2819 | 4.5001 | -1.6449 |
| 0.9 | 1.5540 | -0.7119 | 1.7861 | -0.7962 | 2.1144 | -0.9049 | 2.6588 | -1.0534 | 3.8415 | -1.2819 | 4.5001 | -1.6449 |
| 1.0 | 1.6559 | -0.7501 | 1.7861 | -0.7962 | 2.1144 | -0.9049 | 2.6588 | -1.0534 | 3.8415 | -1.2819 | 4.5001 | -1.6449 |
| 1.1 | 1.7559 | -0.7863 | 1.7861 | -0.7962 | 2.1144 | -0.9049 | 2.6588 | -1.0534 | 3.8415 | -1.2819 | 4.5001 | -1.6449 |
| 1.2 | 1.8544 | -0.8204 | | | 2.1144 | -0.9049 | 2.6588 | -1.0534 | 3.8415 | -1.2819 | 4.5001 | -1.6449 |
| 1.3 | 1.9523 | -0.8529 | | | 2.1144 | -0.9049 | 2.6588 | -1.0534 | 3.8415 | -1.2819 | 4.5001 | -1.6449 |
| 1.4 | 2.0487 | -0.8840 | | | 2.1144 | -0.9049 | 2.6588 | -1.0534 | 3.8415 | -1.2819 | 4.5001 | -1.6449 |
| 1.5 | 2.1442 | -0.9138 | | | | | 2.6588 | -1.0534 | 3.8415 | -1.2819 | 4.5001 | -1.6449 |
| 1.6 | 2.2392 | -0.9421 | | | | | 2.6588 | -1.0534 | 3.8415 | -1.2819 | 4.5001 | -1.6449 |
| 1.7 | 2.3341 | -0.9691 | | | | | 2.6588 | -1.0534 | 3.8415 | -1.2819 | 4.5001 | -1.6449 |
| 1.8 | 2.4290 | -0.9952 | | | | | 2.6588 | -1.0534 | 3.8415 | -1.2819 | 4.5001 | -1.6449 |
| 1.9 | 2.5239 | -1.0202 | | | | | 2.6588 | -1.0534 | 3.8415 | -1.2819 | 4.5001 | -1.6449 |
| 2.0 | 2.6185 | -1.0439 | | | | | 2.6588 | -1.0534 | 3.8415 | -1.2819 | 4.5001 | -1.6449 |
| 2.2 | 2.8094 | -1.0886 | | | | | | | 3.8415 | -1.2819 | 4.5001 | -1.6449 |
| 2.4 | 2.9990 | -1.1302 | | | | | | | 3.8415 | -1.2819 | 4.5001 | -1.6449 |
| 2.6 | 3.1898 | -1.1683 | | | | | | | 3.8415 | -1.2819 | 4.5001 | -1.6449 |
| 2.8 | 3.3798 | -1.2042 | | | | | | | 3.8415 | -1.2819 | 4.5001 | -1.6449 |
| 3.0 | 3.5726 | -1.2376 | | | | | | | 3.8415 | -1.2819 | 4.5001 | -1.6449 |
| 3.2 | 3.7652 | -1.2690 | | | | | | | 3.8415 | -1.2819 | 4.5001 | -1.6449 |
| 3.4 | 3.9599 | -1.2978 | | | | | | | | | 4.5001 | -1.6449 |
| 3.6 | 4.1491 | -1.3251 | | | | | | | | | 4.5001 | -1.6449 |
| 3.8 | 4.3447 | -1.3513 | | | | | | | | | 4.5001 | -1.6449 |
| 4.0 | 4.5236 | -1.3758 | | | | | | | | | 4.5001 | -1.6449 |
| 4.5 | 5.0355 | -1.4319 | | | | | | | | | 4.8570 | -1.6449 |
| 5.0 | 5.2948 | -1.4818 | | | | | | | | | 5.2948 | -1.6449 |
| 5.5 | 5.2948 | -1.5271 | | | | | | | | | 8.5362 | -1.6449 |
| 6.0 | 5.2948 | -1.5677 | | | | | | | | | 8.5362 | -1.6449 |
| 7.0 | 8.5362 | -1.6393 | | | | | | | | | 8.5362 | -1.6449 |
| 8.0 | 8.5362 | -1.7007 | | | | | | | | | | |
| 9.0 | 8.5362 | -1.7543 | | | | | | | | | | |
| 10.0 | 8.5362 | -1.8018 | | | | | | | | | | |

**Table 1** $t_1^*$ and $t_2^*$ for given M and K

our model (LTC) and the GP and SP models.

Although the GP model uses maximization of the expected profit as the objective, it can be easily verified that the model is equivalent to a special case of our model. In the GP model, it is assumed that the selling price is $A$ per conforming item. Let $P(X; u, U)$ be the profit of selling an item with content $X$ and let $\overline{P}(\mu, U)$ be its expected value. For each accepted item, the profit is the difference between $A$ and $cX$. If an item is rejected, the re-processing process continues until the item is finally accepted. Thus, the profit of a rejected item is equal to the difference of $\overline{P}\ (\mu, U)$ and $R$. The profit function of the model is given as follows:

$$P(X; \mu, U) = \begin{cases} A - cX & \text{if} L \leq X \leq U \\ \overline{P}(\mu, U) - R & \text{otherwise} \end{cases} \quad (7.9)$$

It can be shown that

$$\overline{P}(\mu, U) = A - \overline{UC}(\mu, U), \quad (7.10)$$

where $\overline{UC}(\mu, U)$ is the objective function of the LTC model. Since $A$ is a constant, the GP model and the LTC model are equivalent if the capacity constraint is ignored.

The SP model maximizes the per-filling (per-item) profit rather than the per-accepted-item profit. The cost $R'$ is composed of not only the unit reprocessing cost $R$ but also the capacity opportunity cost for applying another filling attempt to a nonconforming or an overfilled item. The per-item profit, denoted as $P'(X; \mu, U)$, is given as follows:

$$P'(X; \mu, U) = \begin{cases} A - cX & \text{if} L \leq X \leq U \\ -R' & \text{otherwise} \end{cases} \quad (7.11)$$

The expected profit per fill attempt is given as

$$\overline{P}'(\mu, U) = (A - c\mu + R')\,[\Phi(t_1) - \Phi(t_2)] - R' + c\sigma\,[\phi(t_1) - \phi(t_2)] \quad (7.12)$$

If we let $R = R'$ , it is easy to verify that $\overline{P}'(\mu, U) = \overline{P}(\mu, U)/p$. Consequently, $\overline{P}'(\mu, U)$ is the expected profit per filling attempt. It was shown that the optimal upper limit satisfies the following equation

$$cU^* - R' = A \quad (7.13)$$

# 4 NUMERICAL RESULTS

In this section, we provide an example of a situation, and then use this example to compare the three models under selected values of demand rate, per-item reprocessing cost, and process variance.

**Example**

Consider a container-filling process with a lower specification limit $L = 3$ ounces. The production rate is 1,000 items per hour, and the variance of the process is $(0.05)^2$ ounces. The unit material cost is \$ 1.0 per ounce, and the cost of reprocessing an item is \$ 0.05. The demand rate is 850 items per hour. Our computational results show that $t_1^*$ and $t_2^*$ are 2.659 and $-1.053$, respectively. As a result, $\mu^*$ and $U^*$ are 3.053 ounces and 3.186 ounces, respectively, which results in a production yield rate of 850 items per hour. The cost of producing an accepted item is \$3.074.

The solution of the GP model under this situation can be obtained by using the table given in Golhar and Pollock (1988). The optimal process mean and upper limit are 3.038 ounces and 3.120 ounces, respectively. The shortage is 125 items per hour, or 14.71% of the demand. Furthermore, the optimal solution of the SP model is closer to our model: the optimal process mean and upper limit are 3.045 ounces and 3.150 ounces, respectively. The shortage is 52 items per hour, or 6.12% of the demand.

In the remainder of this section, a sensitivity analysis is used to study the effects of the following three parameters on the optimal solutions to the three models: $D, R$ and $\sigma$.

**Effect of D**

Table 2 gives the optimal solutions of the three models for selected values of $D$, ranging from 725 items to 990 items per hour. The results show that, when the demand rate is at 725 items per hour, the solution of our model and that of the GP model are identical. This suggests that the demand rate is relatively much lower than the process capacity. Therefore, the producer can use excess capacity to reprocess the nonconforming items and overfilled items. This is also true for the SP model. Notice that, when the process capacity is much larger than the demand rate, the process should not be considered a bottleneck. Therefore, the solution given by the SP model should not be used under this situation.

Table 2 shows that, when the demand increases, both the process mean and the upper limit of our model increase, resulting in a larger yield rate in order to meet the demand. Note also that, because both the GP and SP models do not consider demand in their formulation, their solutions do not change as the demand changes. As the demand increases, we observe that the larger shortages in both the GP and SP models increase. It appears that the SP model is less sensitive to the change in demand, mainly because the process is closer to being considered a bottleneck when the demand increases.

Golhar and Pollock (1988) used an effective way to evaluate a solution by comparing the expected filling result with the ideal one. Note that the material cost in the ideal situation is $cL$, because every item is filled with exactly $L$ units in every filling attempt. This situation can be achieved only when the process variance is extremely small. We define the per-item expected excess cost $\overline{E}$ as the difference between the expected unit cost $\overline{UC}^*$ and the lower bound $cL$:

$$\overline{E} = \overline{UC}^* - cL \tag{7.14}$$

The definition of $\overline{E}$ is noteworthy, since $\overline{E}$ can be regarded as a function of the conforming rate $p(= D/r)$. From Table 2, we see that, as long as $p$ is larger than 0.725, $\overline{E}$ increases as $D$ increases. This indicates that, when the demand is high, in order to satisfy the capacity constraint, the producer must loose production efficiency by filling more contents into a container to increase the yield rate, which leads to a higher expected unit cost of producing an accepted item. Moreover, although the expected unit reprocessing cost decreases because of a larger conforming rate, $\overline{UC}^*$ increases because of the relatively larger material cost. This fact suggests that, to meet a higher conforming rate in a capacitated situation, the cost effect of using a larger process mean is larger than that of a smaller reprocessing cost.

### Effect of R

Table 3 gives the optimal solutions of the three models for selected value of $R$, ranging from \$ 0.005 per item to \$ 0.250 per item. As $R$ increases, reprocessing becomes more costly. To satisfy the demand, however, it is necessary to maintain the same production yield rate. As indicated in the table, both the process

| D | The LTC Model | | | | The GP Model | | | | The SP Model | | | |
|---|---|---|---|---|---|---|---|---|---|---|---|---|
| | $\mu^*$ | $U^*$ | $p$ | $\overline{E}$ | $\mu^*$ | $U^*$ | $p$ | % Shortage | $\mu^*$ | $U^*$ | $p$ | % Shortage |
| 725 | 3.0375 | 3.1203 | 0.725 | 0.0703 | 3.0375 | 3.1203 | 0.725 | 0 | 3.0450 | 3.15 | 0.798 | 0 |
| 750 | 3.0398 | 3.1291 | 0.705 | 0.0704 | 3.0375 | 3.1203 | 0.725 | 3.33 | 3.0450 | 3.15 | 0.798 | 0 |
| 775 | 3.0424 | 3.1391 | 0.775 | 0.0708 | 3.0375 | 3.1203 | 0.725 | 6.45 | 3.0450 | 3.15 | 0.798 | 0 |
| 800 | 3.0452 | 3.1510 | 0.800 | 0.0716 | 3.0375 | 3.1203 | 0.725 | 9.38 | 3.0450 | 3.15 | 0.798 | 0.25 |
| 825 | 3.0486 | 3.1659 | 0.825 | 0.0727 | 3.0375 | 3.1203 | 0.725 | 12.12 | 3.0450 | 3.15 | 0.798 | 3.27 |
| 850 | 3.0527 | 3.1857 | 0.850 | 0.0743 | 3.0375 | 3.1203 | 0.725 | 14.71 | 3.0450 | 3.15 | 0.798 | 6.12 |
| 875 | 3.0577 | 3.2148 | 0.875 | 0.0764 | 3.0375 | 3.1203 | 0.725 | 17.14 | 3.0450 | 3.15 | 0.798 | 8.80 |
| 900 | 3.0641 | 3.2581 | 0.900 | 0.0794 | 3.0375 | 3.1203 | 0.725 | 19.44 | 3.0450 | 3.15 | 0.798 | 11.33 |
| 925 | 3.0720 | 3.2820 | 0.925 | 0.0837 | 3.0375 | 3.1203 | 0.725 | 21.62 | 3.0450 | 3.15 | 0.798 | 13.73 |
| 950 | 3.0823 | 3.3212 | 0.950 | 0.0903 | 3.0375 | 3.1203 | 0.725 | 23.68 | 3.0450 | 3.15 | 0.798 | 16.00 |
| 975 | 3.0980 | 3.3580 | 0.975 | 0.1023 | 3.0375 | 3.1203 | 0.725 | 25.64 | 3.0450 | 3.15 | 0.798 | 18.15 |
| 990 | 3.1163 | 3.4113 | 0.990 | 0.1182 | 3.0375 | 3.1203 | 0.725 | 26.78 | 3.0450 | 3.15 | 0.798 | 19.39 |

**Table 2** Effects of D

mean and the upper limit increase in order to increase the production yield rate.

It is interesting to observe that, when R increases, the production yield rate of both the GP and SP models increases. Furthermore, in both models, a larger process mean and a larger upper limit are used as $R$ increases. As a result, the shortage decreases as $R$ increases. In reality, it seems unreasonable that the process yield rate is dictated by the reprocessing cost. This suggests that these two models should be used with caution.

**Effect of $\sigma$**

It is well known that the performance of a process can be improved by reducing its inherent variation Deming (1986), Taguchi (1978). For a given process mean, a small process standard deviation implies a higher process yield rate. On the other hand, to maintain the same yield rate, the process mean can be set lower when it is smaller. In this situation, the material requirement is reduced and the production yield rate may be improved at the same time. To demonstrate the effect of the process standard deviation on the optimal solution, the optimal solutions for some selected values of $\lambda$, ranging from 0.5 to 0.01 are reported in Table 4.

As expected, the efficiency of the three models improves. It was found that both $\mu^*$ and and $U^*$ decrease for both our model and the GP model, resulting simultaneously in a smaller material cost and a larger production yield rate. This result is also observed for the SP model, except that $U^*$ remains the same in this model. Furthermore, when $\sigma$ is smaller than 1/40, both our model and the GP model have excess capacities. As a result, the solutions to these two models are identical when $\sigma$ is smaller than 1/40. When $\sigma$ is smaller than 1/20, the SP model has excess capacities. These results suggest the importance of improving the process by reducing process variance in controlling production cost and increasing production outputs.

# 5 DISCUSSION

The container-filling industry is one of the process industries characterized by high-volume production, a small number of standard products, and the use of

| R | The LTC Model | | | | The GP Model | | | | The SP Model | | | |
|---|---|---|---|---|---|---|---|---|---|---|---|---|
| | $\mu^*$ | $U^*$ | $p$ | $\overline{E}$ | $\mu^*$ | $U^*$ | $p$ | % Shortage | $\mu^*$ | $U^*$ | $p$ | % Shortage |
| 0.005 | 3.0527 | 3.1845 | 0.850 | 0.0663 | 3.0118 | 3.0357 | 0.277 | 67.41 | 3.0333 | 3.105 | 0.671 | 21.06 |
| 0.010 | 3.0527 | 3.1848 | 0.850 | 0.0672 | 3.0167 | 3.0508 | 0.383 | 54.94 | 3.0347 | 3.110 | 0.690 | 18.82 |
| 0.025 | 3.0527 | 3.1852 | 0.850 | 0.0699 | 3.0265 | 3.0821 | 0.569 | 33.06 | 3.0388 | 3.125 | 0.739 | 13.06 |
| 0.050 | 3.0527 | 3.1857 | 0.850 | 0.0743 | 3.0375 | 3.1203 | 0.725 | 14.71 | 3.0450 | 3.150 | 0.798 | 6.12 |
| 0.075 | 3.0527 | 3.1861 | 0.850 | 0.0787 | 3.0457 | 3.1529 | 0.804 | 5.41 | 3.0506 | 3.175 | 0.838 | 1.41 |
| 0.100 | 3.0527 | 3.1861 | 0.850 | 0.0831 | 3.0522 | 3.1832 | 0.847 | 0.35 | 3.0554 | 3.200 | 0.864 | 0 |
| 0.125 | 3.0575 | 3.2121 | 0.874 | 0.0871 | 3.0575 | 3.2125 | 0.874 | 0 | 3.0596 | 3.225 | 0.883 | 0 |
| 0.150 | 3.0619 | 3.2405 | 0.892 | 0.0904 | 3.0619 | 3.2405 | 0.892 | 0 | 3.0632 | 3.250 | 0.897 | 0 |
| 0.175 | 3.0656 | 3.2661 | 0.905 | 0.0932 | 3.0656 | 3.2706 | 0.905 | 0 | 3.0664 | 3.275 | 0.901 | 0 |
| 0.200 | 3.0688 | 3.2957 | 0.916 | 0.0957 | 3.0688 | 3.2957 | 0.916 | 0 | 3.0693 | 3.300 | 0.917 | 0 |
| 0.225 | 3.0716 | 3.3229 | 0.924 | 0.0979 | 3.0716 | 3.3229 | 0.924 | 0 | 3.0718 | 3.325 | 0.925 | 0 |
| 0.250 | 3.0741 | 3.3499 | 0.931 | 0.0998 | 3.0471 | 3.3499 | 0.931 | 0 | 3.0741 | 3.350 | 0.931 | 0 |

**Table 3** Effects of R

| $\sigma$ | The LTC Model | | | | The GP Model | | | | The SP Model | | | |
|---|---|---|---|---|---|---|---|---|---|---|---|---|
| | $\mu^*$ | $U^*$ | $p$ | $E$ | $\mu^*$ | $U^*$ | $p$ | % Shortage | $\mu^*$ | $U^*$ | $p$ | % Shortage |
| 1/2 | 3.5267 | 4.8559 | 0.850 | 0.6634 | 3.1179 | 3.3562 | 0.277 | 67.41 | 3.0500 | 3.15 | 0.119 | 86.00 |
| 1/4 | 3.2634 | 3.9279 | 0.850 | 0.3361 | 3.0835 | 3.2539 | 0.383 | 54.94 | 3.0498 | 3.15 | 0.235 | 72.35 |
| 1/10 | 3.1054 | 3.3707 | 0.850 | 0.1397 | 3.0530 | 3.1641 | 0.569 | 33.06 | 3.0487 | 3.15 | 0.531 | 37.53 |
| 1/20 | 3.0527 | 3.1857 | 0.850 | 0.0743 | 3.0375 | 3.1203 | 0.725 | 14.71 | 3.0450 | 3.15 | 0.798 | 6.12 |
| 1/30 | 3.0351 | 3.1240 | 0.850 | 0.0525 | 3.0305 | 3.1020 | 0.804 | 5.41 | 3.0397 | 3.15 | 0.883 | 0 |
| 1/40 | 3.0263 | 3.0930 | 0.850 | 0.0416 | 3.0261 | 3.0916 | 0.847 | 0.35 | 3.0346 | 3.15 | 0.917 | 0 |
| 1/50 | 3.0230 | 3.0849 | 0.874 | 0.0348 | 3.0230 | 3.0850 | 0.874 | 0 | 3.0345 | 3.15 | 0.936 | 0 |
| 1/60 | 3.0206 | 3.0802 | 0.892 | 0.0301 | 3.0206 | 3.0802 | 0.892 | 0 | 3.0272 | 3.15 | 0.949 | 0 |
| 1/70 | 3.0187 | 3.0760 | 0.905 | 0.0266 | 3.0187 | 3.0773 | 0.905 | 0 | 3.0245 | 3.15 | 0.957 | 0 |
| 1/80 | 3.0172 | 3.0739 | 0.916 | 0.0239 | 3.0172 | 3.0739 | 0.916 | 0 | 3.0224 | 3.15 | 0.963 | 0 |
| 1/90 | 3.0159 | 3.0718 | 0.924 | 0.0217 | 3.0159 | 3.0716 | 0.924 | 0 | 3.0206 | 3.15 | 0.968 | 0 |
| 1/100 | 3.0148 | 3.0700 | 0.931 | 0.0200 | 3.0148 | 3.0700 | 0.931 | 0 | 3.0191 | 3.15 | 0.972 | 0 |

**Table 4** Effects of $\sigma$

expensive production equipment. In demand forecasting and capacity planning, producers tend to match the capacity and the demand to achieve both maximum capacity utilization and customer satisfaction. This situation is the source of motivation for this paper.

Demand satisfaction with capacity constraints in the process industries can be found in many situations. A typical example can be found in the economic lot scheduling problem (ELSP), in which it is assumed that demand and capacity are known and are constant over an infinite time horizon, back order is not allowed, and multi-products are produced by a single machine Elmaghraby (1978). The solution algorithm is usually to find the total cycle time and the sub-cycle time for each product. The solutions must be feasible so that the production facilities are not overloaded and the demands of each item are satisfied during its product cycle time. Our model can be used to determine the production strategy for ensuring that the output during the given sub-cycle time meets the required demand.

# REFERENCES

[1] Al-Sultan, K. S, "An Algorithm for the Determination of the Optimal Target Values for Two Machines in Series with Quality Sampling Plans", *International Journal of Production Research,* 12(1), pp 37-45, 1994.

[2] Al-Sultan, K. S. and M. A. Al-Fawzan, "Variance Reduction in a Process with Random Linear Drift", Accepted for publicaton in *International Journal of Production Research*, 1996.

[3] Al-Sultan, K. S. and M. A. Rahim, "Economic Selection of Process Parameters: A Literature Survey", Working paper, Department of Systems Engineering, King Fahd University of Petroleum and Minerals, 1994.

[4] Al-Sultan, K. S. and M. F. S. Pulak, "Process Improvement by Variance Reduction for a Single Filling Operation with Rectifying Inspection", Accepted for publicaton in *Production Planning and Control*, 1996.

[5] Bettes, D.C., "Finding an Optimal Target Value in Relation to a Fixed Lower Limit and an Arbitrary Upper Limit", *Applied Statistics,* 11, pp 202-210, 1962.

[6] Bisgaard, S., W.G. Hunter, and L. Pallesen, "Economic Selection of Quality of Manufactured Product", *Technometrics,* 26, pp 9-18, 1984.

[7] Boucher, T. O. and M. Jafari, "The Optimum Target Value for Single Filling Operations with Quality Sampling Plans", *Journal of Quality Technology,* 23, pp 44-47, 1991.

[8] Craig, R. J., "Normal Family Distribution Functions: FORTRAN and Basic Programs", *Journal of Quality Technology,* 16, pp 232-236, 1984.

[9] Carlsson, O., "Determining the Most Profitable Process Level for a Production Process Under Different Sales Conditions", *Journal of Quality Technology,* 23, pp 44-47, 1984.

[10] Carlsson, O., "Economic Selection of a Process Level under Acceptance Sampling by Variables", *Engineering Costs and Production Economics,* 16, pp 69-78, 1989.

[11] Deming, W. E., *Out of the Crisis*, Cambridge, MA: MIT Press, 1986.

[12] Elmaghraby, S.E., "The Economic Lot Scheduling Problem (ELSP): Review and Extensions", *Management Science,* 24, pp 587-598, 1978.

[13] Golhar, D.Y., "Determination of the Best Mean Contents for a 'Canning Problem'", *Journal of Quality Technology,* 19, pp 82-84, 1987.

[14] Golhar, D.Y., "Computation of the Optimal Process Mean and the Upper Limit for a Canning Problem", *Journal of Quality Technology,* 20, pp 193-195, 1988.

[15] Golhar, D.Y., and S.M. Pollock, "Determination of the Optimal Process Mean and the Upper Limit for a Canning Problem", *Journal of Quality Technology,* 20, pp 188-192, 1988.

[16] Hunter, W. G., and C. P. Kartha, "Determining the Most Profitable Target Value for a Production Process", *Journal of Quality Technology,* 9, pp 176-181, 1977.

[17] Nelson, L. S., "Best Target Value for a Production Process", *Journal of Quality Technology,* 10, pp 88-89, 1978.

[18] Pulak, M. F. S. and K. S. Al-Sultan, "On the Optimum Targeting for a Single Filling Operation with Rectifying Inspection", Accepted for publicaton in *Omega*, 1996a.

[19] Pulak, M. F. S. and K. S. Al-Sultan, "A Computer Package for Process Mean Targeting", Accepted for publicaton in *Journal of Quality Technology*, 1996b.

[20] Schmidt, R.L., and P. E. Pfeifer, "Economic Selection of the Mean and Upper Limit for a Canning Problem with Limited Capacity", *Journal of Quality Technology,* 23, pp 312-317, 1991.

[21] Taguchi, G., *Introduction to Quality Evaluation and Quality Control,* Tokyo, Japan: Japanese Standards Association, 1978.

[22] Tang, K., and J. Tang, "Design of Screening Procedures: A Review", *Journal of Quality Technology,* 26, pp 209-226, 1994.

# 8

# OPTIMAL TARGET VALUES IN MULTIPLE CRITERIA ECONOMIC SELECTION MODELS

**O. Carlsson**

*ESA, Department of Statistics,*
*University of Örebro,*
*S-70182,*
*Sweden.*

## ABSTRACT

Increased demands for quality by customers force producers to reduce the proportion of products with unacceptable values of specified quality characteristics. In this paper a system of equations is derived for calculating the optimal target values in multivariate economic selection models when the customer's quality specifications are given as discrete and continuous open intervals e. g., "the larger the better". Simplified approaches are also studied, which makes it easier to quantify the economic impact of e.g., a reduction of variability in quality characteristics of interest. Examples from the pulp and paper industry are given.

**Key words:** process optimization, process control, quality selection

## 1 INTRODUCTION

The variability of actual quality characteristics can almost always, at least in the beginning of a quality improvement program, be reduced by better ways of running the process, but such a reduction cannot go on forever. Sooner or later, such reduction processes start costing money, and the producer has to determine whether or not to continue with further reduction in the process variation. When facing such a problem economic selection models can often be used.

The quality specification of the multivariate process level $\mu = (\mu_1, \mu_2, ..., \mu_p)'$ can be formulated so that $\mu_k$ either attains a certain value or belongs to a given interval, where $\mu_k$ is the expected value of the $k$ th quality characteristic, $k = 1, 2, ..., p$. The first type of specification is especially common in the manufacturing industries where it is usually the midpoint of some specification interval. The parameter $\mu_k$ is then fixed and the producer has only the variability as an action parameter. However, in the pulp and paper industries and many other process industries, quality specifications of the process levels $\mu_k, k = 1, 2, ..., p$, often belong to open-ended intervals with either an upper or a lower bound, e.g., brightness, bursting strength, bending stiffness, ply bond and Cobb. For such situations, the process levels also become important action parameters because changes in the process levels almost inevitably affect the production costs.

An economic selection model consists of three main factors: economy, production and purchaser's quality requirements. The economy factor includes prices and costs in a broad sense, while production includes distribution, variability, dependence/independence, specification levels and process control. The purchaser's quality requirements factor needs some more consideration, because the relation between the producer and the purchaser is the object of a change. Traditionally, the purchasers ensured quality of delivered lots by means of acceptance sampling by attribute or sometimes by sampling by variables. Nowadays, however, certification programs such as ISO 9000 and an increasing cooperation and trust between the supplier and the purchaser imply that the quality assurance can already be performed on the production line. Further, the purchaser's quality requirements have increased dramatically. For instance, a common quality requirement today is that the simple capability index , $C_p$, defined as $C_p = (USL - LSL)/6\sigma$ (USL and LSL are the upper and lower specification levels) should exceed 1.33, and the probability of a non-conforming unit becomes $32 \times 10^{-6}$ when $\mu = (USL + LSL)/2$. Some companies are even more restrictive. Motorola demands under somewhat different conditions that $C_p \geq 2$, which with their interpretation corresponds to a fraction of $3.4 \times 10^{-6}$ nonconformities. The increase in quality requirements is also reflected in the number of non-conforming items allowed in a lot or during a production period; zero nonconformities is the goal.

Most studies in economic selection share the assumption that the quality characteristic is independently and identically univariate normally distributed and that the quality verification is done either by total inspection or by sampling inspection by variables, see for instance Springer (1951), Bisgaard et al.(1984),

Golhar (1987) and Carlsson and Rydin (1993). Later Arcelus and Rahim (1991) and (1994) and Carlsson (1992) studied some bivariate models.

In this paper, a multivariate economic selection model is studied under the assumption of total inspection or acceptance sampling by attribute, i.e., the produced items are sorted as conforming and as non-conforming items according to a multivariate quality characteristic. The model covers situations where the quality verification is done either by the producer or by the purchaser on process data or, in the traditional manner, by the purchaser on sampled data. The variable production cost functions are general but the first derivatives are assumed to exist. Exact and approximate models and their corresponding systems of equations of partial derivatives, to derive the optimal process levels, are studied. In general, the systems of equations consist of dependent equations. But it is shown by examples that the systems of dependent equations can be successfully reduced to a system of independent equations when the expected number of non-conforming items is small, which implies that the optimization can be done independently. Sometimes even the solutions are explicit, which makes it easy to study the economic impact of changes in different parameters, especially variability in the quality characteristics.

# 2 EXACT AND APPROXIMATE MODELS

## 2.1 Incomes and Costs

Suppose that a lot or the result of a production period, below lot, consists of $n$ independent units and that each unit consists of $p$ independent parts. The outcome can be represented by an $n \times p$ matrix of $\mathbf{X}$ random vectors $(\mathbf{X_1}, \mathbf{X_2}, ..., \mathbf{X_p})$ where $\mathbf{X_k} = (X_{1k}, X_{2k}, ..., X_{nk})$ and assume that $E(X_{ik}) = \mu_k, k = 1, ..., p$, exists and denote the vector of expected values with $\mu$, where $\mu = (\mu_1, \mu_2, ..., \mu_p)$. Assume that the net revenue is $a_1$ for an item belonging to an accepted lot or period; otherwise it is $a_2$. The variable production costs can be modelled in different ways. For a detailed discussion see e.g., Bisgaard et al.(1984) or Carlsson and Rydin (1993). Let the production cost functions be general with the restriction that the first derivatives are assumed to exist. The study has to be separated in two cases. Firstly, the variable production costs are functions of the process levels, i.e., $c(\mu) = n \sum_p^{k=1} c_k(\mu_k)$. Such cost functions can be used when the producer's costs are related to the ***input*** in a production process. An example of such production process is the bleaching of wood pulp, where the variable production costs consist of the input of energy and chemicals required to reach an intended particular process level, e.g., brightness ISO%. Secondly, the production costs per lot are a function of the *output* of the production process, $\mathbf{X}$. The production costs are now random variables and can be written $c(\mathbf{X}) = \sum_{i=1}^{n} \sum_{k=1}^{p} c_k(X_{ik})$. Such production cost functions can be applied, for instance, in the steel industry, where the quality characteristics are some resulting geometric measures.

# 3 PROBABILITY DISTRIBUTIONS

Suppose that the $k$th part within the $i$th unit has the same distribution $F_k$, whose first moment exists, and assume that $A_k$ are given intervals on the $x_k$ axis and denote $P(X_{ik} \subset A_k) = p_k$, and , $q_k = 1 - p_k, k = 1, ..., p$. Let $D_{ik}$ be an indicator variable associated with each $X_{ik}$ such that

$$D_{ik} = \begin{cases} 0 & \text{if } X_{ik} \subset A_k \\ 1 & \text{otherwise} \end{cases} \tag{8.1}$$

$i = 1, 2, ...., n, k = 1, 2, ..., p$. Further, let $D_i = \sum_{k=1}^{p} D_{ik}$ and $D = \sum_{i=1}^{n} D_i$. Then, $D = 0$ when a lot has zero nonconformities.

Closed intervals are excluded from this study because of the common presence of mid-interval target values. The study is restricted to open intervals of the type "larger is better", (Taguchi 1986) i.e., $X_{ik} > l_k$, where $l_k$ are prescribed lower specification levels, $k = 1, 2, ..., p$. It can be noted that restrictions on the process levels of the form $\mu_k \geq T_k$ are permitted in the model, where $T_k$ are prescribed lower process levels or target values, $k = 1, 2, ..., p$. Changes in what follows when some of the requirements are the type "smaller is better" are obvious. The definition (8.1) can now be rewritten as

$$D_{ik} = \begin{cases} 0 & \text{if } X_{ik} \geq l_k \\ 1 & \text{otherwise} \end{cases} \tag{8.2}$$

$i = 1, 2, ..., n, k = 1, 2, ..., p$. The probability $q_k$ can be written as $q_k = F_k(l_k) = F_k$ and the exact probability of zero nonconformities becomes

$$P(D = 0) = \prod_{k=1}^{p} (1 - F_k)^n \tag{8.3}$$

The distribution of $D$ for $d, d = 0, 1, ...,$ can be easily calculated by using the following theorem of Feller (1957). If $D_{ik}$ are independent, $\max(q_k)$ tends to zero and $n\sum_{k=1}^{p} q_k = \lambda$ remains constant when $n$ goes to infinity, $i = 1, 2, ..., n, k = 1, 2, ..., p$, then $D$ is approximately Poisson distributed with parameter $\lambda$.

The distribution function for the number of nonconformities in a lot is then

$$P(D \leq d) = \sum_{j=0}^{d}(1/j!)\left(n\sum_{k=1}^{p} F_k\right)^j \exp\left(-n\sum_{k=1}^{p} F_k\right) \tag{8.4}$$

and it can also be written as an incomplete Γ-function i.e.,

$$P(D \leq d) = \int_R (z^d/d!)\exp(-z)dz \tag{8.5}$$

where $R$ is the interval $(n\sum_{k=1}^{p} F_k, \infty)$. For the case of zero nonconformities the distribution function (8.4) reduces to

$$P(D = 0) = \exp\left(-n\sum_{k=1}^{p} F_k\right) \tag{8.6}$$

Further, for small values of $n\sum_{k=1}^{p} F_k$ the probability of zero nonconformities (8.6) can be further approximated by the first order Taylor expansion

$$P(D = 0) = 1 - n\sum_{k=1}^{p} F_k \tag{8.7}$$

## 3.1 Different Models

If a lot is accepted when the number of nonconformities is less than or equal to $d$, the expected net income, $ENI(\mathbf{X})$, based on the Poisson approximation, becomes

$$\begin{aligned} ENI(\mathbf{X}) &= n\left\{a_2 + (a_1 - a_2)\left[\sum_{j=0}^{d}(1/j!)\left(n\sum_{k=1}^{p} F_k\right)^j \exp\left(-n\sum_{k=1}^{p} F_k\right)\right]\right\} \\ &\quad -n\left\{\sum_{k=1}^{p} c_k(\mu_k)\right\} \end{aligned} \tag{8.8}$$

in the *input* case and

$$ENI(\mathbf{X}) = n\left\{a_2 + (a_1 - a_2)\left[\sum_{j=0}^{d}(1/j!)\left(n\sum_{k=1}^{p}F_k\right)^j \exp\left(-n\sum_{k=1}^{p}F_k\right)\right]\right\}$$
$$-n\sum_{i=1}^{n}\sum_{k=1}^{p}E\left[c_k(X_{ik})\right] \quad (8.9)$$

in the *output* case.

The derivatives of (8.8) and (8.9) with respect to $\mu_k, k = 1, 2, ..., p$, are easily calculated when using the integral (8.5), and they are identical except for the last term. For example, the derivatives of (8.8) form the system of equations

$$-n(a_1 - a_2)(\partial F_k/\partial \mu_k)\left[(1/d!)\left(n\sum_{k=1}^{p}F_k\right)^d \exp\left(-n\sum_{k=1}^{p}F_k\right)\right] = \partial c_k(\mu_k)/\partial \mu_k$$

which for the zero nonconformities case simplifies to

$$-n(a_1 - a_2)(\partial F_k/\partial \mu_k)\exp\left(-n\sum_{k=1}^{p}F_k\right) = \partial c_k(\mu_k)/\partial \mu_k, \; k = 1, 2, ..., p \quad (8.10)$$

The expected net incomes (8.8) and (8.9) can be approximated with the approximate expected net income, $ANI(X)$, for the zero nonconformities case, when applying the Taylor expansion (8.7). The approximate net income, $ANI(\mathbf{X})$, simplifies for the *input* case to

$$ANI(\mathbf{X}) = n\left[a_1 - (a_1 - a_2)n\sum_{k=1}^{p}F_k - \sum_{k=1}^{p}c_k(\mu_k)\right] \quad (8.11)$$

and for the *output* case to

$$ANI(\mathbf{X}) = n\left[a_1 - (a_1 - a_2)n\sum_{i=1}^{p}F_k - \sum_{k=1}^{n}\sum_{k=1}^{p}E(c_k(X_{ik}))\right] \quad (8.12)$$

As above, the derivatives of (8.11) and (8.12) with respect to $\mu_k, 1, 2, ..., p$, are identical except for the last term, and the derivatives of (8.11) give the system of equations

$$-n(a_1 - a_2)(\partial F_k/\partial \mu_k) = \partial c_k(\mu_k)/\partial \mu_k \quad (8.13)$$

The equations in the system (8.13), $\partial ANI(\mathbf{X})/\partial\mu_k = 0$, differ from the corresponding equations in the system (8.10), $\partial ENI(\mathbf{X})/\partial\mu_k = 0, k = 1, 2, ..., p$, by not containing the exponential term, $\exp(-n\sum_{k=1}^{p} F_k)$. Then, for both the *input* and the *output* case the system of equations based on the Taylor approximation (8.13) consists of $p$ independent equations, which implies that the optimization can be done independently for each quality characteristic.

# 4 EXAMPLES

Two examples are given. The first is to show the simplicity, and the second the accuracy, of proposed approximations.

**Example 1**

Arcelus and Rahim (1991) studied an economic selection model with two independent quality characteristics under item-by-item inspection. One quality characteristic was continuous and followed a normal distribution, $X_1 \sim N(\mu, \sigma^2)$, and the other was an attribute variable and Poisson distributed, $X_2 \sim Po(\lambda)$. They derived and solved numerically the system of equations for the optimal target values. In an example, the continuous quality characteristic, $X_1$, was the weight of a paper roll with the lower specification limit $l_1 = 525$ lb per roll. The attribute quality characteristic, $X_2$, was the number of brownish spots with the upper specification limit $l_2 \leq 3$. Their model contained four different net income functions depending on the outcome. Here, we restrict the study to only two outcomes and corresponding net incomes. The producer's net income is $a_1$ for an accepted roll and $a_2$ for a rejected roll, and a roll is accepted when both $X_1 \geq 525$ and $X_2 \leq 3$, otherwise it is rejected. The variable part of the production cost function for $X_1$ was assumed to be linear and of the output type, $c_1(X_1) = c_1 \cdot (X_1 - l_1)$, and the production cost function for $X_2$ was assumed to be exponential and the input type, $c_1(X_2) = c_2 \exp(l_2 - \lambda)$. In this case $\mathbf{X}$ is a $1 \times 2$ matrix. The expected net income for the whole based on the Taylor approximation, (8.11) and (8.12) , can be written as

$$\begin{aligned} ANI(\mathbf{X}) &= a_1 - (a_1 - a_2)\left\{\Phi\left((l_1 - \mu)/\sigma\right) + \sum_{j=1}^{l_2} \exp(-\lambda/k!)\right\} \\ &\quad -c_1 \cdot (\mu - l_1) - c_2 \exp\cdot(l_2 - \lambda) \qquad (8.14) \end{aligned}$$

The corresponding equations become

$$\begin{aligned} \phi\left[(l_1-\mu)/\sigma\right] &= c_1\sigma/(a_1-a_2) \\ \lambda^{l_2}\exp(-\lambda)/l_2 &= c_2\exp(l_2-\lambda) \end{aligned}$$

and they have the solution

$$\begin{aligned} (\mu_{opt}-l_1)/\sigma &= \left\{-\ln\left[2\pi\left(c_1\sigma/(a_1-a_2)\right)^2\right]\right\}^{1/2} \\ \lambda_{opt} &= e(c_2 l_2!)^{1/2} \end{aligned}$$

**Example 2**

Carlsson and Rydin (1993) studied an economic selection model for bleaching of pulp. A sample of 25 rolls was taken from a lot of 250 rolls, and the only quality characteristic of interest, $X$, brightness ISO%, was found to be $N(\mu, .2025)$. The customer's quality requirement was given as the lower capability index, $\hat{C}_{pk} = \left(\overline{X} - l_1\right)\sqrt{n}/3S \geq 1.33$, at the lower specification limit $l_1 = 85\%$ ISO. The net income for each roll belonging to an accepted or to an unaccepted lot was 700SEK and 300SEK, respectively. Further, the main variable production costs consisted of the *input* of different chemicals and could be properly described with the exponential function $c(\mu) = 147 \cdot \exp(.1772(\mu - 85))$.

$$i = 1, 2, ..., 25$$

Assume that another customer instead demands that all $X_i \geq 85\% ISO$, $i = 1, 2, ..., 25$, which is equivalent to zero nonconformities. In this case the number of nonconformities in the sample, $D$, is distributed as $bin\,[25, \Phi\left((85-\mu)/.45\right)]$. The exact expected net income, founded on (8.3), $EXNI(\mathbf{X})$, not earlier defined in this paper, becomes

$$\begin{aligned} EXNI(\mathbf{X}) &= 250\{300 + 400\left[1 - \Phi\left((85-\mu)/.45\right)\right]^{25} \\ &\quad -147\exp\left(.1732(\mu - 85)\right)\} \end{aligned}$$

where $\mathbf{X}$ here is a $25 \times 1$ matrix. Numerical calculations give $\mu_{opt} = 86.50\%$ISO, $EXNI_{opt}(\mathbf{X}) = 126,280$ SEK and the probability of rejection becomes .0108. The expected net income, based on the Poisson approximation (8.6), $ENI(\mathbf{X})$, becomes

$$\begin{aligned} ENI(\mathbf{X}) &= 250\{300 + 400\exp\left[-25\Phi\left((85-\mu)/.45\right)\right] \\ &\quad -147\exp\left(.1732(\mu - 85)\right)\} \end{aligned}$$

and $\mu_{opt} = 86.50\%ISO$, $ENI_{opt}(\mathbf{X}) = 126,280$ SEK and the probability of rejection becomes .0108.

The expected net income for the approximate model founded on the Taylor expansion (8.7) can be written as

$$ANI(\mathbf{X}) = 250\,\{300 + 400\,[1 - 25\Phi\,((85 - \mu)/.45)] - 147\exp\left(.1732(\mu - 85)\right)\}$$

and $\mu_{opt} = 86.50\%ISO$, $ANI_{opt}(\mathbf{X}) = 126,275$ SEK and the probability of rejection becomes .0108.

# REFERENCES


[1] Arcelus, F.J. and M.A. Rahim, "Joint Determination of Optimum Variable and Attribute Target Means", *Naval Research Logistics*, 38, pp 851-864, 1991.

[2] Arcelus, F.J. and M.A. Rahim, "Simultaneous Economic Selection of Variables and an Attribute Target Mean", *Journal of Quality Technology*, 26, pp 125-133, 1994.

[3] Bisgaard, S., W.G. Hunter, and L. Pallesen, "Economic Selection of Quality of Manufactured Products", *Technometric*, 26, pp 9-18, 1984.

[4] Carlson, O., "Determining the Most Profitable Process Level Under Different Sales Conditions", *Journal of Quality Technology*, 16, pp 44-49, 1984.

[5] Carlson, O., "Quality Selection of a Two-dimensional Process Level under Single Acceptance Sampling by Variables", *International Journal of Production Economics*, 27, pp 43-55, 1992.

[6] Carlsson, O. and S. Rydin, "Quality Selection under Sampling Inspection with an Exponential Production Cost Function", *International Statistical Review*, 61, pp 109-119, 1993.

[7] Feller, W., *An Introduction to Probability Theory and Its Applications*, 2, Ed., Wiley, New York, 1957.

[8] Golhar, D.Y., "Determination of the Best Mean Contents for a Canning Problem", *Journal of Quality Technology*, 19, pp 82-84, 1987.

[9] Springer, C.H., "A Method for Determining the Most Economic Position of a Process Mean", *Industrial Quality Control*, 8, pp 36-39, 1951.

[10] Taguchi, G. *Introduction to Quality Engineering*, Asian Productivity Organization, Tokyo, Japan, 1986.

# 9

# UNIFORMITY OF PRODUCTION VS. CONFORMANCE TO SPECIFICATIONS IN THE CANNING PROBLEM

**F. J. Arcelus**

*Faculty of Administration,*
*University of New Brunswick,*
*Fredericton, New Brunswick,*
*Canada E3B 5A3.*

## ABSTRACT

This paper analyses an issue of great practical importance for many production processes, namely how to coordinate the apparently contradictory goals of producing not only in accordance to specifications but also with as much uniformity as possible in the characteristic of interest. The primary objective is to assess the viability of combining the twin quality objectives of minimizing rejection rates and maximizing the uniformity of production of the resulting items. The basic import of the study is that, when flexibility in setting specification limits and non-uniformity penalties exists, optimal results can be obtained which yield approximately the same profit per unit as that associated with the traditional within-specifications policy, while at the same time providing lower rejection rates and decreases in process variability.

**Key Words** Quality, Variance Reduction, Taguchi Penalty Function, Conformance -to-Specifications, Optimization

## 1 INTRODUCTION

In presenting their case against the zero-defects approach to quality control, Taguchi and Clausing (1990, p. 68) state:

> "Robustness derives from consistency. Where deviation is consistent, adjustment to the target is possible; catastrophic stack-up is more

likely from scattered deviation within specifications than from consistent deviation outside. This regard for consistency, for being on target, has a fascinating and practical application."

Consistency or uniformity of production is a widely accepted objective of any quality management program. Consistent performance adjusted to target or uniformity of production around some given socially ideal value forms the cornerstone of the Taguchi approach (e.g., Ross (1988); Taguchi (1985, 1986); Taguchi, et al.(1988)). Its predominance in the industrial world over the traditional conformance to specifications is unquestionable (e.g., Montgomery (1992)).

And yet, conformance to specifications is still an important objective, specially in the presence of waste and/or poor design and/or arbitrariness or excessively narrow quality specifications (e.g., Lee and Woo (1989)) and/or too high process variance, often in spite of great efforts to reduce it (e.g., McClish (1983, 1985)). Under these conditions, if revenue is to be generated, some degree of consistency may have to be sacrificed, in favor of producing within specifications. Hence, some trade-off between the two objectives has to be contemplated (e.g., Arcelus and Rahim (1996); Kackar (1986); Singpurwalla (1992)). In this area, neither the Taguchi nor the consumer/producer risk approaches provide much guidance (e.g., Easterling, et al.(1991); Singpurwalla, 1992).

It is the purpose of this paper to study such a trade-off. Our points of departure are the canning problem of Golhar and Pollock (1988; 1992) and the work of Arcelus and Rahim (1996) towards the modeling of this trade-off. To that effect, the paper is organized as follows. The next section presents the original formulation of the canning problem, as in Golhar and Pollock (1988) and the various variations needed to model the trade-off between objectives. Included here are those in Arcelus and Rahim (1996) and a new model, designed to strengthen the methodology needed to evaluate the trade-off. This is followed by the derivation of the optimality conditions and by an assessment of their differences across models. Numerical examples will be used throughout to highlight the main issues in the comparative analysis. A Conclusions section completes the paper.

# 2 SIMULTANEOUS MODELLING OF UNIFORMITY AND CONSISTENCY

In light of the objectives of the paper, this section presents and contrasts three models. The first, reproduces the canning problem of Golhar and Pollock (1988), as an example of a model emphasizing conformance to specifications. The other two introduce the notion of conformity. One considers a unique target for both objectives. The other considers the effect of different targets.

## 2.1 Model 1- Conformance-to-Specifications Model (M1)

The M1 of Golhar and Pollock (1988) considers the familiar scenario of a can being filled with an expensive ingredient. For the can to be acceptable, its weight, $X$, must fall within predetermined specifications, with the lower and upper limits denoted by $L$ and $U$ ounces, respectively. The random variable, $X$ is assumed to be normally distributed, with a mean of $\mu$ and a standard deviation of $\sigma$. Each acceptable can sells for a price of $A$ and incurs in a production cost of $c$ per ounce. Those which are not, are reprocessed at an additional cost of $R$ per can, defined as the rejection penalty. Its profit function, $P_1$, may be written as follows:

$$\begin{aligned} P_1 &= A - cX_1, && \text{if } L \leq X_1 \leq U_1 \\ &= E_1 - R, && \text{otherwise} \end{aligned} \qquad (9.1)$$

where $E_1$ represents the expected profit per can, which, given the normality assumptions with respect to $X_1$, can be written as :

$$E_1 = A - R[1/(\Delta\Phi_1) - 1] - C(\mu_1 - \sigma\Delta\phi_1/\Delta\Phi_1)$$

where

$$\begin{aligned} \Delta\phi_1 &= \phi(u_1) - \phi(l_1) \\ \Delta\Phi_1 &= \Phi(u_1) - \Phi(l_1) \\ l_1 &= (L - \mu_1)/\sigma \\ u_1 &= (U_1 - \mu_1)/\sigma \end{aligned} \qquad (9.2)$$

$\Phi$ and $\phi$ are the normal cumulative distribution function and its probability density function, respectively. The objective is to find the values of the standard normal variates, $l_1$ and $u_1$, evaluated at the respective specification limits, a fixed $L$ and a variable $U_1$, so as to maximize the expected profit per part, as defined in (2.2). Note that the real variables of interest are $\mu_1$ and $U_1$, but for analytical convenience $l_1$ and $u_1$ are used. A similar approach is used for the other models.

## 2.2 Model 2 - Unified Conformance-Uniformity-Target Model (M2)

The M1 of (2.1) considers conformance to specifications as the main objective. Hence, the target value which optimizes the manufacturer's production/rejection trade-off is optimal. On the other hand, if uniformity is the only issue, Taguchi's socially ideal value is in order. Consensus exists only when one objective is paramount.

The controversy arises when both objectives are important and hence need to be considered simultaneously. In this case, the salient issue is how to model the manufacturer/consumer trade-off that results from incorporating the profit function into both objectives. Arcelus and Rahim (1996) justify the need for a combined approach in terms of the need to account for two sets of costs, as follows:

> "One set deals with the manufacturing costs of meeting the conformance to specifications objective. The other set includes the cost to society or the consumer's costs and are related to the uniformity-of-production objective. When the process variance is small enough to render the former objective relatively unimportant, then only the uniformity objective remains salient and the Taguchi formulation attempts to deal with this problem. However, if conformance is still an issue, both objectives are salient, in which case both should be represented in the objective function."

The main problem in combining the two objectives lies in deciding whether two target values, one for each objective, should be used or whether a combined target is more appropriate. Convincing arguments exist on either side. On the one hand, it may be argued that Taguchi's socially ideal value approach, with its presumption of negligible nonconforming costs, is not adequate, because it makes no provision for the inclusion, when needed, of the traditional manufacturer's production/rejection trade-off, as does the $M1$ of (2.1). The $M2$ of this section is intended to evaluate such an approach. It is based on the proposition that neither objective should be used, because of the need for the simultaneous satisfaction of both objectives. Instead, the optimal target should correspond to neither and should represent a balance between the two criteria. The case for the other approach and its corresponding model appears in the next section.

The $M2$ of Arcelus and Rahim (1996) adds a penalty to the profit, $P_2$, of a can proportional to the square deviation of its weight from the combined target. This uniformity penalty factor, denoted by $a_2$, is computed as the ratio of the average loss, $R$, to the square of the tolerance, $(u_2 - l_2)^2$ (e.g., Arcelus and Rahim, 1996; Kackar, 1986; Taguchi, 1985). Then, the profit per can, $P_2$, and its corresponding expected profit per can, $E_2$, are given by

$$\begin{aligned} P_2 &= A - cX_2, -a_2(X_2 - \mu_2)^2, && \text{if } L \le X_2 \le U_2 \\ &= E_2 - R, && \text{otherwise} \end{aligned}$$

where

$$E_2 = E_1 - a_2\sigma^2 \left\{1 - \left[u_2\phi(u_2) - l_2\phi(l_2)\right]/\Delta\Phi\right\}$$

and

$$\begin{aligned} \Delta\phi_2 &= \phi(u_2) - \phi(l_2) \\ \Delta\Phi_2 &= \Phi(u_2) - \Phi(l_2) \\ l_2 &= (L - \mu_2)/\sigma \\ u_2 &= (U_2 - \mu_2)/\sigma \\ a_2 &= 4R/\left[(u_2 - l_2)^2\sigma^2\right] \end{aligned} \tag{9.3}$$

As before, the decision variables are the two normal variates, $l_2$ and $u_2$, evaluated at the respective specification limits, a fixed $L$ and a variable $U_2$ and the objective is the maximization of the expected profit, $E_2$.

## 2.3 Model 3- Independent Conformance-Uniformity-Target Model (M3)

The other side of the controversy on whether a combined target or two independent ones should be used hinges upon the proposition that the Taguchi quadratic loss concept is based on the idea that any deviation away from some socially ideal value should be penalized, not just deviations which fall outside specification limits. The important thing to note is that the Taguchi philosophy is not so much concerned with uniformity of production per se, but rather uniformity of production around some given socially ideal value. Hence, the $M3$ of this section uses two target values. The first, denoted here by $\mu_3$, represents the process mean for which optimality results from trade-offs between production costs outside and inside the specification limits. The other, $T$, represents the given socially ideal weight of the can. Otherwise, the function form of the profit, $P_3$, and its corresponding expected profit, $E_3$, are derived as in (2.3) above, with the change in subscript from 2 to 3, when appropriate , i.e.,

$$\begin{aligned} P_3 &= A - cX_3 - a_3(X_3 - T)^2, && \text{if } L \leq X_3 \leq U_3 \\ &= E_3 - R, && \text{otherwise} \end{aligned}$$

where

$$E_3 = E_2 - a_3\sigma^2 t_3 \left(2\Delta\phi_3 - t_3\Delta\Phi_3\right)/\Delta\Phi_3$$

and

$$\begin{aligned} \Delta\phi_3 &= \phi(u_3) - \phi(l_3) \\ \Delta\Phi_3 &= \Phi(u_3) - \Phi(l_3) \\ l_3 &= (L - \mu_3)/\sigma \\ u_3 &= (U_3 - \mu_3)/\sigma \\ a_3 &= 4R/\left[(u_3 - l_3)^2\sigma^2\right] \\ t_3 &= (T - \mu_3)/\sigma \end{aligned} \tag{9.4}$$

As before, the decision variables are the two normal variates, $l_3$ and $u_3$ , evaluated at the respective specification limits, a fixed $L$ and a variable $U_3$ and the objective is the maximization of the expected profit, $E_3$.

The Lagrangian function and the optimality conditions for all three models appear in Appendix A. To simplify notation, subscripts are dropped unless needed. A numerical example is discussed next. All computations are done with the Optimization Toolbox of MATLAB (1992).

# 3 EMPIRICAL RESULTS

To facilitate comparability of results, the industrial canning problem used by Golhar and Pollock (1988) and Arcelus and Rahim (1996) is also used here. Unless specifically stated, the basic values of the parameters are:

$$(R, C, L, \sigma) = (\$0.2, \$0.5, 3.0\,\text{ounces}, 0.4\,\text{ounces}) \qquad (9.5)$$

The value of $A$ is not needed, since, as shown in Appendix A, it does not appear in the optimality conditions. Hence, the optimal values of the objective functions are given in terms of $A - E_m$, $m = 1, 2, 3$. The term $A - E_m$ denotes the optimal break-even revenues (BERs), obtained from minimizing $A - E_m$ or its equivalent from maximizing $E_m$, as given in Appendix A.

The primary purpose of this section is to highlight the impact of the various uniformity/conformity policies, underlying each model on the optimality conditions. The discussion of the results is based upon graphical illustrations of the various sensitivity analyses conducted. To simplify the exposition, the mathematical development does not convey any additional insight and hence it is omitted to simplify the exposition. From Appendix A, it is clear that there are four process constants of interest, namely $R, c, \sigma$ and $T$. To that effect, the analysis starts with the effect of $R$ and $T$ on the BERs, for various values of the rejection penalty, $R$, higher and lower than the production cost, $c$, per can and for various values of $T$. The line with $T = 0$ corresponds to the $M1$ model of Figure 1 with asterisks (*). $M2$, with straight line, depicts the effect of $R$, when the target value, $T$, is also equal to the average weight, $\mu_2$. As stated before, the optimal average weight of a can in $M2$ does not correspond to that of $\mu_1$. Rather, it results from a compromise between the two objectives, conformity and uniformity. The next three are associated with $M3$ and relate to three different socially ideal values of the weight. In practice, if the socially expected weight of a can is listed at $T$ ounces, it is expected that $L$ be set at a somewhat lower value, representing the minimum weight acceptable. In Figure 1, the $L - T$, or its equivalent $(L - T)/\sigma$, is set at three possible values. They are the minimum possible, $M3(T = L)$, included here as a lower bound for $T - L$ and two more realistic cases, $M3(T = L + .05)$ and $M3(T = L + .1)$ representing differences in $T - L$ of .05 and of .1, respectively.

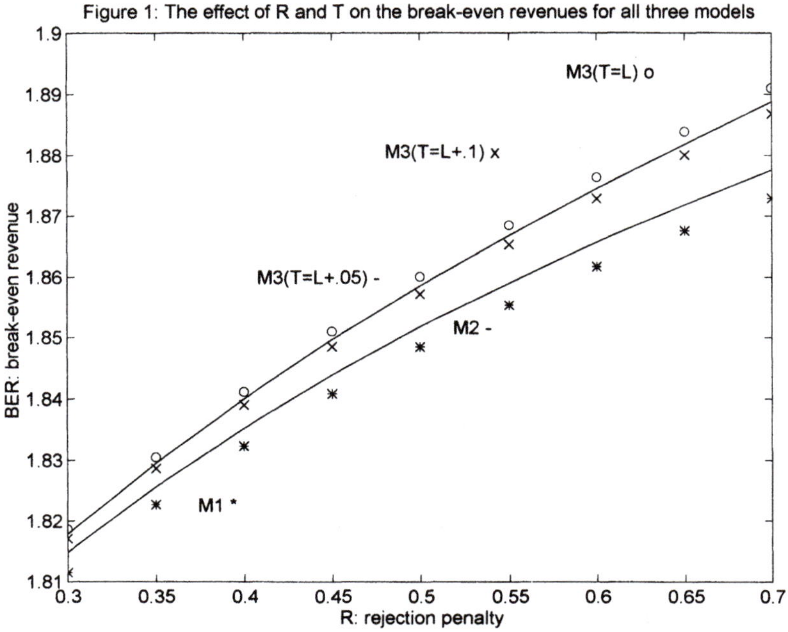


Even a cursory look at Figure 1 and at the underlying data suggests the following observations. The first and most obvious is that increases in $R$ lead to decreases in the profitability of each can or, equivalently, to increases in its BER. Within this context, $M1$ yields the most profitable policy, since the manufacturer has to be concerned with only one objective, namely conformance to specifications. $M2$ comes a close second, since both objectives can be combined into one by the selection of a can weight which also represents the target weight, even if the latter may represent the socially ideal one. Once $T$ is set independently, then the closer it is set to $L$, the less profitable is the policy. This is to be expected, since the need to keep the weights as close as possible to a $T - L$ increases the probability of rejection and thus of incurring the rejection penalty. Finally, it should be observed that the differences among the BERs are relatively small, not only for the example presented in this paper, but for others not reported here. This confirms once again the claim that increases in quality do not necessarily have to be expensive if appropriate policies are set in place. It also confirms one of the results of Arcelus and Rahim (1996) to the effect that Golhar and Pollock (1988)'s policy of setting flexible upper limits are tantamount to increasing the uniformity of the can weights.

Figure 2 depicts the effect of fluctuations in $R$ and/or $T$ in an important quality indicator for the conformance objective, namely the probability that the weight of a given can falls between a given lower limit $L$ and its optimally obtained upper limit, $U^*$. For $M1$, the smaller the rejection penalty, $R$, the lower the incentive to produce within the limits and hence the higher the probability of rejection. The addition of the uniformity incentive in $M2$, with its associated new penalty, increases the probability of acceptance. As $R$ increases, both policies yield almost equal probabilities. For reasons stated above, increases in $R$ lead to increase in the acceptance probabilities. However, the rate of increase is in inverse relation to the magnitude of the $T - L$ difference. However, in our extensive computational experience, no substantial differences have been observed on the basis of fluctuations in the value of $T - L$.

Figure 2: The effect of R and T on the probability of acceptance

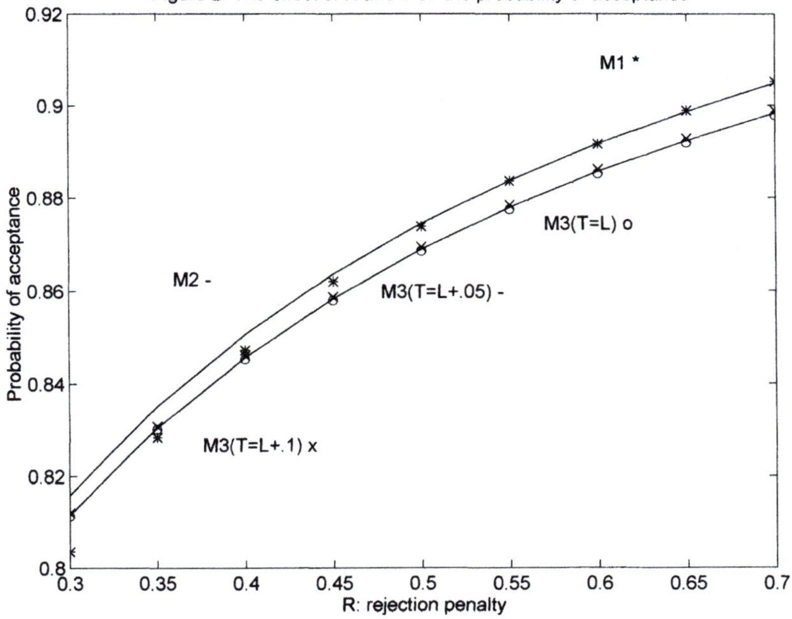


Figures 3, 4 and 5 illustrate the impact of $R$ and $T$ on the optimal mean weight of the can, $\mu^*$, the optimal standardized lower limit, $l^*$ and the optimal standardized target value, $t^* = (\mu^* - T)/\sigma$. $M1$ and $M2$, for reasons similar to those alluded to earlier, yield the highest mean weight, $\mu^*$ (Figure 3). This is as it should be. Given the higher priority allotted to the conformance objective, it is expected that, in the absence of a non-uniformity penalty independent of the mean weight, the optimal average weight would be set further away from the mean than in the case of $M3$. This can also be seen in the larger negative values of the standardized lower limit (Figure 4). The three $M3$ cases result in lower average weights, since the uniformity objective requires values closer to the lower limit. How well the uniformity objective may be met is illustrated in Figure 5. As expected, the closer $T$ is to the lower limit, the further apart from each other are the uniformity and conformance target values of $\mu^*$ and $T$. Nevertheless, as in the cases of Figures 1 and 2, the latter shows once again very low variation. Further discussion on this point is provided in the Conclusions section.

Figure 3: The effect of R and T on the optimal mean can weight

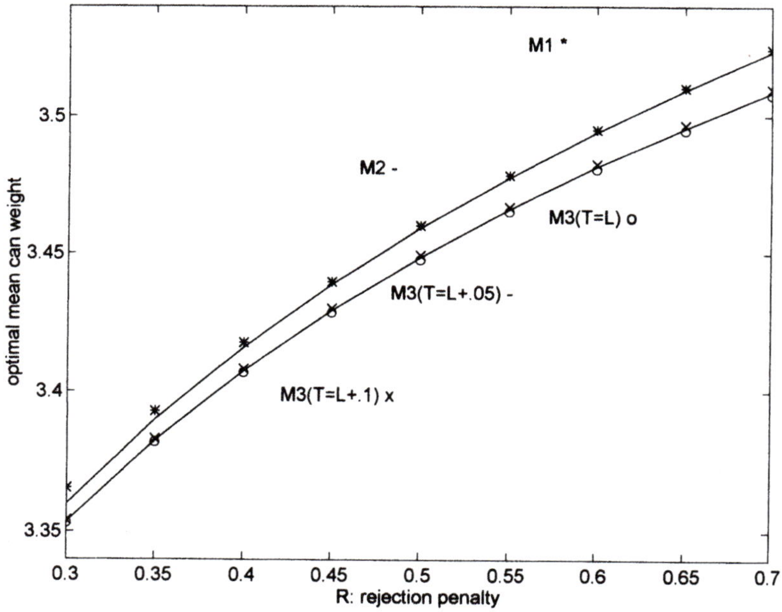


The influence of changes in $\sigma$ on the optimal policies is straightforward to discern and hence no graphical illustratiions are provided. Higher variability results in lower profits, higher rejection rates, less uniformity and larger mean can weights. Finally, the impact of the unit production cost, $c$, is of two types.

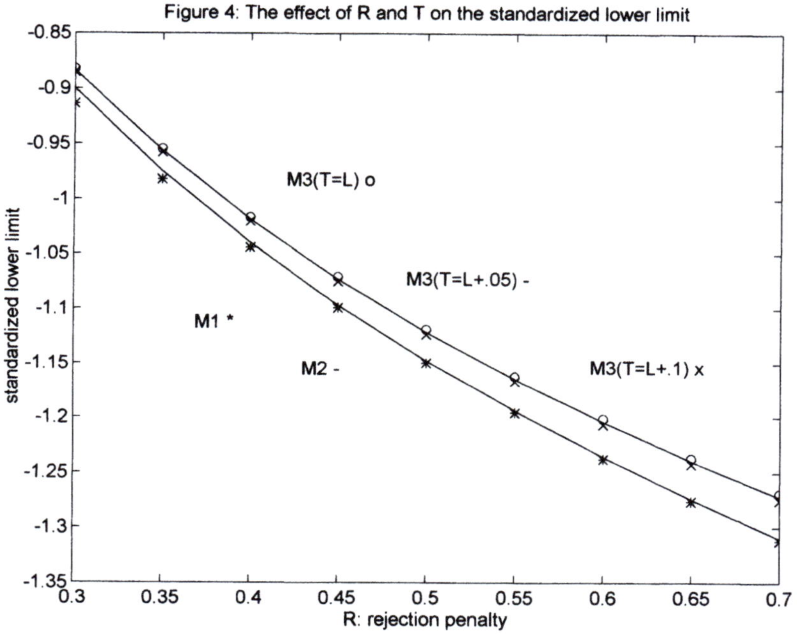


First, larger production costs lead to lower profits and to an increase in the relative importance of the conformity objective vis a vis that of production uniformity. The second hinges upon its magnitude relative to that of the rejection penalty, as has been illustrated in Figures 1-5.

# 4 SOME CONCLUDING COMMENTS

This paper has endeavored to analyse an issue of great practical importance for many production processes, namely how to coordinate the apparently contradictory goals of producing not only in accordance to specifications but also with as much uniformity as possible in the characteristic of interest. To that effect three models have been evaluated, each emphasising one of the issues of the paper. The first, M1, considers the conformace-to-specifications objective. M2 and M3 incorporate the uniformity objective into the formulation. The difference between these two models is the target value to be selected for the satisfaction of such an objective. M3 adopts Taguchi's standard "societal" ap-

Figure 5: The effect of R and T on the standardized target value

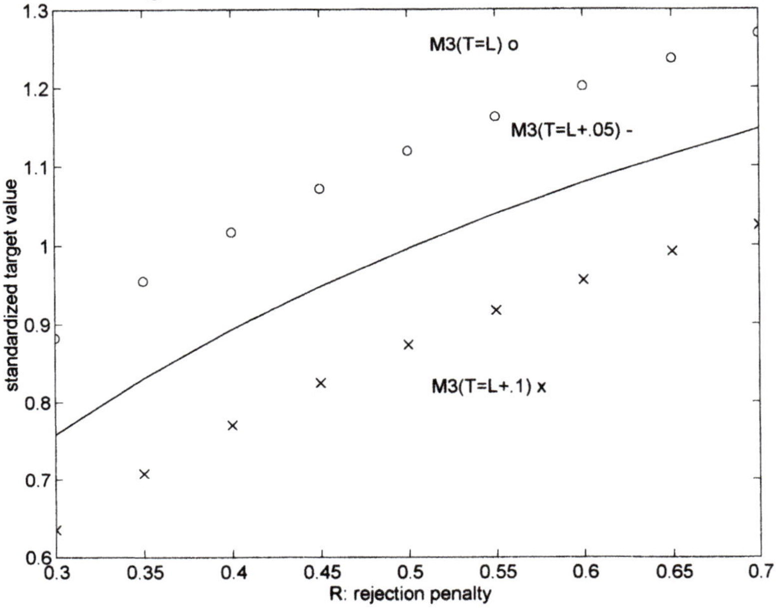


proach, with uniformity target value being a parameter of the model, outside the control of the producer. On the other hand, M2 seeks a target value that simultaneously optimizes both objectives, even it may differ from that of M3. All three address a different concern and the rationale for their inclusion in the paper is presented in the appropriate sections.

M2 and M3 include both goals in the objective function and attempt to optimize them simultaneously. Such an approach implies equal importance assigned to uniformity and conformity to specifications. Several alternatives may also be considered in future research endeavors. Two spring immediately to mind, given their industrial applicability. The conformity objective of this paper assigns equal importance to underweight (under L) and overweight (over U) items, by rejecting any can whose weight falls outside these limits and then assigning a common penalty, R, for the reprocessing of these cans, regardless of the reasons for original rejection. This may not be an appropriate objective in all two-sided cases. Often, underweight or, in general, falling below the lower limit, implies rejection of the item, with the high opportunity costs, associated with the loss of consumer goodwill. At the same time, going over the upper limit carries a different set of penalties mostly associated with giveaway costs. For these cases, two possible types of models may be considered. One assigns different importance to each goal represented in the objective function, by optimizing a weighted average of the objectives, with the determination of the weights left at the discretion of the user (e.g., Arcelus and Banerjee (1987)). The second alternative consists of optimizing a particular objective, subject to at least a minimum level of satisfaction for the second, expressed in Melloy (1991) as the maximum acceptable risk of noncompliance. Whereas the second approach appears easier to implement, given the well-known difficulties in estimating weights, the determination of the more appropriate technique for each application remains an empirical question, which only further research may be able to ascertain.

The basic import of the study is clear from the graphical illustrations. The example used here plus numerous others studied provide clear indication of consistency of policies regardless of which technical objective has preference, if any. The primary implication of this result is quite attractive for the practitioner. When flexibility in setting specification limits and non-uniformity penalties exists, optimal results can be obtained which yield approximately the same profit per unit than as associated with the traditional within-specifications policy, while at the same time providing lower rejection rates and decreases in process variability.

Obviously, a study of this type has to be validated through the use of a plethora of examples for many different production processes under a variety of objectives. But the thrust of the argument is straightforward. Flexibility in parameter setting in order to allow the process to adjust optimally the penalties associated with the two types of non-compliance appears the best policy.

# Acknowledgment

Financial assistance for the completion of this research from the Natural Sciences and Engineering Research Council of Canada is gratefully acknowledged.

# REFERENCES


[1] Arcelus, F.J. and P.K. Banerjee, "Optimal Production Plan in a Tool Wear Process with Rewards for Acceptable Undersized and Oversized Parts," *Engineering Costs and Production Economics*, 11(1), pp 13-19, 1987.

[2] Arcelus, F.J. and M.A. Rahim, "Reducing Performance Variation in the Canning Problem, *European Journal of Operational Research*, accepted, 1996.

[3] Easterling, R.G., M.E. Johnson, T.R. Bement and C.J. Nachtsheim, "Statistical Tolerance Based on Consumer's Risk Considerations," *Journal of Quality Technology*, 23, pp 1-11, 1991.

[4] Golhar, D.Y. and S.M. Pollock, "Determination of the Optimal Process Mean and the Upper Limit for a Canning Problem,", *Journal of Quality Technology*, 20, pp 188-192, 1988.

[5] Golhar, D.Y. and S.M. Pollock, "Cost Savings Due to Variance Reduction in a Canning Process," *IIE Transactions*, 24, pp 89-92, 1992.

[6] Kackar, R.N., "Taguchi's Quality Philosophy : Analysis and Commentary," *Quality Progress*, December 21-29, 1986.

[7] Lee, W.J. and T.C. Woo, "Optimum Selection of Discrete Tolerance," *Transactions of ASME, Journal of Mechanisms, Transmissions and Automation in Design*, 111(2), pp 243-251, 1989.

[8] MATLAB, *Optimization Toolbox*, The Mathworks Inc., Natick, MA.

[9] McClish, W.P., "Statistical Process Control - An Eye Opener," *Proceedings of Ceramic Engineering and Science*, pp 557-563, 1983.

[10] McClish, W.P., "Use of Statistical Controls in Production," *Proceedings of Ceramic Engineering and Science*, pp 165-171, 1985.

[11] Melloy, B.J., "Determining the Optimal Process Mean and Screening Limits for Packages Subject to Compliance Testing," *Journal of Quality Technology*, 23(4), pp 318-323, 1991.

[12] Montgomery, D.C., "The Use of Statistical Process Control and Design of Experiments in Product and Process Improvement," *IIE Transactions*, 24, pp 4-17, 1992.

[13] Ross, P.J., *Taguchi Techniques for Quality Engineering*, New York, McGraw Hill, 1988.

[14] Singpurwalla, N.D., "A Bayesian Perspective on Taguchi's Approach to Quality Engineering and Tolerance Design," *IIE Transactions*, 24, pp 18-27, 1992.

[15] Taguchi, G. "Quality Engineering in Japan," *Communications in Statistics : Theory and Methods*, 14, pp 2785-2801, 1985.

[16] Taguchi, G. *Introduction Quality Engineering*, Asian Productivity Organization, Tokyo, 1986.

[17] Taguchi, G., and D. Clausing, "Robust Quality," *Harvard Business Review*, 68, pp 65-75, 1990.

[18] Taguchi, G., E.A. Elsayed and T. Hsiang, *Quality Engineering in Production Systems*, McGraw Hill, New York, 1988.

# APPENDIX A

## OPTIMALITY CONDITIONS

Lagrangian Function: Min $Z = E - \sum_{j=1}^{3} \lambda_j y_j$, for $m = 1, 2, 3$

Optimality Conditions
$$\begin{array}{lll} y_1 & l - (L - \mu)/\sigma & = 0, \text{ for } m = 1,2,3 \\ y_2 & u - (U - \mu)/\sigma & = 0, \text{ for } m = 1,2,3 \\ y_3 & a - 4R/[(u-l)^2\sigma^2] & = 0, \text{ for } m = 1,2 \end{array}$$

$E'_u - 2aE'_a h_m/(u - l) = 0$ and $E'_l + h_m E'_u + c\sigma = 0$
where

$$\begin{aligned} h_m &= 0, && \text{for } m = 1 \\ &= 1, && \text{for } m = 2,3 \\ E'_l &= -\phi_l \left[R - c\sigma(l\Delta\Phi + \Delta\phi)\right]/\Delta\Phi^2 \text{ for } m = 1 \\ &= E'_{l_1} - a\sigma^2 \left\{\Delta\Phi(1 - l^2) - [u_2\phi(u_2) - l_2\phi(l_2)]\right\}, \text{ for } m = 2 \\ &= E'_{l_2} - 2a\sigma^2 t\phi_l(t\Delta\Phi + \Delta\phi)/\Delta\Phi^2 \text{ for } m = 3 \end{aligned}$$

$$
\begin{aligned}
E'_u &= \phi_u \left[R - c\sigma(u\Delta\Phi + \Delta\phi)\right] / \Delta\Phi^2 \text{ for } m = 1 \\
&= E'_{u_1} + a\sigma^2 \left\{\Delta\Phi(1 - u^2) - [u_2\phi(u_2) - l_2\phi(l_2)]\right\}, \text{ for } m = 2 \\
&= E'_{u_2} - 2a\sigma^2 t\phi_u(t\Delta\Phi + \Delta\phi)/\Delta\Phi^2 \text{ for } m = 3 \\
E'_a &= \text{ for } m = 1 \\
&= \sigma^2 \left\{[u_2\phi(u_2) - l_2\phi(l_2)] - \Delta\Phi\right\} / \Delta\Phi \text{ for } m = 2 \\
&= E'_{a_2} + \sigma^2 t\phi - t\Delta\Phi)/\Delta\Phi \text{ for } m = 3
\end{aligned}
$$

# PART IV

## OPTIMAL SETUP, CONTROL, MONITORING AND TESTING

# 10

# A STEPWISE-OPTIMAL PROCEDURE FOR SETTING MACHINES AND ADJUSTING PROCESSES

**B. J. Melloy, M. A. Coffin and P. C. Kiessler**

*College of Engineering and Science,*
*Clemson University,*
*South Carolina,*
*USA.*

## ABSTRACT

Machine setting is one of the predominant assignable causes of quality variation in a production process. Therefore, a systematic procedure for setting a machine to target is essential, especially in advance of short production runs. Grubbs has developed such a procedure; it yields a final setting which is a minimum variance unbiased estimator of the target. Nevertheless, the intermediate settings are not explicitly considered, and as such, any economic losses or other consequences which may ensue are neglected. Hence, the objective of this research is to develop a supplementary procedure that optimizes the intermediate settings while maintaining the desirable properties of the final setting.

**Key words:** machine centering, machine setup, process adjustment, optimal adjustment

## 1 INTRODUCTION

Certain processes are setup dominant; that is, the effect of setup on the quality characteristics dominates the other process variables (Juran and Gryna (1980)). When this is the case, it is imperative to center the setup before production starts (Juran and Gryna (1980)). Examples of such processes are punching, drilling, cutting to length, broaching, die cutting, die drawing, molding, coil winding, labeling, sheet-metal bending, flame cutting, heat sealing, printing, and presswork. In fact, some processes are "...so highly reproducible that if the setup is correct the lot will be correct" (Juran and Gryna (1980)). Similarly, in the context of using Shewhart control charts to monitor processes, Montgomery (1991), lists three primary sources of special causes of variation; here again, "improperly adjusted machines" is cited.

Moreover, both the trend in many industries towards smaller volume production schedules and the increasing use of just-in-time inventory systems result in smaller lot sizes with correspondingly shorter production runs (Cullen (1989)). Naturally, shorter production[1] runs require more frequent machine changeovers (and accompanying setups) to accommodate the large variety of jobs which commonly use the same process or equipment. In such an environment, accurate setups are even more critical because in many instances the "... job is completed before the sequence of "analysis, feedback, and corrective action" can be completed" (Seder (1988)).

Process setting or adjustment represents one dimension of the overall setup problem. Other aspects of the problem that have been examined include setup simplification (Granger (1989); Noaker (1995)), setup time reduction (DeGarmo, Black and Kosher (1996)), setup scheduling (Ladany (1994); Pugh (1987); Pugh (1988)), the effects of loss (due to setup) on production scheduling (de Matta and Guignard (1994)), and determining the optimal process setting (Ladany (1994); Ladany (1995); Ladany and Ben-Arieh (1990); Pugh (1987); Pugh (1988)). However in these studies, the investigators either have not addressed setup adjustment, or have assumed that the process is initially on target, which in practice is rarely the case (Mackertich (1990)). Investigators that have considered this problem usually employ experimental methods, under conditions where the relationship(s) between the product characteristic(s) and the process setting(s) are unknown (Bhogesara, Nunn and McCarthy (1995); Parikh, Quilty and Gardiner (1991)).

Setup adjustment has been described as "...an often neglected aspect of process design." Consequently, setup may often be performed on a "cut-and-try basis" (Juran and Gryna (1980)). Grubbs (1983), however, has developed a systematic analytical procedure for setting machines and adjusting processes. This procedure prescribes a series of adjustments which are based upon a product characteristic (e.g., a dimension) of consecutively produced groups of items. That is, the product characteristic of interest is measured, and the machine is subsequently adjusted according to a formula based on the difference between the average measure and the target. (Naturally, this requires knowledge of the

[1] Shewhart control charts are best suited for application in high volume manufacturing. Thus, the trend in industry towards short-run production has led to the development of statistical process control methods expressly for this environment (e.g., see Farnum (1992)).

relationship between the product characteristic and the process setting[2].) This procedure is designed to bring the mean of the process on target, while compensating for both the intrinsic measurement error and process variability. In fact, the procedure yields a final adjusted setting that is a minimum variance unbiased estimator of the target.

Nevertheless, while the final setting is explicitly considered, the intermediate settings and product characteristics are not. The latter could certainly be a concern, for example, in the event that out-of-spec products have to be discarded, or if the items are difficult, time-consuming, or expensive to rework. Accordingly, the objective of this research will be to develop an adjustment strategy that will optimize the intermediate settings and/or product characteristics within the framework of the existing procedure, while maintaining the desirable properties of the final setting.

# 2 MODEL DEFINITION

The values of the product characteristic are determined by the position of the mean, and any deviations therefrom which may occur at that point in time; that is

$$Y_{ij} = X_{ij} + U_i \tag{10.1}$$

for $i = 1, 2, \ldots, n$, and $j = 1, 2, \ldots, m$, where $Y_{ij}$ denotes the characteristic value of the $j$-th item of the $i$-th group; $U_i$ denotes the process mean value prior to the $i$-th adjustment; and $X_{ij}$ denotes the deviation of the value of the $j$-th item of the $i$-th group from the mean. The deviations from the mean are assumed to have an expected value of zero, and a standard deviation equal to $\sigma_X$.

In certain instances, the value of the product characteristic may be obscured to some degree by measurement error. This error is reflected in the basic construct

[2] Juran and Gryna (1980) cite several ways that this knowledge can be secured: "from the planners, who provide information relating process variables to product characteristics"; "from cut-and-try experience by the operator"; and "from the fact that the units of measure for product and process are identical." Another alternative would be to pursue an inverse regression approach. Such an approach is appropriate here since the objective is to estimate the value of the independent variable (the process setting) corresponding to a measured value of the dependent variable (the product characteristic). This common situation is referred to as the calibration problem. Draper and Smith (1981), for example, describe inverse prediction procedures for both linear and nonlinear models; eighteen additional references on the subject are also provided in this source.

(as defined by Equation (10.1)) thus:

$$O_{ij} = Y_{ij} + E_{ij} \tag{10.2}$$

for $i = 1, 2, \ldots, n$, and $j = 1, 2, \ldots, m$, where $O_{ij}$ denotes the measure (i.e., the observed value) of the $j$-th item of the $i$-th group; and $E_{ij}$ denotes the error associated with the measure of the $j$-th item of the $i$-th group. The measurement method or device is assumed to be unbiased (that is, there is no systematic error), with a precision of $\sigma_E$.

Lastly, Grubbs' (1983) derivation required the assumption that the deviations, as well as the measurement errors, were uncorrelated. Secondly, it was also implicitly assumed therein that $X_{1j}$ and $E_{1j}$, for $j = 1, 2, \ldots m$, were not correlated with $U_1$. Thirdly, it was stated that the deviations and measurement errors were assumed to be mutually independent. Herein, the second assumption will remain in effect. However, the first and third assumptions will be replaced by the following single assumption: the collection of random variables, $\{X_{hj}, E_{ij}$, for $h = 1, 2, \ldots, n$, $i = 1, 2, \ldots, n$, and $j = 1, 2, ..., m\}$, may be correlated for $h = i$, but are otherwise uncorrelated.

Regardless of which set of assumptions is considered, it necessarily follows that the process mean and the deviations therefrom are uncorrelated. (This result will become more apparent when Theorem 1 is subsequently presented in Section 4). Nevertheless, the latter assumption has noteworthy implications which distinguish it from those it replaces. First, in practice, it is relatively more straightforward to demonstrate a lack of correlation than it is to establish statistical independence. Moreover, it will not be necessary to separately characterize the measurement errors and process deviations, which may also be difficult in practice. Rather, it is only necessary to characterize the sum, or, in other words, the *apparent* deviations from the mean, which can be readily observed.

# 3 GRUBBS' ADJUSTMENT PROCEDURE

Grubbs' (1983) adjustment procedure involves measuring (a particular characteristic of) the members of a series of $n$ groups of $m$ items, and successively adjusting the machine by a fraction of the difference between the average group measure and the target. (The word "group" is used here in a generic sense, as the adjustments may be based on the measures of individual items.) First, the average measure of the $i$-th group, $\overline{O}_i$, is obtained by averaging the $m$

measurements of the individual items as follows:

$$\overline{O}_i = \sum_{j=1}^{m} O_{ij}/m, \tag{10.3}$$

for $i = 1, 2, \ldots, n$.

Subsequently, the machine is adjusted by a fraction of the difference between the average measure, $\overline{O}_i$, and the target value, $t$; that is

$$A_i = k_i(\overline{O}_i - t), \tag{10.4}$$

for $i = 1, 2, \ldots, n$, where $k_i$ is the constant that represents the degree of fractional correction.

The values of the corrections are determined such that the adjustment procedure will yield a final adjusted mean setting that is a minimum variance unbiased estimator of the target. Specifically, the equation for the optimal correction is

$$k^* = 1/i, \tag{10.5}$$

for $i = 1, 2, \ldots, n$ (refer to Appendix A for further detail).

Lastly, Grubbs (1983) stated that the adjustment procedure may continue until such time as "...the standard deviation of the final adjusted level is suitably small." Clearly, the latter "...can be made arbitrarily small by a suitable choice of the number of items which are to be measured and used in the adjustment process." Nevertheless, it should be noted that unless

$$mn \geq \sigma^2_{X+E}/(\sigma^2_D + \mu^2_D) \tag{10.6}$$

the mean square error of the final adjusted setting will be greater than that of the original setting, since

$$MSE[U_{n+1}] = E[(U_{n+1} - t)^2] = (E[U_{n+1}] - t)^2 + V[U_{n+1}] = (\sigma^2_{X+E})/mn, \tag{10.7}$$

due to Equations (10.9) and (A.8), and

$$MSE[U_1] = \mu^2_D + \sigma^2_D, \tag{10.8}$$

due to Equation (A.1), where $\mu_D$ and $\sigma^2_D$ denote the mean and variance of the initial displacement, respectively. In other words, unless the number of items to be employed ($mn$) equals or exceeds the minimum specified in Equation (10.6), the benefit of proceeding with the adjustment procedure would be arguable.

# 4 OPTIMIZATION OF INTERMEDIATE SETTINGS AND PRODUCTS

The equation for the final adjusted mean setting, $U_{n+1}$, can be found by substituting the result for the $i$-th optimal correction (10.5), into the equation for the process mean setting after the $i$-th adjustment (A.3), via mathematical induction:

**Theorem 1**

$$U_{n+1} = t - (1/mn)\sum_{i=1}^{n}\sum_{j=1}^{m}(X_{ij} + E_{ij})$$

for $n = 1, 2, \ldots,$

**Proof**

Refer to Appendix B

From this result, it is easily verified both that the final setting is an unbiased estimator of the target; that is

$$E[U_{n+1}] = t, \tag{10.9}$$

and that, indeed, $V[U_{n+1}] = (\sigma^2_{X+E})/mn$, as indicated by Equation (A.8).

More importantly, Theorem 1 also reveals that although the final setting is dependent on the collective number of items ($mn$), it is not dependent upon the group size selected when the collective number of items is fixed. Thus, the opportunity exists to optimize both the intermediate states of the process and the attributes of the items produced during the adjustment procedure by manipulating the group size (and thereby the number of adjustments, or vice versa), without any possibility of adversely affecting the final mean setting.

First, for the set of intermediate mean settings, $\{U_1, U_2, \ldots, U_n\}$, the measure of optimality employed will be the average mean square error, hereafter denoted by $\tau(n)$; thus

$$\tau(n) \equiv E\left[\sum_{i=1}^{n}(U_i - t)^2/n\right]$$

$$\begin{aligned} &= \left(V[U_1] + (E[U_1]-t)^2\right)/n + (1/n)\delta(n)\sum_{i=2}^{n}\left(V[U_i] + (E[U_i]-t)^2\right) \\ &= (\sigma_D^2 + \mu_D^2)/n + \delta(n)\left[(\sigma_{X+E}^2)/mn\right]\sum_{i=1}^{n-1}(1/i) \qquad (10.10) \end{aligned}$$

with

$$\delta(n) = \begin{cases} 0 & \text{if } n = 1 \\ 1 & \text{if } n > 1 \end{cases}$$

due to Equations (10.9), (A.1), and (A.8).

A similar measure, in the form of the expected value of the average squared distance from target, will be employed for the product characteristics. That is, for the set of intermediate product characteristics,

$$\{Y_{11}, Y_{12}, \ldots, Y_{1m}, Y_{21}, Y_{22}, \ldots, Y_{2m}, \ldots, Y_{n1}, Y_{n2}, \ldots, Y_{nm}\},$$

$$\begin{aligned} E\left[\sum_{i=1}^{n}\sum_{j=1}^{m}(Y_{ij}-t)^2/mn\right] &= E\left[\sum_{i=1}^{n}\sum_{j=1}^{m}(X_{ij}+U_i-t)^2/mn\right] \\ &= \left[\sum_{i=1}^{n}\sum_{j=1}^{m}\left[E(U_i-t)^2 + 2E(U_i-t)E[X_{ij}]\right.\right. \\ &\qquad \left.\left. +E[X_{ij}^2]\right]/mn\right] \\ &= \left[\sum_{i=1}^{n}\sum_{j=1}^{m}\left[E(U_i-t)^2 + \sigma_X^2\right]/mn\right] \\ &= \tau(n) + \sigma_X^2 \qquad (10.11) \end{aligned}$$

due to Equations (10.1) and (10.10) and the assumption that the process settings and deviations therefrom are uncorrelated.

Observe that the rightmost term in Equation (10.10) reflects the variability introduced by the adjustments, which increases as the number of adjustments is increased. The other term is indicative of both the variability and the bias due to the initial random displacement, the effect of which diminishes as the number of adjustments increases, in contrast to the rightmost term. Thus a trade-off is evident as the number of adjustments is increased (or decreased). Lastly, the rightmost term in Equation (10.11) represents the inherent variation of the process (or machine) at a given mean setting; thus this term does not appear in Equation (10.10). Moreover, since this term is a constant, the value of $n$ which minimizes this equation will minimize (10.10) (and vice versa).

The objective, then, is to determine the value of $n$ that minimizes Equation (10.10), subject to the restrictions that $(mn)$ is fixed and that $n$ is an integer-multiple of $m$. Toward this end, it can be shown that $\tau$ has a unique minimum at

$$n' = \lceil \rho(mn) \rceil, \tag{10.12}$$

for $\rho(mn) \in [1, mn]$, where

$$\rho(mn) = (\sigma_D^2 + \mu_D^2)/(\sigma_{X+E}^2/mn), \tag{10.13}$$

and $\lceil \rho(mn) \rceil$ denotes the greatest integer less than or equal to $\rho(mn)$. If $\rho(mn)$ is not an integer-multiple of $m$, then the optimal solution, $n^*$, will be an adjacent integer-multiple of $m$, since $\tau$ is uni-modal on the domain of interest.

In the event that $\rho(mn) \notin [1, mn]$, then one of either two outcomes will result, both of which require further consideration. First, when $\rho(mn) < 1$, no (nonzero) solution exists. This result confirms the earlier statement regarding the questionable benefit of undertaking the adjustment procedure when $MSE[U_1] < MSE[U_{n+1}]$. The reason for this becomes apparent in view of the fact that $\rho(mn)$ may be alternately expressed as

$$\rho(mn) = MSE[U_1]/MSE[U_{n+1}], \tag{10.14}$$

due to Equations (10.7) and (10.8). Secondly, when $\rho(mn) > mn$, $n^* = mn$. This result is also intuitively appealing, since it would clearly be advantageous to adjust at the earliest opportunity when $MSE[U_1] >> MSE[U_{n+1}]$.

To illustrate further, the optimal number of adjustments for procedures ranging in the number of items used from two to eight are listed in Table 1. Observe that the results therein are consistent with the observations above; that is, $n^*$ is small when $\rho(mn)$ is small, and becomes larger as $\rho(mn)$ increases. Finally, it appears from Table 1 that in general, the proposed optimization procedure is not sensitive to small errors in estimation.

# 5 SUMMARY

Grubbs' (1983) procedure for setting machines and processes was considered once again herein. This procedure yields a final setting which is a minimum

| $mn$ | Interval for $\rho(mn)$ | $n^*$ |
|---|---|---|
| 2 | $[1.0, 2.0]$ | 1 |
| | $[2.0, \infty)$ | 2 |
| 3 | $[1.0, 2.25]$ | 1 |
| | $[2.25, \infty)$ | 3 |
| 4 | $[1.0, 2.0]$ | 1 |
| | $[2.0, 3.3\overline{3}]$ | 2 |
| | $[3.3\overline{3}, \infty)$ | 4 |
| 5 | $[1.0, 2.60417^1]$ | 1 |
| | $[2.60417, \infty)$ | 5 |
| 6 | $[1.0, 2.0]$ | 1 |
| | $[2.0, 3.0]$ | 2 |
| | $[3.0, 4.7]$ | 3 |
| | $[4.7, \infty)$ | 6 |
| 7 | $[1.0, 2.88611^2]$ | 1 |
| | $[2.88611, \infty)$ | 7 |
| 8 | $[1.0, 2.0]$ | 1 |
| | $[2.0, 3.3\overline{3}]$ | 2 |
| | $[3.3\overline{3}, 6.07619^3]$ | 4 |
| | $[6.07619, \infty)$ | 8 |

[1] 125/48
[2] 1039/360
[3] 638/105

**Table 1** The optimal number of adjustments as a function of $mn$ and $\rho(mn)$.

variance unbiased estimator of the target. Moreover, the intermediate settings possess the same desirable properties as the final setting.

It was demonstrated that while the final setting is dependent on the collective number of observations used in the adjustment process, it is nonetheless not a function of the group (sample) size. Conversely, the intermediate settings are dependent on the sample size, and as a consequence, these settings, and the products thereof, are not necessarily optimal (despite the desirable properties of the former).

Thus, a supplementary procedure was developed which determines the sample size so as to optimize the intermediate settings and products. This procedure is performed within the framework of Grubbs' original procedure, and as such, the desirable properties of both the intermediate and final settings are maintained. Although the proposed procedure requires the estimation of parameters beyond that of the original procedure, it is premised on a milder set of assumptions which are both more amenable to the estimation of these parameters and easier to verify in practice. Furthermore, there is evidence to suggest that the procedure is robust to estimation error.

## REFERENCES

[1] Bhogesara, Anil R., Robert E. Nunn, and Stephen P. McCarthy, "Injection Molding Manufacturing Productivity - Machine Setup," *Proceedings of the 53rd Annual Technical Conference*, Boston, Massachusetts, pp 576-580, 1995.

[2] Cullen, C. C., "SPC for Short Production Runs," *Proceedings of the Quality in Electronics Meeting*, San Jose, California, pp 147-150, 1989.

[3] DeGarmo, E. P., J T. Black, and R. Kosher, *Materials and Processes in Manufacturing*, Eighth Edition, Prentice Hall, Englewood Cliffs, New Jersey, 1996.

[4] de Matta, Renato and Monique Guignard, "Studying the Effects of Production Loss Due to Setup in Dynamic production Scheduling," *European Journal of Operational Research* 72(1), pp 62-73, 1994.

[5] Draper, N. R. and H. Smith, *Applied Regression Analysis*, Second Edition, John Wiley & Sons, New York, 1981.

[6] Farnum, N. R., "Control Charts for Short Runs: Nonconstant Process and Measurement Error," *Journal of Quality Technology*, 24(3), pp 138-144, 1992.

[7] Granger, C., "Benefits of Single Set-up Production," *Machinery and Production Engineering* 147, pp 24-35, 1989.

[8] Grubbs, F. E., "An Optimum Procedure for Setting Machines or Adjusting Processes." *Journal of Quality Technology* 15(4), pp 186-189, 1983.

[9] Juran, J. M., and F.M. Jr. Gryna, *Quality Planning and Analysis*, Second Edition, McGraw-Hill Book Company, New York, 1980.

[10] Ladany, Shaul P., "Optimal Combined Set-up and Calibration Policy," *International Journal of Advanced Manufacturing Technology*, 9(2), pp 134-140, 1994.

[11] Ladany, Shaul P., "Optimal Set-up of a Manufacturing Process with Unequal Revenue from Oversized and Undersized Items," *Proceedings of the 1995 IEEE Annual International Engineering Management Conference, Singapore, Singapore*, pp 428-432, 1995.

[12] Ladany, S. P. and D. Ben-Arieh, "Optimal Industrial Robot-Calibration Policy," *International Journal of Advanced Manufacturing Technology*, 5, pp 345-357, 1990.

[13] Mackertich, Neal A., "Precontrol vs. Control Charting: A Critical Comparison," *Quality Engineering*, 2(3), pp 253-260, 1990.

[14] Montgomery, D. C., *Introduction to Statistical Quality Control*, Second Edition, John Wiley & Sons, New York, 1991.

[15] Noaker, Paula M., "Simplifying Setup," *Manufacturing Engineering*, 115(1), pp 35-39, 1995.

[16] Parikh, Mayank R., William F. Jr. Quilty, and Keith M. Gardiner, "SPC and Setup Analysis for Screen Printed Thick Films," *IEEE Transactions on Components, Hybrids, and Manufacturing Technology*, 14(3), pp 493-498, 1991.

[17] Pugh, G. Allen, "The Most Economic Setting for a Uniformly Shifting Process," *Proceedings of the 9th Annual Conference on Computers and Industrial Engineering*, Atlanta, Georgia, pp 381-385, 1987.

[18] Pugh, G. Allen, "An Algorithm for Economically Setting a Uniformly Shifting Process," *Computers and Industrial Engineering*, 14(3), pp 237-240, 1988.

[19] Seder, Leonard A., "Job Shop Industries," in *Juran's Quality Control Handbook*, Fourth Edition, Edited by J. M. Juran and Frank M. Gryna, McGraw Hill, New York, 1988.

# APPENDIX A

Initially, the process mean setting has a value of:

$$U_1 = t + D, \tag{A.1}$$

where $D$ represents the initial displacement, relative to the target value. It is assumed that this displacement is random (and of course, unknown), with mean $\mu_D$ and variance $\sigma_D^2$. (Grubbs (1983) considered two variations of this fundamental construct; one wherein the initial displacement was treated as a constant, and the other in which the displacement was considered to be a random variable with mean zero. Instances of random displacement, without the restriction that the mean is equal to zero, are of interest herein.)

Subsequent to the $i$-th adjustment, the mean setting becomes

$$U_{i+1} = U_i - A_i, \tag{A.2}$$

for $i = 1, 2, \ldots, n$. Proceeding recursively, it can be demonstrated that the expression for the process mean setting after the $i$-th adjustment is

$$U_{i+1} = t + \prod_{j=1}^{i}(1 - k_j)D - \sum_{j=1}^{i} k_j(1/m)\sum_{j=1}^{m}(X_{ij} + E_{ij}) \prod_{l=j+1}^{i} (1 - k_l), \tag{A.3}$$

for $i = 1, 2, \ldots, n$, due to Equations (A.1), (A.2), (10.1), (10.2), and (10.4), where it is specified that

$$\prod_{l=i+1}^{i} (1 - k_l) = 1 \tag{A.4}$$

(adapted from Grubbs, 1983). Correspondingly, the expected value and variance of the $(i + 1)$-st mean setting can be shown to be

$$E[U_{i+1}] = t + \prod_{j=1}^{i}(1 - k_j)E[D], \tag{A.5}$$

and

$$V[U_{i+1}] = \prod_{j=1}^{i}(1-k_j)^2\sigma_D^2 + [(\sigma_{X+E}^2)/m]\sum_{j=1}^{i} k_j^2 \prod_{l=j+1}^{i}(1-k_l)^2, \qquad \text{(A.6)}$$

respectively, for $i = 1, 2, \ldots, n$ (adapted from Grubbs, 1983).

In order to obtain a minimum variance unbiased estimator of the target, Equation (A.6) is minimized subject to the constraint that:

$$\prod_{j=1}^{i}(1-k_j) = 0. \qquad \text{(A.7)}$$

The solution of this mathematical program yields (10.5) as the equation for the optimal correction.

(Grubbs actually obtained Equation (10.5) as the solution to the model with a constant initial displacement. Under conditions consistent with this assumption, Equation (A.5) would fundamentally have the same form, and as such, (A.7) would still be imposed. On the other hand, Equation (A.6) would be reduced to

$$V[U_{i+1}] = [(\sigma_{X+E}^2)/m]\sum_{j=1}^{i} k_j^2 \prod_{l=j+1}^{i}(1-k_l)^2.$$

Nevertheless, since

$$\prod_{j=1}^{i}(1-k_j)^2\sigma_D^2 \geq 0,$$

and since

$$\prod_{j=1}^{i}(1-k_j^*)^2\sigma_D^2 = 0,$$

it follows, therefore, that Equation (10.5) also minimizes (A.6).)

Hence, when Equation (10.5) is substituted into (A.6),

$$V^*[U_{i+1}] = (\sigma_{X+E}^2)/mi \qquad \text{(A.8)}$$

is obtained as the equation for the minimal variance of the $(i+1)$-st mean setting, for $i = 1, 2, \ldots, n$ (adapted from Grubbs, 1983). (It is noteworthy that each of the *adjusted* mean settings (i.e., excepting the original one) is a minimum variance unbiased estimator of the target. Thus in this regard,

Grubbs' procedure resembles a "greedy" algorithm. That is, an algorithm wherein each stage (adjustment) is optimized without regard to the others. Of course in general, such an approach would yield a suboptimal solution to the unified series of stages (adjustments).)

Finally, inspection of Equation (A.5) reveals that when $E[D] = 0$, the imposition of Equation (A.7) is no longer required to guarantee that the estimator will be unbiased. Under this specific condition, the solution of the unconstrained mathematical program yields another (considerably more complicated) equation for the optimal correction, which, when substituted into Equation (A.6), yields a result (somewhat) less than that of (A.8). Nevertheless, the "...accuracy on every occasion or for each machine an (sic) [on] practical grounds may be nearly as precise ...In fact, it is probably true that the required precision can be accounted for easily by proper choice of ..." the number of items used in the adjustment process, "...thus making the difference ...somewhat academic" (Grubbs, 1983). Therefore, (10.5) will be employed as the equation for the correction, irrespective of the value of $E[D]$.

# APPENDIX B

**Theorem 1**

$$U_{n+1} = t - (1/mn)\sum_{i=1}^{n}\sum_{j=1}^{m}(X_{ij} + E_{ij})$$

for $n = 1, 2, \ldots$.

**Proof**

First, suppose that the following supposition for $U_r$ is true:

$$U_r = t - [1/m(r-1)]\sum_{i=1}^{r-1}\sum_{j=1}^{m}(X_{ij} + E_{ij}) = t - [1/(r-1)]\sum_{i=1}^{r-1}(\overline{X}_{ij} + \overline{E}_{ij}).$$

Secondly, $U_{r+1}$ will be expanded as

$$\begin{aligned} U_{r+1} &= U_r - A_r = U_r - k_r^*(\overline{O}_r - t) = U_r - (1/r)(U_r + \overline{X}_i + \overline{E}_i - t) \\ &= U_r[(r-1)/r] - (1/r)(\overline{X}_i + \overline{E}_i - t), \end{aligned}$$

through the use of Equations (10.1), (10.3), (10.4), (10.5) and (A.2). Then, proceeding with the expansion of $U_{r+1}$ by employing the supposition above for $U_r$ yields

$$\begin{aligned} U_{r+1} &= t[(r-1)/r] - [1/(r-1)]\sum_{i=1}^{r-1}(\overline{X}_{ij} + \overline{E}_{ij})[(r-1)/r] - (1/r)(\overline{X}_i \\ &\quad + \overline{E}_i - t) \\ &= t - (1/r)\sum_{i=1}^{r}(\overline{X}_{ij} + \overline{E}_{ij}). \end{aligned}$$

Hence, since setting $i = 1$ in Equation (A.3) demonstrates that the supposition also holds for $r = 2$:

$$U_2 = t - \overline{X}_{ij} - \overline{E}_{ij},$$

the stated theorem is true.

# 11

# SHIFT DETECTIONS OF PROCESS MEAN USING REGRESSION AND CROSS-CORRELATION ANALYSES

**E. A. Elsayed[1], M. Gultekin[1] and J. H. Byun[2]**

[1]*Department of Industrial Engineering,*
*Rutgers University,*
*P.O. Box 909,*
*Piscataway, NJ 08855-0909*
*USA.*

[2]*Department of Industrial Engineering,*
*Gyeongsang National University,*
*Chinju, Gyeongnam 660-701,*
*Korea.*

## ABSTRACT

Process monitoring and adjustment is one of the main activities of the on-line quality control. It involves the detection of unacceptable deviations in the product quality characteristics or levels of process parameters and the adjustments required in order to minimize these deviations. In this chapter, we propose methods based on simple linear and weighted least squares regression techniques and cross-correlation values to detect gradual or sudden shift in the process mean. We evaluate the performance of the proposed methods in terms of delay in shift detection and the number of false alarms signaled. The methods are compared with those previously developed in the literature such as the Shewhart chart and are shown to be effective in detecting shift in the process mean when some conditions are satisfied.

**Key words:** gradual shift, regression, cross-correlation

## 1 INTRODUCTION

The broad purpose of the overall quality system is to produce units that are robust to all noise factors. Robustness implies that the product's functional characteristics are not sensitive to variation caused by noise factors Taguchi et al.(1989). In order to achieve robustness, quality control and engineering efforts must begin in the product design phase and be continued through production engineering and production operation phases. We refer to the quality control activities during the product planning, design and production engineering phases as *off-line* quality control, whereas the quality control activities during actual production are referred to as *on-line* quality control.

The off-line quality control focuses on the determination of the optimal levels of product characteristics and the optimal parameters of the production processes by using design of experiments. The on-line quality control focuses on the minimization of the deviations of the product characteristics and the parameters of the production processes from the optimal target levels. In other words, both off-line and on-line quality control seek the optimization of the product quality and the parameters of the production processes. In this chapter, we present approaches that detect shifts in the process mean and product quality when they occur in order to minimize losses due to deviations from the targets.

One of the most important problems in industry which relates to on-line quality engineering is the shift in the process mean. The shift of the process mean may occur in different ways. The mean may increase/decrease suddenly and may stay at the new value or it may increase/decrease gradually in a linear or a nonlinear fashion. This, of course, depends on the factors causing the shift.

In this chapter, we consider a process in which the mean increases or decreases gradually in a linear fashion. The wear-out of cutting tools, gradual clogging of filling nozzles in a filling process or operator fatigue are examples of such a kind of shift. The shift in the process mean may cause significant losses in product quality and the producer's future sales. Therefore, it is important to detect the shift of the process mean as soon as it occurs and provide corrective actions to eliminate or minimize future occurrences of similar shifts. In this chapter, we present new methods based on regression techniques and cross-correlation to detect the shift in the process mean. The proposed methods are effective and easy to implement.

In Section 2, we give a brief discussion of the shift detection techniques that are commonly used in industry. In Section 3, we discuss the shift of process means and describe the models that form the basis of our study. We propose new methods for detecting the shift in the process mean in Section 4. We compare the performance of the proposed methods with individual Shewhart chart and present our analysis and results in Section 5.

# 2 APPROACHES FOR SHIFT DETECTION

Control charts, Bayesian techniques, time series analysis and filtering methods are frequently used to detect shift in the process mean as early as possible and to provide information about the shift. We present these techniques and discuss their limitations and strengths as given below.

## 2.1 Control Charts

Control charts are used to ensure that parts are produced with minimum deviations from the target values. They include Shewhart, Cumulative Sum (CUSUM), Exponentially Weighted Moving Average (EWMA) and multivariate control charts.

### *Shewhart Control Chart*

The Shewhart control chart is usually designed in such a way that false alarms are minimized. In doing so, the chart becomes insensitive to small changes in the process mean. One way to overcome this disadvantage is to use supplementary runs rules (Nelson (1984, 1985)). Wheeler (1983), derives the power function for different combinations of the four main runs rules and estimates the probabilities of Type I error under different conditions. Davis and Woodall (1988), compare supplementary runs rules and the trend rule (signal when some number of successive points all increase or if the same number of successive points all decrease in value) and conclude that using the trend rule does not provide additional advantage when the runs rules are already being used. A Markov chain approach in Champ and Woodall (1987), extends the $\overline{x}$ control chart with supplementary runs rules. Another modification of the Shewhart chart to improve its ability to detect shift of process means is to vary sampling intervals and/or sample sizes (Prabhu et al.(1994), Reynolds et al.(1988)). One important disadvantage associated with these modified approaches is the increase in sampling cost.

### *CUSUM Control Chart*

Johnson and Leone (1962a, 1962b, 1962c) show the mathematical development of the CUSUM control charts for sample mean, variance and range for Poisson and binomial random variables. The CUSUM chart detects small shifts in the process mean (shifts less than one standard deviation away from the mean) more rapidly than the Shewhart chart but the ability of the Shewhart chart in detecting large shifts of the process mean is better than that of the CUSUM chart (Lucas (1976). One of the approaches that can increase the power of CUSUM chart in detecting the shift of the process mean is to use a parabolic-mask (Lucas (1973)). However, designing such a chart is not as easy as designing a CUSUM chart with V-mask. Lucas (1982), suggests a combined Shewhart-CUSUM quality control scheme. Starting the CUSUM chart at some non-zero value (Lucas and Crosier (1982)) improves its ability to detect shifts, especially when the process is out-of-control in the early stages.

## *EWMA Control Chart*

The application of the EWMA control chart is suggested for processes where repeated sampling is not possible or appropriate (Wortham and Ringer (1971)). The EWMA chart performs like a Shewhart control chart as the weighting factor, $\lambda$, becomes closer to one and it performs like a CUSUM control chart as $\lambda$ becomes closer to zero (Hunter (1986)). Lucas and Saccucci (1990), evaluate the run-length properties of EWMA control schemes by representing the EWMA statistic as a continuous-state Markov chain. The EWMA control chart with parabolic control limits discussed by Holmes and Mergen (1992), performs better than EWMA with parallel limits in terms of the average run length (ARL) values. Domangue and Patch (1991), propose an omnibus EWMA chart which is based on the exponentiation of the absolute value of the standardized sample mean of the observation. Omnibus EWMA chart can be said to perform better in detecting the shift in process variance than the shift in the process mean.

## *Multivariate Control Charts*

Hotelling introduced the $T^2$ control chart as a technique for monitoring the overall quality of several characteristics. Alt (1985), discusses the multivariate control charts for the process mean and dispersion. One drawback of Hotelling's $T^2$ chart is the difficulty in determining the quality characteristic that causes the shift. Murphy (1987), develops some methods to select the out-of-control variable. To improve the response time of the $T^2$ chart, Holmes and Mergen (1993), introduce the Mean Square Successive Differences (MSSD) method to better estimate the variance-covariance matrix. Wierda (1994), proposes a step-down procedure instead of the $T^2$ control chart when information is available on how often the characteristics are affected by a special cause of variation.

## 2.2 The Bayesian Technique

The Bayesian technique is a recursive method that can be applied to many discrete or continuous distributions. Yousry et al.(1991), and Sturm et al.(1991), develop models using an empirical Bayesian method to monitor and analyze the process data. Although a comparison is not made, since the estimate for the process parameter is updated each time, it is reasonable to expect that this technique performs better than the control chart techniques discussed earlier in this chapter.

## 2.3 Time Series Analysis

The autocorrelation structure of the data is captured by using time series models when the observations are dependent. Yourstone and Montgomery (1989), show that the Shewhart chart performs worse in terms of detection of the process shift and the number of false alarms when applied to the correlated data. The same authors present an application of the group autocorrelation control chart (GACC) which is applied to the residuals of an ARIMA model (Yourstone and Montgomery (1989)). Guidelines for adjusting the control limits of $\bar{x}$, $S$, $R$ and $S^2$ charts for correlated samples are given in Vasilopoulos and Stamboulis (1978), and Yang and Hancock (1990). Alwan and Roberts (1988), discuss the common-cause and special-cause control charts. The special-cause chart in which the residuals are used in a standard control chart is found to be very effective in identifying irregular data points. Wardell et al.(1994), study the run length distribution of the special-cause charts and conclude that these charts are not suitable to use for positively correlated processes. They are very effective in detecting large shifts and in processes where the observations are negatively correlated. Bagshaw and Johnson (1975), study the effect of serial correlation on the performance of the CUSUM charts and they show that the run length of the CUSUM chart depends on the correlation structure of the observations (Johnson and Bagshaw (1974)). Yashchin (1993), transforms the sequence of serially correlated observations to an independently and identically distributed sequence, which leads to a practically acceptable approximation.

## 2.4 Filtering

One of the most commonly used filtering approaches is Kalman filtering which estimates the state vector of a dynamic system from noisy observations. It can be regarded as a Bayesian-like technique (Meinhold and Singpurwalla (1983)).

Phadke (1981), uses Kalman filtering to obtain the best estimate of the true defective index and its confidence interval. Downing et al.(1980), compare Kalman filtering with the CUSUM control chart.

# 3 SHIFT OF PROCESS MEAN

There are many methods for controlling product quality during production cycles. Inspection of products during manufacturing, employment of diagnostic and adjustment processes, improvement of production processes, and the use of automatic control systems are some of the methods used. In this chapter, we monitor the mean of the product quality in order to detect gradual shifts in its value.

The problem under study deals with a production process where the mean of the product characteristic being monitored increases gradually in a linear fashion. Observations collected at equal time intervals directly from the process are assumed to be identically, independently and normally distributed with a known mean, $\mu$, and a known variance, $\sigma^2$. At the beginning, the process is assumed to be in control. Then the process mean starts to shift linearly and the observation at time $t_s$ is the first observation affected by this shift. The shifted mean at time $t$, $\mu(t)$, is defined as

$$\mu(t) = \mu + (t - (t_s - 1))tan\theta$$

where $\theta$ is the angle that $\mu(t)$ makes with $\mu$ as shown in Figure 1.

As shown in the figure, the observations oscillate around a straight line with slope equals zero until time $t_s - 1$, and thereafter around a straight line with a slope of $tan\theta$. In order to detect the change in the mean of the process, the observations are fitted to a straight line and the change in the slope of this line is observed. This is accomplished by one or more of the models presented below.

### Models

The models for shift detection are based on fitting the observations to a line using three different methods as described below:

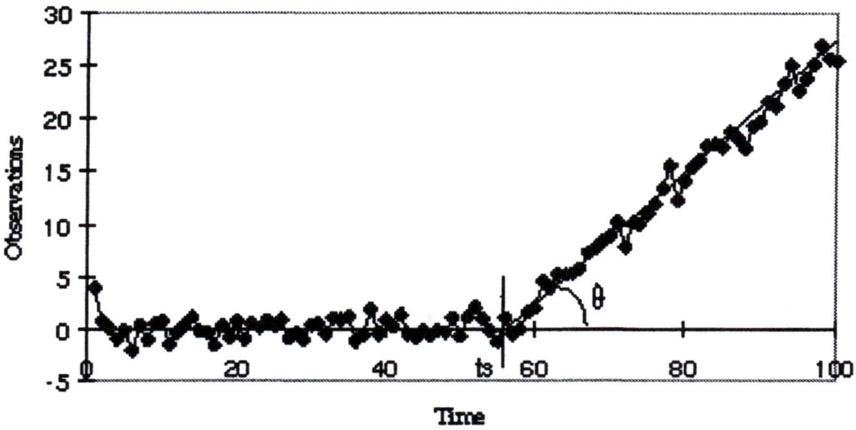


Figure 1. Observations versus time.

- The observations from time $t = 1$ to time $t = T$ (the current time) are fitted to a line using simple linear regression. (The models with the prefix REG are of this kind.)
- The most recent thirty observations are fitted to a line using simple linear regression. Thirty observations are chosen in order to ensure normality. (The models with the prefix REG30 are of this kind.)
- The most recent thirty observations are fitted to a line using weighted least squares regression. (The models with the prefix WLSR30 are of this kind.) The weights are assigned to the observations in such a way that the most recent observation has the highest weight:

$$w_t = \lambda^{(30-t)}, \ t = 1, 2, ..., 30$$

where $\lambda$ is the weighting factor and $w_t$ is the weight assigned to the observation at time $T+t-30$ ($T$ being the current time). Extensive simulation results show that higher $\lambda$ values cause early detection of shifts with fewer false alarms. It is found that the appropriate value of $\lambda$ is 0.9.

# 4 METHODS FOR SHIFT DETECTION

In this section, we present four new methods for detecting shifts in the process (or product) mean. These methods are based on the models given in Section 3. The methods are described below.

The size of the window length may have an effect on the performance of the proposed methods and it may need further investigation.

## 4.1 Application of $3\sigma_s$ Limits

The estimated slope of a straight line fitted to normally distributed observations is also normally distributed. The slope estimate in our study is normally distributed with mean equals to zero. At every new observation, the slope of the fitted line is estimated, the standard deviation of the slope estimate ($\sigma_s$) is calculated and $3\sigma_s$ limits are built around this slope estimate. An out-of-control signal is given if the estimate falls outside these limits.

We apply the $3\sigma_s$ limits to the slope estimates obtained by REG, REG30 and WLSR30. When the $3\sigma_s$ limits are applied, we refer to these models as REG-$3\sigma$, REG30-$3\sigma$ and WLSR30-$3\sigma$, respectively.

It must be noted that the control limits are static for REG30-$3\sigma$ and WLSR30-$3\sigma$, while in the REG-$3\sigma$ model they are dynamic. The reason for this is that in REG-$3\sigma$ the variance of slope estimates changes each time a new observation is added, but it stays constant for both REG30-$3\sigma$ and WLSR30-$3\sigma$ models.

## 4.2 One Step Ahead Predictions

In this method, we use the model REG (when this model is used in prediction we refer to it as REG-PRED) to detect the shift in the process mean. One step ahead prediction is made and $100(1-\alpha)\%$ prediction control limits are built around it. If the observed value is outside this prediction control limit, an out-of-control signal is given. The $\alpha$ confidence is taken as 0.0027 which corresponds to the 3 standard deviation limits.

## 4.3 Application of $T^2$ Chart

$T^2$ chart is used to simultaneously monitor the behavior of quality characteristics that are jointly distributed as a multivariate normal. In the models where the most recent thirty observations are used to fit a line, namely, REG30 and WLSR30, a moving window can be observed where the time axis, varying from 1 to 30, is kept fixed. The window moves by one unit each time a new observation is added and the oldest observation is deleted. Thus 29 of the observations shown in the window are the same observations that appeared in the previous window. As a result, the slope estimate at time $t$ is correlated with and up to the slope estimates at time $t \pm 29$.

We use the two most recent slope estimates, which in this case form a bivariate normal distribution, obtained by REG30. Each time a new slope estimate is obtained, the $T^2$ statistic is calculated and an out-of-control signal is given if the calculated $T^2$ value falls outside the control limits.

The $T^2$ statistic is defined as

$$T^2 = (\underline{\hat{\beta}_1}(t))^T \Sigma^{-1} (\underline{\hat{\beta}_1}(t))$$

where $(\underline{\hat{\beta}_1}(t))^T$ is the transpose of the vector, $\underline{\hat{\beta}_1}(t)$, formed by the two slope estimates calculated at times $t$ and $t-1$. $\Sigma^{-1}$ is the inverse of the covariance matrix of the slope estimates, $\Sigma$. The $(ii)$th element of the covariance matrix gives the variance of the slope estimate while the $(ij)$th element gives the covariance between the slope estimates calculated at times $t-1$ and $t$, respectively.

Note that the slope estimates are considered individually and the target mean for the slope estimates is zero. $T^2$ statistic is distributed according to a Chi-square distribution with 2 degrees of freedom and we use $\alpha = 0.005$ for the evaluation of Type I error $(\chi^2_{2,\alpha})$. Thus, the control limits for the $T^2$ chart are as follows:

$$\text{Upper Control Limit} = \chi^2_{2,0.005}$$

$$\text{Lower Control Limit} = 0$$

## 4.4 Cross-Correlation

Cross-correlation is a measure of how well a series of events correlates with another series of events. The cross-correlation functions of two general stochastic processes $x(t)$ and $z(t)$ are defined as

$$R_{xz}(t_1, t_2) = E[x(t_1)z(t_2)]$$

$$R_{zx}(t_1, t_2) = E[z(t_1)x(t_2)]$$

where $E[\,]$ is the expectation operator.

The time dependent cross-correlation functions are defined as (McConnel (1995)):

$$R_{xz}(t, \tau) = \lim_{T\to\infty} \frac{1}{2T} \int_{-T}^{T} x(t)z(t+\tau)\mathrm{dt} \qquad (11.1)$$

$$R_{zx}(t, \tau) = \lim_{T\to\infty} \frac{1}{2T} \int_{-T}^{T} z(t)x(t+\tau)\mathrm{dt} \qquad (11.2)$$

where $\tau = t_2 - t_1$ is the time lag between the two processes. Since our objective is to detect the shift of the process mean as soon as it occurs, we assume the time lag between the two processes to be zero. In this way, when a shift occurs, the effect of the deviation in the observations and the slope estimates is reflected sooner by the cross-correlation function. We consider the slope estimates obtained from the model REG30 ($\hat{\beta}(t)$) and the observations ($x(t)$) to be two stochastic processes and define $R_{x\hat{\beta}}(t, \tau)$ as the cross-correlation between $x(t)$ and $\hat{\beta}(t)$ calculated at time $t$. Since the observations are taken at discrete time intervals, we adapt Equation (1) as

$$R_{x\hat{\beta}}(t, 0) = \frac{1}{T}\Sigma_{i=t-T+1}^{t} x(i)\hat{\beta}(i) \qquad (11.3)$$

where $T$ is the length of the time interval in which the cross-correlation value is calculated.

An out-of-control signal is given when the calculated cross-correlation value is outside the upper or lower control limits given by:

$$\text{Upper Control Limit} = \text{mean} + k\frac{\overline{R}}{d_2} \qquad (11.4)$$

$$\text{Lower Control Limit} = \text{mean } -k\frac{\overline{R}}{d_2} \tag{11.5}$$

where
$k$ is a constant,
$\overline{R}$ is the mean of the ranges of the cross-correlation values calculated from each subgroup of size $n$,
$\overline{R}/d_2$ is the estimate of the standard deviation of the $R_{x\hat{\beta}}(t)$.

# 5 EXPERIMENTS, ANALYSIS AND RESULTS

Five hundred observations are randomly generated from a normal distribution with mean zero and a standard deviation ($\sigma$) that varies from 0.1 to 5.0. The linear shift in the process mean starts at observation 349, i.e., $t_s$=350. The values of the angle $\theta$ vary from 1 to 45 degrees. Each run is replicated 25 times.

The performance measure is the delay in detecting the shift in the process mean. The performance of the proposed models is compared for all $\sigma/\theta$ values. The other performance measure considered is the cumulative number of false alarms (signaling shift when there is no shift) over the 25 runs. The performance of the proposed models is compared with that of the Shewhart chart when applied directly to the observations from the process.

As shown in Figures 2 and 3, the Shewhart chart outperforms REG-3$\sigma$, REG30-3$\sigma$ and WLSR30-3$\sigma$ for most of the $\sigma/\theta$ values. But WLSR30-3$\sigma$ shows better performance when $\sigma/\theta$ is small. WLSR30-3$\sigma$ and REG30-3$\sigma$, when used together, detect shifts sooner than Shewhart for $\sigma/\theta$ values that are approximately between 0.02 and 0.4. The performance of REG-3$\sigma$ for $1.25 < \sigma/\theta < 2.50$ is quite noticeable. The slope of the line used in REG-3$\sigma$ better reflects the mean of the process as new observations are added, but when a shift occurs the line fails to immediately adjust itself accordingly. When the shift is not too small or too large, the fitted line would detect the shift in the process mean before any observed value falls outside the control limits of the Shewhart chart. Table 1 shows that the REG-3$\sigma$ method provides the least number of false alarms.

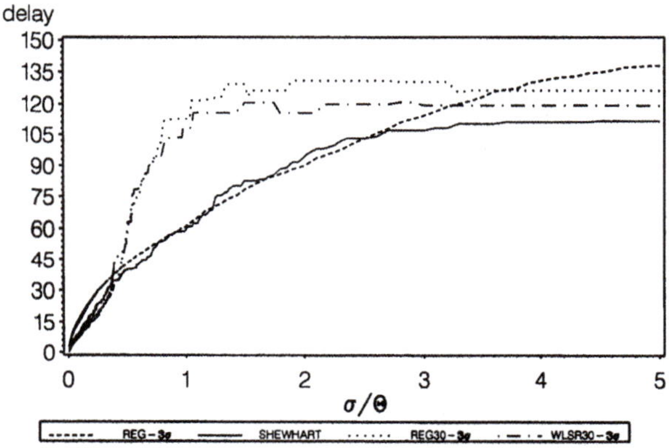


Figure 2. Comparison of the performance of REG $-3\sigma$, REG30 $-3\sigma$, WLSR30 $-3\sigma$ methods and the SHEWHART chart

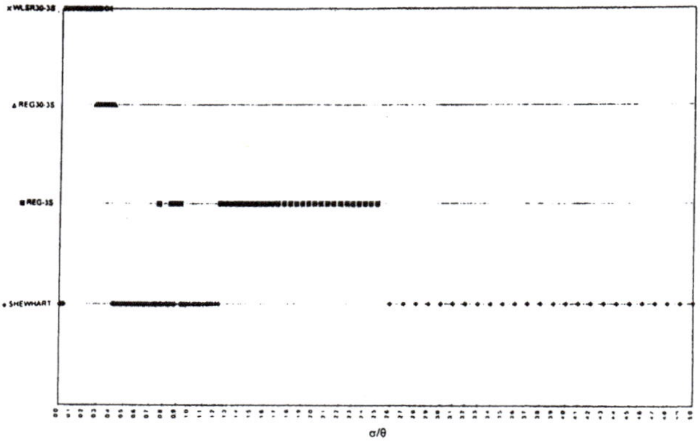

Figure 3: The method with the shortest delay for a given σ/θ ratio.

We then compare the performance of REG-PRED with that of the Shewhart chart. The plots of the performance are given in Figure 4. It is seen that the Shewhart chart always detects shifts sooner than REG-PRED. This is partly due to the fact that the prediction control limits are larger than the limits of the Shewhart chart. The number of false alarms given by REG-PRED is less

| Method | Number of false alarms (in 25 runs) |
|---|---|
| Shewhart | 20 |
| REG30-3$\sigma$ | 23 |
| REG-3$\sigma$ | 0 |
| WLSR30-3$\sigma$ | 24 |

**Table 1** Number of false alarms for the REG30-3$\sigma$, REG-3$\sigma$, and WLSR30-3$\sigma$ methods and the Shewhart chart

than the ones given by the Shewhart chart (Table 2). The reason that the values in Table 2 are close to each other is that the charts start at time 120, which is a large enough time for the prediction interval to become narrower and closer to the limits of the Shewhart chart.

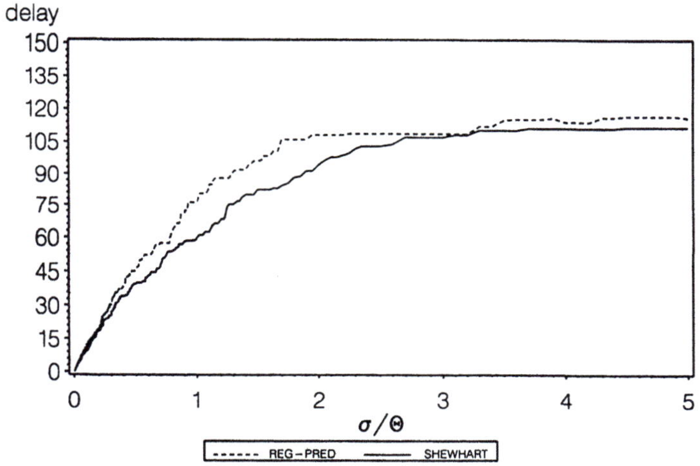


Figure 4. Comparison of the performance of the REG – PRED and the SHEWHART chart

The performance of the Shewhart chart is significant when compared to that of the REG30-$T^2$ model (Figure 5). In detecting shifts where $\sigma/\theta$ is less than 0.3, the two models show almost the same performance. Moreover, the numbers of

| Method | Number of false alarms (in 25 runs) |
|---|---|
| Shewhart | 20 |
| REG-PRED | 17 |

**Table 2** Number of false alarms for the REG-PRED method and the Shewhart chart

false alarms signaled by the two methods are close to each other as shown in Table 3.

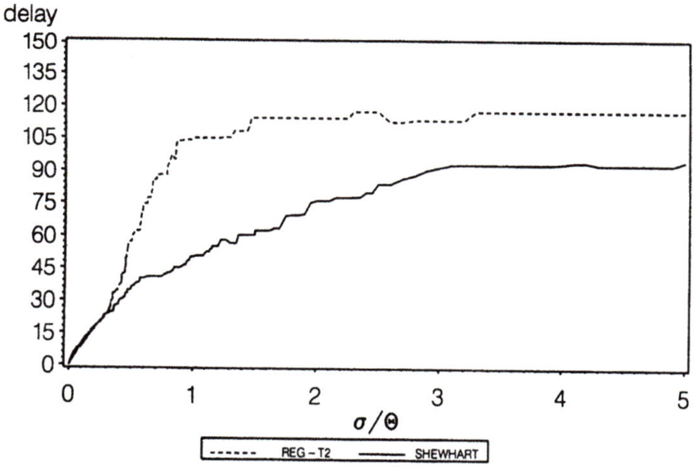


Figure 5. Comparison of the performance of the REG $-$ 30T$^2$ method and the SHEWHART chart

In order to be able to evaluate the performance of the cross-correlation method, the optimal values for the parameters are calculated. During the simulation, the first hundred $R_{x\hat{\beta}}(t,0)$ values as given in Equation (3) are used to estimate the mean and the standard deviation of the cross-correlation values. The $R_{x\hat{\beta}}(t,0)$'s are treated as independent from each other while calculating $\overline{R}/d_2$ needed for

| Method | Number of false alarms (in 25 runs) |
|---|---|
| REG30-$T^2$ | 34 |
| Shewhart | 41 |

**Table 3** Number of false alarms for the REG30-$T^2$ method and the Shewhart chart

the control limits of Equations (4) and (5). Since they are correlated, we underestimate the standard deviation and therefore they need to be adjusted. As $k$, given in Equations (4) and (5), increases, the control limits become wider and the number of false alarms decreases, but delay in shift detection increases. In order to have a rational comparison, we keep the number of false alarms signaled in the Shewhart chart and the cross-correlation method close to each other and take $k = 6$. Then we compare their performance in terms of delay.

The length of $T$, the time interval of interest, has a direct effect on the estimated value of the cross-correlation. The size of the subgroup, $n$, to estimate the standard deviation has also an effect on the delay and the number of false alarms. As the subgroup size increases, the range in each subgroup increases and the control limits become wider causing a delay in shift detection.

By keeping the upper and lower control limits for the cross-correlation data constant, we conducted the experiments shown in Table 4 to investigate the effect of $T$ and $n$ on the ability to detect the shift in the process mean. It can be seen that as $T$ increases, the delay decreases, but the number of false alarms increases. If the number of false alarms is kept constant for all $T$ values, it is seen that long time periods result in making the cross-correlations insensitive to shifts or changes in the stochastic processes $x(t)$ and $\hat{\beta}(t)$. In turn, a shift in the process mean becomes difficult to detect. On the other hand, short time periods react to shifts or changes sooner. In this case, the shift in the process mean becomes more evident and is easier to detect. A similar argument is valid for the subgroup size, $n$. In Table 4, it can be observed that as $n$ increases, delay also increases, and the number of false alarms increases. If the number of false alarms is kept constant, then the delay increases as $n$ decreases.

The improvement obtained in delay by increasing $T$ from 1 to 2 is higher than that obtained by increasing $T$ from 2 to 3 or 3 to 4. An opposite behavior can

| | $\sigma/\theta = 0.1$ | | | | $\sigma/\theta = 2$ | | | | $\sigma/\theta = 5$ | | | |
|---|---|---|---|---|---|---|---|---|---|---|---|---|
| | $T = 1$ | $T = 2$ | $T = 3$ | $T = 4$ | $T = 1$ | $T = 2$ | $T = 3$ | $T = 4$ | $T = 1$ | $T = 2$ | $T = 3$ | $T = 4$ |
| $n = 5$ | 10.08 | 9.52 | 9.20 | 9.04 | 105.32 | 83.68 | 74.76 | 71.72 | 103.12 | 99.92 | 99.60 | 92.44 |
| | (15) | (34) | (48) | (62) | (15) | (34) | (48) | (62) | (15) | (34) | (48) | (62) |
| $n = 10$ | 10.16 | 9.60 | 9.68 | 9.72 | 107.64 | 92.68 | 85.56 | 78.84 | 104.80 | 103.88 | 109.96 | 107.00 |
| | (12) | (20) | (31) | (40) | (12) | (20) | (31) | (40) | (12) | (20) | (31) | (40) |
| $n = 25$ | 10.80 | 10.08 | 9.88 | 9.96 | 114.08 | 97.12 | 96.72 | 99.12 | 124.88 | 111.80 | 113.04 | 114.68 |
| | (5) | (10) | (18) | (29) | (5) | (10) | (18) | (29) | (5) | (10) | (18) | (29) |

**Table 4** Effect of $T$ and $n$ on the delay and the number of false alarms (shown in parenthesis). The upper values in rows for $n = 5$, 10 and 25 represent the number of observations before the shift is detected and the lower values represent the cumulative number of false alarms

| Method | Number of false alarms (in 25 runs) |
|---|---|
| Cross-Correlation | 19 |
| Shewhart | 20 |

**Table 5** Number of false alarms for Shewhart and cross-correlation method

be observed in the case of false alarms. A reasonable number of false alarms and delay, $T = 2$ and $n = 10$, enable us to detect the shift in the process mean sooner than other values.

The cross-correlation method for $T = 2$ and $n = 10$ outperforms the Shewhart chart for approximately all ranges of $\sigma/\theta$. The performance plots are shown in Figure 6d. The numbers of false alarms signaled by the two methods are almost the same as shown in Table 5. The success of the cross-correlation method is explained by the simultaneous increase/decrease in the observations and the slope of the fitted line when there is a shift. The performance of the cross-correlation method in terms of the delay in shift detection is calculated for $T = 1$ and 2, and the upper and lower control limit values are calculated for $n = 5$, 10 and 25. These are also presented in Figure 6.

# 6 CONCLUSIONS

In this chapter, we propose methods for shift detection of process means based on regression techniques (REG-3$\sigma$, REG30-3$\sigma$, WLSR30-3$\sigma$, REG-PRED and REG30-$T^2$) and the cross-correlation method. We compare the performance of these methods with that of the Shewhart chart applied directly to the observations. The performance criteria are the delay in detecting the shift in the process mean and the number of false alarms signaled.

REG-PRED and REG30-$T^2$ fail to detect the shift in the process mean sooner than the other methods. The remaining three regression methods, REG-3$\sigma$, REG30-3$\sigma$ and WLSR30-3$\sigma$, are effective in detecting the shift sooner than the Shewhart chart for certain $\sigma/\theta$ values.

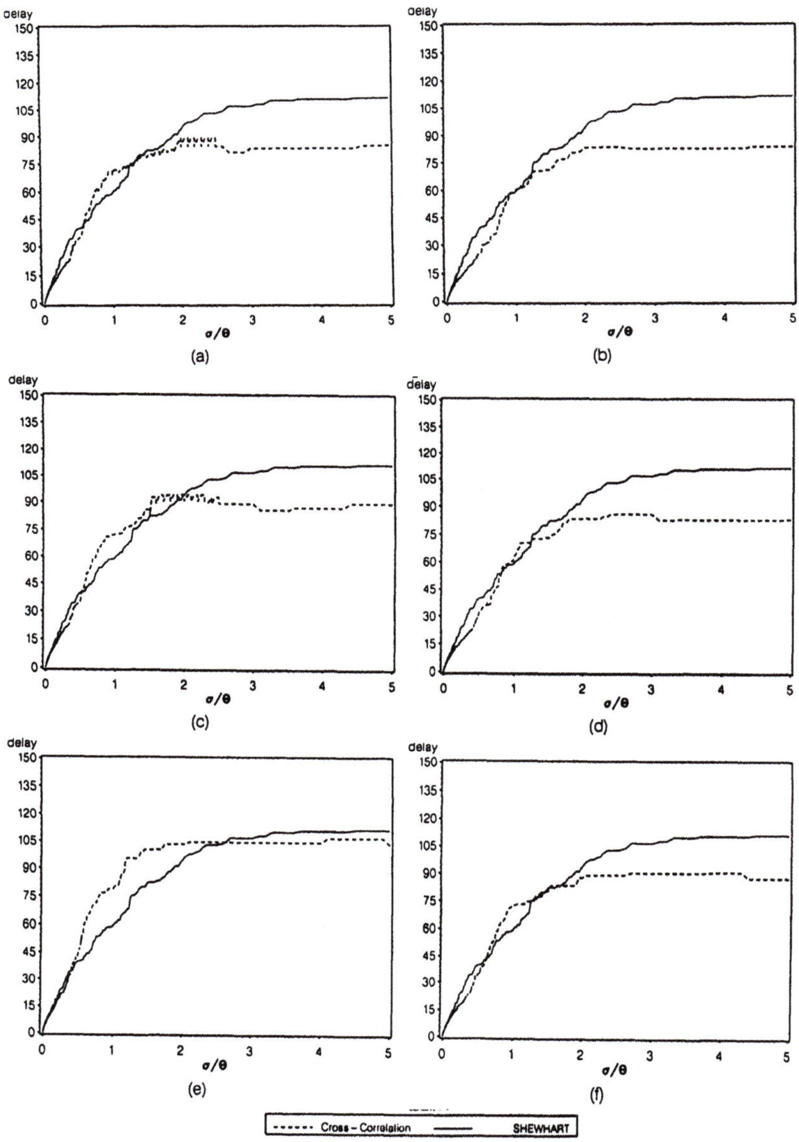


Figure 6. Comparison of the performance of the SHEWHART chart and the cross-correlation method (a) T=1, n=5 (b) T=2, n=5 (c) T=1, n=10 (d) T=2, n=10 (e) T=1, n=25 (f) T=2, n=25

When there is a small change in the process mean, a CUSUM chart is more successful than a Shewhart chart in detecting this change whereas the Shewhart chart is more effective in detecting large shifts. In this study we show that very large shifts in the process mean can be signaled by the REG-3$\sigma$, REG30-3$\sigma$ and WLSR30-3$\sigma$ sooner than the Shewhart chart. There is no significant difference in the number of false alarms signaled by these methods. It is suggested that these methods be used in combination with a Shewhart chart in order to improve shift detection.

Among all methods, the cross-correlation method shows the best performance in terms of detecting the shift. We see that except for a small interval, the cross-correlation method outperforms the Shewhart chart. This method is effective and practical in detecting the linear shifts in the process mean as early as possible.

# REFERENCES

[1] Alt, F. B., "Multivariate Quality Control", *Encyclopedia of Statistical Sciences*, edited by N. L. Johnson and S. Kotz, 6, pp 110-122, 1985.

[2] Alwan, L. C. and H.V. Roberts, "Time-Series Modeling for Statistical Process Control," *Journal of Business and Economic Statistics*, 6(1), pp 87-95, 1988.

[3] Bagshaw, M. and R. A. Johnson, "The Effect of Serial Correlation on the Performance of CUSUM Tests II," *Technometrics*, Vol. 17(1), pp 73-80, 1975.

[4] Champ, C.W. and W.H. Woodall, "Exact Results for Shewhart Control Charts With Supplementary Runs Rules," *Technometrics*, 29(4), pp 393-399, 1987.

[5] Davis, R.B. and W.H. Woodall, "Performance of the Control Chart Trend Rule Under Linear Shift," *Journal of Quality Technology*, 20(4), pp 260-262, 1988.

[6] Domangue, R. and S.C. Patch, "Some Omnibus Exponentially Weighted Moving Average Statistical Process Monitoring Schemes," *Technometrics*, 33(3), pp 299-313, 1991.

[7] Downing, D. J., D.H. Pike, and G.W. Morrison, "Application of the Kalman Filter to Inventory Control," *Technometrics*, 22(1), pp 17-22, 1980.

[8] Holmes, D.S. and A.E. Mergen, "Parabolic Control Limits for the Exponentially Weighted Moving Average Control Charts," *Quality Engineering*, 4(4), pp 487-495, 1992.

[9] Holmes, D.S. and A.E. Mergen, "Improving the Performance of the $T^2$ Control Chart," *Quality Engineering*, 5(4), pp 619-625, 1993.

[10] Hunter, S.J., "The Exponentially Weighted Moving Average", *Journal of Quality Technology*, 18(4), pp 203-209, 1986.

[11] Johnson, N. L. and F. C. Leone, "Cumulative Sum Control Charts, Mathematical Principles Applied to their Construction and Use, Part I," *Industrial Quality Control*, 19, pp 15-21, 1962a.

[12] Johnson, N. L. and F. C. Leone, "Cumulative Sum Control Charts, Mathematical Principles Applied to their Construction and Use, Part II," *Industrial Quality Control*, 19, pp 29-36, 1962b.

[13] Johnson, N. L. and F. C. Leone, "Cumulative Sum Control Charts, Mathematical Principles Applied to their Construction and Use, Part III," *Industrial Quality Control*, 19, pp 22-28, 1962c.

[14] Johnson, R. A. and M. Bagshaw, "The Effect of Serial Correlation on the Performance of CUSUM Tests I," *Technometrics*, 16(1), pp 103-112, 1974.

[15] Lucas, J.M., "A Modified 'V' Mask Control Scheme," *Technometrics*, 15(4), pp 833-847, 1973.

[16] Lucas, J.M., "The Design and Use of V-Mask Control Schemes," *Journal of Quality Technology*, 8(1), pp 1-12, 1976.

[17] Lucas, J.M., "Combined Shewhart-CUSUM Quality Control Schemes," *Journal of Quality Technology*, 14(2), pp 51-59, 1982.

[18] Lucas, J.M. and R.B. Crosier, "Fast Initial Response for CUSUM Quality-Control Schemes: Give Your CUSUM a Head Start," *Technometrics*, 24(3), pp 199-205, 1982.

[19] Lucas, J.M. and M. S. Saccucci, "Exponentially Weighted Moving Average Control Schemes: Properties and Enhancements," *Technometrics*, 32(1), pp 1-12, 1990.

[20] McConnell, K. G., *Vibration Testing: Theory and Practice*, John Wiley and Sons, Inc., New York, 1995.

[21] Meinhold, R.J. and N.D. Singpurwalla, "Understanding the Kalman Filter," *The American Statistican*, 37(2), pp 123-127, 1983.

[22] Murphy, B. J., "Selecting Out of Control Variables with the $T^2$ Multivariate Quality Control Procedure," *The Statistican*, 36, pp 571-581, 1987.

[23] Nelson, L.S., "The Shewhart Control Chart-Tests for Special Causes," *Journal of Quality Technology*, 16(4), pp 237-239, 1984.

[24] Nelson, L.S., "Interpreting Shewhart Control Charts," *Journal of Quality Technology*, 17(2), pp 114-116, 1985.

[25] Phadke, M.S., "Quality Audit Using Adaptive Kalman Filtering," *ASQC Quality Congress Transactions*, pp 1045-1052, 1981.

[26] Prabhu, S.S., D.C. Montgomery, and G.C. Runger, "A Combined Adaptive Sample Size and Sampling Interval Control Scheme," *Journal of Quality Technology*, 26(3), pp 164-176, 1994.

[27] Reynolds, M.R. JR., R.W.Amin, J.C. Arnold, and J.A. Nachlas, "Charts With Variable Sampling Intervals', *Technometrics*, 30(2), pp 181-191, 1988.

[28] Sturm, G.W., C.J. Feltz, and M.A. Yousry, "An Empirical Bayes Strategy for Analysing Manufacturing Data in Real Time," *Quality and Reliability Engineering International*, 7(3), pp 159-167, 1991.

[29] Taguchi, G., E. A. Elsayed, and T. Hsiang, *Quality Engineering in Production Systems*, McGraw-Hill Book Company, New York, 1989.

[30] Vasilopoulos, A. V. and A. P. Stamboulis, "Modification of Control Chart Limits in the Presence of Data Correlation," *Journal of Quality Technology*, 10(1), pp 20-30, 1978.

[31] Wardell, Don G., H. Moskowitz and R. D. Plante, "Run-Length Distributions of Special-Cause Control Charts for Correlated Processes," *Technometrics*, 36(1), pp 3-17, 1994.

[32] Western Electric Company, Inc., *Statistical Quality Control Handbook*, Western Electric, New York, 1956.

[33] Wheeler, D.J., "Detecting a Shift in Process Average: Tables of the Power Function for $\bar{x}$ Charts," *Journal of Quality Technology*, 15(4), pp 155-170, 1983.

[34] Wierda, S. J., "Multivariate Step-down Control Charts for the Mean," unpublished paper presented at the 38th Fall Technical Conference, 1994.

[35] Wortham, A.W. and L.J. Ringer, "Control Via Exponential Smoothing," *The Logistics Review*, 7(32), pp 33-40, 1971.

[36] Yang, K. and W. M. Hancock, "Statistical Quality Control for Correlated Samples," *International Journal of Production Research*, 28(3), pp 595-608, 1990.

[37] Yashchin, E., "Performance of CUSUM Control Schemes for Serially Correlated Observations," *Technometrics*, 35(1), pp 37-52, 1993.

[38] Yourstone, S.A. and D.C. Montgomery, "A Time-Series Approach to Discrete Real-Time Process Quality Control," *Quality and Reliability Engineering International*, 5, pp 309-317, 1989.

[39] Yourstone, S.A. and D.C. Montgomery, "Detection of Process Upsets-Sample Autocorrelation Control Chart and Group Autocorrelation Control Chart Applications," *Quality and Reliability Engineering International*, 7, pp 133-140, 1991.

[40] Yousry, M.A., G.W. Sturm, C.J. Feltz, and R. Noorossana "Process Monitoring in Real Time: Empirical Bayes Approach- Discrete Case," *Quality and Reliability Engineering International*, 7(3), pp 123-132, 1991.

# 12

# OPTIMAL CONTROL AND MONITORING OF DETERIORATING PRODUCTION PROCESSES

**J. Yang and V. Makis**

*Department of Mechanical and Industrial Engineering,*
*University of Toronto,*
*Toronto, Ontario,*
*Canada M5S 1A4.*

## ABSTRACT

For processes subject to deterioration, process control, generally referred to as EPC (engineering process control) or APC (automatic process control), should be applied in conjunction with the process monitoring, SPC (statistical process control), to identify the occurrence of assignable causes and try to eliminate them. However, a certain degree of autocorrelation is present in almost all real processes which has a major impact on the performance of classical control charts. To improve this performance, we propose a new statistic for the monitoring of the process mean to compensate for the lost portion of the process deviation due to the autocorrelation. The classical Shewhart chart is applied to this statistic. Using the Markov chain approach, we obtained explicit formulas for the run length distribution, average run length and the standard deviation of the run length.

**Key words:** tool-wear process, random shocks, autocorrelation, optimal control, residual response, process monitoring, average run length

## 1 INTRODUCTION

In many industrial processes, the mean shifts during production and if it is not reset after a long period of operation, a large proportion of nonconforming items are produced. Examples of such processes are metal machining operations involving milling machines, drilling machines, lathes, etc., where the shift in the mean value of the quality characteristic is primarily due to tool-wear. The tool-wear problem is well-known in quality control and has been addressed by several authors.

Manuele (1945) suggested using a control chart to decide when to reset the process with the linear trend. Assuming the process mean shifts in the positive direction, his method consists in setting the center line of the $\overline{x}$-chart $3\sigma$ units above the lower specification limit and making the adjustment to the process when the $\overline{x}$ graph first crosses the line $3\sigma$ units below the upper specification limit. This method is applicable only to wide specification ranges, say at least $9\sigma$. The first formulation of the tool-wear problem as an optimization problem is due to Gibra (1967). He presented a model combining resetting cost and penalty cost for nonconforming parts and derived an equation that can be solved graphically to obtain the optimal resetting point. He assumed that the number of parts produced can be treated as approximately continuous.

Drezner and Wesolowsky (1989) considered a similar model with the economic loss defined as a quadratic function of the deviation of the quality characteristic from its target value. They considered a constant tool-wear rate and derived a formula for the expected average cost per unit time and minimized an approximation to this objective function. An exact analysis under more general assumptions was done by Makis (1996). He showed that the optimal initial setting of the process mean and the optimal resetting time can be found by solving two nonlinear equations. Arcelus et al.(1982) dealt with the problem of obtaining the optimum production schedule when the quality characteristic of the product is subject to a systematic shift in both the mean and the variance.

Quesenberry (1989) considered a problem of designing a machining compensator for the situation where the adjustment of the process mean is not costly and can be made as often as after each part. In all these papers, the process model was a regression model and it was assumed that the mean of the quality characteristic is a function of the operation time since the last resetting of the process and the mean accumulated wear.

Models of production processes, where the process mean is a random variable determined by the starting level and deterioration incurred since the last reset, have been considered, for example, by Schneider et al.(1990), Drezner and Wesolowsky (1991), Crowder (1992), Jensen and Vardeman (1993) and Makis and Yang (1996). These models are suitable for the production settings where the accumulated wear (or deterioration) of the process can be monitored on line and used for decision making.

For processes subject to deterioration, process control, generally referred to as EPC (engineering process control) or APC (automatic process control), should be applied in conjunction with the process monitoring, SPC (statistical process control), to identify the occurrence of assignable causes and try to eliminate them.

As the tools for process control and process monitoring respectively, EPC and SPC play complementary rather than competitive roles in the continuous quality improvement. This has been observed by MacGregor (1992), Box and Kramer (1992), Montgomery et al.(1994) and other authors. SPC tries to reduce variation by detecting and eliminating special causes, while EPC is used to adjust the process economically, by taking into account different cost factors such as the quality cost and the adjustment cost. Hence, EPC and SPC should be integrated, especially for deteriorating processes. In the integration, the role of SPC is to diagnose if the prescribed process model has been violated, and then to improve the process at the *process level*, whereas EPC is used to adjust the process in the framework of the prescribed model.

SPC charts have been developed and applied, under the i.i.d. assumption, to processes encountered in industrial, economic, environmental and other areas. However, the i.i.d. assumption is hardly valid in practice and a certain degree of autocorrelation is present in almost all real processes.

As noted by Vander Wiel et al.(1992), autocorrelation is not necessarily bad. Autocorrelation means that the process is predictable, and hence, the optimal process control can be applied to reduce predictable variability by adjusting the process.

The presence of autocorrelation has a major impact on the performance of the classical control charts. This has been studied by a number of authors, as early as in the 1970's by Bagshaw and Johnson (1975), and recently by Alwan (1992). To overcome this impact, the usual approach is to fit an adequate stochastic model to the data obtained from the process and then apply a control chart to the residuals rather than to the original data. Berthouex, Hunter and Pallesen (1978) first used this method in their sewage treatment plant monitoring problem. Alwan and Robert (1988) and Harris and Ross (1991) proposed a time series model for the stochastic modelling, and Montgomery and Mastrangelo (1991) considered EWMA as a one-step-ahead predictor to approximate the exact ARIMA approach. Shewhart charts applied to the residuals are also referred to as special-cause charts (SCC).

For an effective monitoring of the process, it is important to study the response of the residuals to the occurrence of special causes.

Wardell et al.(1992) studied how the forecast (and the residual in the case of a special-cause chart) of $ARMA(1,1)$ process would react to a step shift in the process mean, if the process is deterministic, i.e., no noise is present. They showed that the mean value of the residual of the $ARMA(1,1)$ process exponentially decreases to a new non-zero level and used the results to explain the difference in ARLs for several types of SPC charts. Later, Wardell et al.(1994) extended the result to the $ARMA(p,q)$ process and derived the run length distributions for special-cause charts used to monitor the mean of this process.

In this paper, we study the problem of the effective monitoring of a controlled production process subject to variation from both the deterministic tool-wear drift and random shocks. The model of the process was first formulated by Vander Wiel (1991) and later studied by Crowder (1992), Jensen and Vardeman (1993) and Makis and Yang (1996).

The application that originally motivated the development of the model was an operation of a computer-controlled bar lathe that forms a cylindrical metal piece from a metal bar in a turning operation by cutting away the excess metal. Diameters of finished cylinders tend to become larger as the cutting tool wears down. The stochastic shocks can result from small differences in the bar stock properties, environmental condition changes, etc. For a more detailed description, see Jensen and Vardeman (1993).

Crowder (1992) studied the problem of the optimal control of the process with zero drift, symmetric loss function and a target value $T = 0$ over a finite time horizon. Jensen and Vardeman (1993) found an optimal control policy for a model with a deterministic drift and an adjustment error, and Makis and Yang (1996) established the structure of the optimal policy for the asymmetric loss function and a non-zero target value.

The behavior of the residuals of the controlled process was studied in Yang and Makis (1995) where it was found that the residuals behave as a sequence of i.i.d. normal random variables with zero means when only common causes of variation are present, and as a sequence of independent normal random variables with means exponentially decreasing to zero when a step shift occurs. This behavior can be explained by the autocorrelation, through which the effect of special causes is filtered at a constant rate $w$ at each stage, and the filtered portion is transferred into the estimated process means (or predictors). The results are summarized in Section 2.

Due to the exponential decrease of the residual means, the SPC charts applied to the process residuals may not be very effective. To improve the performance of the control charts, we propose in Section 3 a new statistic for the monitoring of the process mean. The idea is to compensate for the lost portion of the process deviation due to the autocorrelation. The classical Shewhart chart is applied to this statistic.

Using the Markov chain approach, we obtain explicit formulas for the run length distribution, average run length (ARL) and the standard deviation of the run length (SDRL).

# 2 MODEL DESCRIPTION AND PROPERTIES OF RESIDUALS

The controlled process studied in this paper is described as follows:

$$\begin{cases} Y_t = \Theta_t + \epsilon_t, \\ \Theta_t = \Theta_{t-1} + u_{t-1} + d + \nu_t, \quad t = 1, 2, \cdots \end{cases} \tag{12.1}$$

where

$\Theta_t$ represents the true process mean at time $t$,
$d$ represents a known nonrandom drift per period,
$u_{t-1}$ represents the amount of adjustment at stage $t-1$. The control action $u_{t-1}$ is determined by the observations obtained up to $t-1$. An example is the control policy presented by Jensen (1989). If no control action is taken, $u_{t-1} \equiv 0$. Otherwise, $u_{t-1} \neq 0$, and the process mean will be adjusted before the next sample is taken.
$Y_t$ represents the measurement obtained on $\Theta_t$,
$\epsilon_t$ represents the measurement error at time $t$ ,
$\nu_t$ represents random shock to the process at time $t$ and it is assumed that $\{\epsilon_t\}$ and $\{\nu_t\}$ are two sequences of independent and identically distributed normal random variables with zero means and variances $\sigma_\epsilon^2$ and $\sigma_\nu^2$, respectively. These two sequences are assumed to be independent of each other and $\sigma_\epsilon^2$ and $\sigma_\nu^2$ are assumed to be known.

The proofs of the theorems presented in this section can be found in Yang and Makis (1995).

Let $Y^t = (Y_t, Y_{t-1}, Y_{t-2}, \cdots)$, $y_t$ be the observed value of $Y_t$, and $y^t = (y_t, y_{t-1}, y_{t-2}, \cdots)$.

To initialize the iterative procedure for estimating $\Theta_t$ for $t = 1, 2, \cdots$ in Equation ( 12.1), we assume that, conditional on $y^0$ (the history of the particular process or a pilot run), $\Theta_0$ is normally distributed with $E(\Theta_0|Y^0 = y^0) = \hat{\Theta}_0$ and $V(\Theta_0|Y^0 = y^0) = q^2$, where

$$q^2 = ((\sigma_\nu^4 + 4\sigma_\epsilon^2\sigma_\nu^2)^{0.5} - \sigma_\nu^2)/2$$

is the steady state variance.

Define

$$\hat{\Theta}_t = E(\Theta_t|Y^t = y^t),$$

and the *process residual* as

$$R_t = Y_t - E(Y_t|Y^{t-1} = y^{t-1}) = Y_t - (\hat{\Theta}_{t-1} + u_{t-1} + d).$$

Then, it can be shown that given $Y^t = y^t$,

$$\hat{\Theta}_t = w \cdot y_t + (1 - w) \cdot (\hat{\Theta}_{t-1} + u_{t-1} + d) \tag{12.2}$$

for $t = 1, 2, \cdots$, where

$$w = (q^2 + \sigma_\nu^2)/(q^2 + \sigma_\nu^2 + \sigma_\epsilon^2) \tag{12.3}$$

We have the following result.

**Theorem 1**

*For Model (Equation ( 12.1)), the process residuals $\{R_t\}$ behave as a sequence of i.i.d. normal random variables, with common mean 0 and variance*

$$s^2 = q^2 + \sigma_\nu^2 + \sigma_\epsilon^2. \tag{12.4}$$

Theorem 1 describes the property of the process residuals when the process is statistically in-control.

Now, suppose that a step shift of size $\Delta$ occurs to the process mean at $t = t_0$, i.e.,

$$\left\{\begin{array}{l} \left\{\begin{array}{l} Y_{t_0} = \Theta_{t_0} + \epsilon_{t_0}, \\ \Theta_{t_0} = \Theta_{t_0-1} + u_{t_0-1} + d + \nu_{t_0} + \Delta, \end{array}\right. \\ \left\{\begin{array}{l} Y_t = \Theta_t + \epsilon_t, \\ \Theta_t = \Theta_{t-1} + u_{t-1} + d + \nu_t, \quad t = 1, 2, \cdots, t_0 - 1, t_0 + 1, t_0 + 2, \cdots \end{array}\right. \end{array}\right. \tag{12.5}$$

Then from Equation ( 12.5), given that $Y^{t_0} = y^{t_0}$, we have

$$\begin{aligned} \hat{\Theta}_{t_0} &= w \cdot y_{t_0} + (1 - w) \cdot (\hat{\Theta}_{t_0-1} + u_{t_0-1} + d + \Delta) \\ &= [w \cdot y_{t_0} + (1 - w) \cdot (\hat{\Theta}_{t_0-1} + u_{t_0-1} + d)] + (1 - w) \cdot \Delta. \end{aligned} \tag{12.6}$$

Let

$$\left\{\begin{array}{ll} \hat{\Theta}_t^* & = \hat{\Theta}_t, \quad t = \cdots, t_0 - 2, t_0 - 1 \\ \hat{\Theta}_t^* & = w \cdot y_t + (1 - w) \cdot (\hat{\Theta}_{t-1}^* + u_{t-1} + d), \quad t = t_0, t_0 + 1, \cdots \end{array}\right. \tag{12.7}$$

and

$$R_t^* = Y_t - (\hat{\Theta}_{t-1}^* + u_{t-1} + d), \quad \text{for } t = \cdots, t_0 - 1, t_0, t_0 + 1, \cdots.$$

Note that when a step shift occurs, $\hat{\Theta}_t^*$ *will not be equal to* $\hat{\Theta}_t$ *any longer.* However, because usually the process deviation cannot be detected as soon as it occurs, we actually use $\hat{\Theta}_t^*$ instead of $\hat{\Theta}_t$ as the process mean estimator, and $R_t^*$ instead of $R_t$ as the process residual, before we detect the occurred deviation and adjust the process.

For model Equation ( 12.5), let

$$r_t = y_t - (\hat{\Theta}_{t-1}^* + u_{t-1} + d)$$

be the observed value of $R_t^*$ and denote $R^t = (R_t^*, R_{t-1}^*, \cdots)$, and $r^t = (r_t, r_{t-1}, \cdots)$.

**Theorem 2**

**Dynamic Response of Residuals to a Step Shift**
*Assume that a step shift of size $\Delta$ occurred to the process mean at $t = t_0$. Then, $\{R_t^*\}$, for $t \geq t_0$, is a sequence of independent normally distributed random variables with the same variance $s^2$, and exponentially decreasing* means.

$$(1 - w)^{t - t_0} \cdot \Delta \tag{12.8}$$

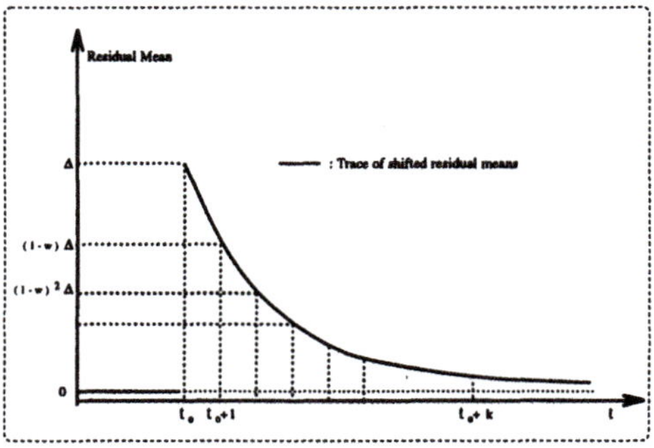


**Figure 1** Dynamic response of the residual mean to a step shift of size Δ

The result in Theorem 2 is illustrated by Figure 1 which shows the behavior of the residual mean after the occurrence of a shift of size Δ. The shift in the mean of the original *autocorrelated* controlled process will be converted to the exponentially decreasing mean (i.e., a ramping shift) in the process $\{R_t^*\}$.

We can see from Equations ( 12.2) and ( 12.8), that the filtering rate $w$ plays an important role in the process mean prediction and in studying the response of the residuals to the external shifts.

We can also see that the results in this section do not depend on the specific form of the control policy $\{u_t\}$. Hence, the monitoring scheme based on the residuals can be used for any control policy adopted.

# 3 PROCESS MONITORING

It follows from Theorem 2, that after the occurrence of a shift, the residual means will be exponentially decreasing. Hence, the control charts applied to the residuals may not be very effective.

To compensate for the lost portion of the process deviation due to the autocorrelation, we propose the following statistic for monitoring the process mean:

$$V_{\tau,t} = w \cdot \sum_{i=1}^{\tau-1} R_{t-i}^* + R_t^*$$

where $R_t^*$ is the process residual at time $t$.

From Theorem 1 and Theorem 2, we have the following result.

**Theorem 3**

**Properties of $V_{\tau,t}$** *When the process is in-control,*

$$E(V_{\tau,t}) = 0$$

*for all $t$. When a shift of size $\Delta$ occurs to the process mean at time $t_0$,*

$$E(V_{\tau,t_0+j}) = E(R_{t_0}^*) = \Delta$$

*for all $j = 0, 1, \cdots, \tau - 1$, and*

$$E(V_{\tau,t+\tau-1}) = E(R_t^*)$$

*for all $t \geq t_0$.*

Theorem 3 indicates that the mean value of $V_{\tau,t}$ is equal to $\Delta$ for $\tau$ steps after the occurrence of the shift and then exponentially decreases to zero. In this paper, we consider a special case $\tau = 2$. Denote

$$V_t = V_{2,t}.$$

Then,

$$Var(V_t) = (1 + w^2) \cdot s^2$$

regardless, the process is in-control or out-of-control.

The $k \cdot \sigma$ Shewhart chart can be applied to the sequence $\{V_t\}$, and the process is assumed to be out-of-control if $|V_t| > k \cdot [Var(V_t)]^{0.5}$.

To derive the formulas for the run length distribution, ARL and SDRL, we consider the process $\{(V_t, R_t^*)\}$ which is a non-homogeneous 2-dimensional continuous-state Markov chain. We will discretize the state space.

Given $k$, the control limit parameter of the Shewhart chart applied to $\{V_t\}$, we divide $[-k \cdot (1+w^2)^{0.5} \cdot s, k \cdot (1+w^2)^{0.5} \cdot s]$ into $(n-1)$ equal subintervals $I_{v_1}, I_{v_2}, \cdots, I_{v_{n-1}}$, and denote $I_{v_n} = (-\infty, -k \cdot (1+w^2)^{0.5} \cdot s) \cup (k \cdot (1+w^2)^{0.5} \cdot s, \infty)$. Let $v_i$ be the middle point of $I_{v_i}$ for $i = 1, 2, \cdots, n-1$. Then, we approximate $V_t$ as $v_i$ if $V_t \in I_{v_i}$, and denote $V_t = v_n$ if $V_t \in I_{v_n}$.

Because of the symmetry, we consider only positive shifts. To discretize the state space for $R_t^*$, we divide $[-3 \cdot s,\ 6 \cdot s]$ into $(m-2)$ subintervals given by $I_{r_1}, I_{r_2}, \cdots, I_{r_{m-2}}$ of equal length of $9 \cdot s/(m-2)$. Denote $I_{r_0} = (-\infty, -3 \cdot s)$ and $I_{r_{m-1}} = (6 \cdot s, \infty)$. Let $r_j$ be the middle point of $I_{r_j}$ for $j = 1, \cdots, m-2$, $r_0 = -3 \cdot s - 0.5 \cdot 9 \cdot s/(m-2)$ and $r_{m-1} = 6 \cdot s + 0.5 \cdot 9 \cdot s/(m-2)$. We approximate $R_t^*$ as $r_j$ if $R_t^* \in I_{r_j}$ for $j = 0, 1, \cdots, m-1$.

Then the discretized state space is

$$S = \{(v, r) | v \in \{v_1, v_2, \cdots, v_{n-1}, v_n\},\ r \in \{r_0, r_1, \cdots, r_{m-1}\}\}$$

where $v_n$ denotes the out-of-control state, which we assume is an absorbing state. Denote

$$P_l(t) = P(R_t^* \in I_{r_l})$$

The transition probability, for $i \neq n$, is given by

$$P(V_{t+1} = v_h, R_{t+1}^* = r_l | V_t = v_i, R_t^* = r_j) = \begin{cases} P_l(t+1) & \mathit{if}\ w \cdot r_j + r_l \in I_{v_h} \\ 0 & \mathit{otherwise} \end{cases}$$

and the transition matrix from stage $t$ to $t+1$ has the form

$$P_{nm \times nm}(t) \equiv \begin{array}{c} \text{“in”} \\ \text{“out”} \end{array} \overset{\begin{array}{cc} \text{“in-control”} & \text{“out-of-control”} \end{array}}{\begin{bmatrix} Q_{(n-1)m \times (n-1)m}(t) & H_{(n-1)m \times m}(t) \\ 0_{m \times (n-1)m} & T_{m \times m}(t) \end{bmatrix}},$$

where $T$ is the identity matrix.

Suppose that a shift occurs to the process immediately after $t = 0$. Then,

$$F_{(n-1)m \times 1}(l) = 1_{(n-1)m \times 1} - [\prod_{t=0}^{l-1} Q_{(n-1)m \times (n-1)m}(t)] \cdot 1_{(n-1)m \times 1}, \quad l = 1, 2, \cdots,$$

where the $i$-th element of $F_{(n-1)m \times 1}(l)$ is the probability that the run length $\leq l$ given that the process was in the $i$-th in-control state at $t = 0$.

Let $A_{(n-1)m \times 1}$ and $VAR_{(n-1)m \times 1}$ denote the conditional ARL and $SDRL^2$ $(n-1)m \times 1$ vectors given the initial state at $t = 0$. To obtain formulas for $A_{(n-1)m \times 1}$ and $VAR_{(n-1)m \times 1}$, we will use the following lemma (for a proof, see Yang and Makis 1996).

**Lemma 1**

If $X$ is a non-negative integer random variable, and $E[X(X+1)\cdots(X+m-1)]$ exists for some integer $m > 0$, then

$$E[X(X+1)\cdots(X+m-1)] = m \cdot \sum_{x=1}^{\infty} x(x+1)\cdots(x+m-2) \cdot P(X \geq x),$$

where $x(x+1)\cdots(x+m-2) \equiv 1$ for $m = 1$.

The formulas are in the the following theorem.

**Theorem 4**

$$A_{(n-1)m\times 1} = \sum_{l=1}^{\infty}[\prod_{t=0}^{l-2} Q_{(n-1)m\times(n-1)m}(t)] \cdot 1_{(n-1)m\times 1},$$

and

$$\begin{aligned} VAR_{(n-1)m\times 1} = {} & 2 \cdot \sum_{l=1}^{\infty} l \cdot [\prod_{t=0}^{l-2} Q_{(n-1)m\times(n-1)m}(t)] \cdot 1_{(n-1)m\times 1} \\ & -A_{(n-1)m\times 1} \odot (A_{(n-1)m\times 1} + 1_{(n-1)m\times 1}), \end{aligned}$$

where $\prod_{t=0}^{-1} Q(t) = I$, the operator $\odot$ multiplies the elements at the same position and the result is a vector of the same length.

Let $\tilde{Q}_{(n-1)m\times(n-1)m}$ be the (row-wise standardized) "in-control" block of the transition matrix when the process is in-control, and $\pi$ be the steady distribution of the discretized process state given that the process is in-control. Then $\pi$ can be obtained as a solution to the following system of equations:

$$\begin{cases} \pi = \tilde{Q}^T_{(n-1)m\times(n-1)m} \cdot \pi \\ \sum \pi_i = 1 \end{cases} \tag{12.9}$$

where $\pi_i$ is the $i$-th element of $\pi$.

Suppose that the probability mass function of the initial state at $t = 0$ is $\pi_i$. Then, by conditioning, we can find the run length distribution, ARL and SDRL. The computational procedure is described in detail in the next section.

# 4 DESCRIPTION OF THE COMPUTATIONAL PROCEDURE AND NUMERICAL EXAMPLE

Assume that a step shift of size $\Delta$ occurred to the process mean at $t = 0$ ($\Delta = 0$ implies there is no shift). Then, from Theorem 2,

$$P_l(t) = P(R_t^* \in I_{r_l}) = \begin{cases} \Phi(-3 - \frac{(1-w)^t \cdot \Delta}{s}) & \text{if } l = 0 \\ \Phi(-3 + \frac{9 \cdot l}{m-2} - \frac{(1-w)^t \cdot \Delta}{s}) & \\ -\Phi(-3 + \frac{9 \cdot (l-1)}{m-2} - \frac{(1-w)^t \cdot \Delta}{s}) & \text{if } l = 1, \cdots, m-2 \\ 1 - \Phi(6 - \frac{(1-w)^t \cdot \Delta}{s}) & \text{if } l = m-1 \end{cases}$$

For the computational purpose, we will be interested in finding the "in-control" portion, $Q$, of the transition matrix $P$. The "in-control" states are

$$\{\{(v_i, r_j), j = 0, 1, \cdots, m-1\}, i = 1, 2, \cdots, n-1\}$$

Let $q_{i',j'}(t)$ be the element of the matrix $Q(t)$, where $0 \leq i', j' \leq (n-1) \cdot m - 1$

We define an $m \times m$ integer matrix, $\mathbf{B}$, as follows.

Let $b_{j_1,j_2}$ be the element of $\mathbf{B}$, where $0 \leq j_1, j_2 \leq m-1$. Put

$$b_{j_1,j_2} = \begin{cases} h, & if\ w \cdot r_{j_1} + r_{j_2} \in I_{v_h} \text{ and } h \neq n \\ 0 & \text{otherwise} \end{cases}$$

For integers $a$ and $b$, let $[a/b]$ denote the integer part of $a/b$, and $rem(a, b) = a - b \cdot [a/b]$. Then, for any $i', j'$ and $t$,

$$q_{i',j'}(t) = \begin{cases} P_{rem(j',m)}(t) & \text{if } b_{rem(i',m),rem(j',m)} = [j'/m] \\ 0 & \text{otherwise} \end{cases}$$

Notice that every $m$ rows of $Q(t)$ are identical. $Q(t)$ is a sparse matrix, and the portion which is not equal to zero is determined by matrix $\mathbf{B}$. The computational advantage of using matrix $\mathbf{B}$ is memory savings and an increased speed of computation, because it can be computed and stored at the very beginning, and then used in each iteration later.

The ARL and SDRL can be computed using the formulas in Theorem 4. When applying the formulas, in each iteration of $l$, only $Q(l-2)$ needs to be computed, the final product matrix and the summation matrix obtained up to $l$ can be stored in two permanently defined matrices, and used in the next iterations. In this way, the required space and memory is kept at minimum, and the computational speed is maximized.

In our programming, we define two discretization parameters $H$ and $K$ for the residuals and V-statistics, respectively. For the residuals, we first divide the interval $[-3s, 3s]$ into $2 \cdot H + 1$ subintervals of equal length and then, due to the positive shift in the process mean, we add $H + 1$ subintervals of the same width to cover the interval $[3s, 6s + \frac{3s}{2 \cdot H+1}]$. The total number of the discretization subintervals for the residuals is $m = 3 \cdot H + 2$. The "in-control" range of the V-statistic, $[-k \cdot s(1+w^2)^{0.5}, k \cdot s(1+w^2)^{0.5}]$ is divided into equal subintervals of length $6s(1+w^2)^{0.5}/(2 \cdot K + 1)$. Hence, the total number of discretization subintervals for the V-statistic is $n = 2 \cdot K + 2$ (one subinterval for the "out-of-control" state).

Denote

$$ARL(L) = \pi' \cdot (\sum_{l=1}^{L} [\prod_{t=0}^{l-2} Q_{(n-1)m \times (n-1)m}(t)] \cdot 1_{(n-1)m \times 1})$$

Obviously, the finer the discretization, i.e., the larger the values of $H$ and $K$, the more accurate the computational results will be on one hand, and the slower the computation on the other hand. Therefore, there is a trade-off between accuracy and speed when choosing the discretization parameters. Our experience with S-Plus programming indicates that $H = 3$ and $K = 3$, i.e., $m = 11$ and $n = 8$ is a proper choice.

**Example 1**

We will show the computation of the ARL for $\Delta = 1$, $k = 3$, $w = 0.5$ and $s = 1$. For the discretization, we have chosen $H = 3$ and $K = 3$, i.e., $m = 11$ and $n = 8$. Table 1 shows the convergence of $\{ARL(L)\}$ and the final result (when $L = \infty$).

| $L$ | 200 | 400 | 600 | 800 | 1000 | 1200 | 1400 | 1600 | 1800 | $\infty$ |
|---|---|---|---|---|---|---|---|---|---|---|
| $ARL(L)$ | 132 | 191 | 218 | 230 | 235 | 238 | 239.4 | 239.9 | 240.1 | 240.3 |

**Table 1** Computation of the ARL

# REFERENCES


[1] Alwan, L. C., and H. V. Roberts, "Time-Series Modeling for Statistical Process Control", *Journal of Business and Economic Statistics*, 6(1), pp 87-95, 1988.

[2] Alwan, L. C., "Effects of Autocorrelation on Control Chart Performance", *Communications in Statistics - Part A: Theory and Methods*, 21(4), pp 1025-1049, 1992.

[3] Arcelus, F. J., P. K. Bannerjee, and R. Chandra, "The Optimal Production Run for a Normally Distributed Quality Characteristic Exhibiting Non-negative Shifts in Process Mean and Variance", *IIE Trans.*, 14, pp 90-98, 1982.

[4] Bagshaw, M., and R. A. Johnson, "The Effect of Serial Correlation on the Performance of CuSum Tests II", *Technometrics*, 17(1), pp 73-80, 1975.

[5] Berthouex, P. M., W. G. Hunter, and L. Pallesen, "Monitoring Sewage Treatment Plants: Some Quality Control Aspects", *Journal of Quality Technology*, 10(4), pp 139-149, 1978.

[6] Box, G., and T. Kramer, "Statistical Process Monitoring and Feedback Adjustment — A Discussion", *Technometrics*, 34(3), pp 251-267, 1992.

[7] Crowder, S. V., "An SPC Model for Short Production Runs: Minimizing Expected Cost", *Technometrics*, 34(1), pp 64-73, 1992.

[8] Drezner, Z., and G. O. Wesolowsky, "Control Limit for a Drifting Process with Quadratic Loss", *International Journal of Production Research*, 27(1), pp 13-20, 1989.

[9] Drezner, Z., and G. O. Wesolowsky, "Optimizing Control Limits under Random Process Shifts and a Quadratic Penalty Function", *Communic. Statist. — Stochastic Models*, 7(3), pp 363-377, 1991.

[10] Gibra, T. N., "Optimal Control of Process Subject to Linear Trends", *J. Indus. Engg.*, pp 35-41, 1967.

[11] Harris, T. J., and W. H. Ross, "Statistical Process Control Procedures for Correlated Observations", *The Canadian Journal of Chemical Engineering*, 69, pp 48-57, 1991.

[12] Jensen, K. L., *Optimal Adjustment in the Presence of Process Drift and Adjustment Error*, Ph.D. Dissertation, Dept. of Statistics, Iowa State University, Ames, Iowa, 1989.

[13] Jensen, K. L., and S. B. Vardeman, "Optimal Adjustment in the Presence of Deterministic Process Drift and Random Adjustment Error", *Technometrics*, 35(4), pp 376-389, 1993.

[14] MacGregor, J. F., Discussion for "Statistical Process Monitoring and Feedback Adjustment — A Discussion" by Box and Kramer, *Technometrics*, 34(3), pp 273-275, 1992.

[15] Makis, V., "Optimal Tool Replacement with Asymmetric Quadratic Loss", *IIE Trans.*, 28(6), pp 463-466, 1996.

[16] Makis, V. and J. Yang, "Optimal Control of a Deteriorating Production Process", presented at *IFORS'96*, Vancouver, Canada, July, 1996.

[17] Manuele, J., "Control Chart for Deteriorating Tool Wear", *Industrial Quality Control*, 1, pp 7-10, 1945.

[18] Montgomery, D.C., J.B. Keats, G.C. Runger, and W.S. Messina, "Integrating Statistical Process Control and Engineering Process Control", *Journal of Quality Technology*, 26(2), pp 79-87, 1994.

[19] Montgomery, D. C., and C. M. Mastrangelo, "Some Statistical Process Control Methods for Autocorrelated Data", *Journal of Quality Technology*, 23(3), pp 179-193, 1991.

[20] Quesenberry, C. P., "An SPC Approach to Compensating a Tool-Wear process", *Journal of Quality Technology*, 20(4), pp 220-229, 1989.

[21] Schneider, H., K. Tang, and C. O'Cinneide "Optimal Control of a Production Process Subject to Random Deterioration", *Operations Research*, 38(6), pp 1116-1122, 1990.

[22] Vander Wiel, S. A., "Optimal Discrete Adjustments for Short Production Runs", Statistical Research Report 101, AT&T Bell Laboratories, Murray Hill, NJ, 1991.

[23] Vander Wiel, S. A., W. T. Tucker, F. W. Faltin, and N. Doganaksov, "Algorithmic Statistical Process Control: Concepts and an Application", *Technometrics*, 34(3), pp 286-297, 1992.

[24] Wardell, D.G., H. Moskowitz, and R.D. Plante, "Control Charts in Presence of Data Correlation", *Management Science*, 38(8), pp 1084-1105, 1992.

[25] Wardell, D.G., H. Moskowitz, and R.D. Plante, "Run-Length Distribution of Special-Cause Control Charts for Correlated Processes", *Technometrics*, 36(1), pp 3-17, 1994.

[26] Yang, J. and V. Makis, "Dynamic Response of Residuals of the Steady Process with Optimal Control and Deterministic Drift", *Working Paper no. 95-12*, Dept. of Industrial Engineering, Univ. of Toronto, 1995.

[27] Yang, J. and V. Makis, "Monitoring Tool-Wear Rate Change in a Controlled Production Process", *The Proceedings of the Fourth International Conference on Automation Technology*, 1, pp 727-734, Hsinchu, Taiwan, July 8-11, 1996.

# 13

# LOT SIZING AND LIFE TESTING FOR QUALITY IMPROVEMENT OF ITEMS SOLD WITH WARRANTY

**I. Djamaludin[1], R.J. Wilson[2] & D.N.P. Murthy[3]**

[1] *Technology Management Centre,*
*The University of Queensland,*
*Brisbane, Qld, 4072,*
*Australia.*

[2] *Department of Mathematics,*
*The University of Queensland,*
*Brisbane, Qld, 4072,*
*Australia.*

[3] *Department of Mechanical Engineering,*
*The University of Queensland,*
*Brisbane, Qld, 4072,*
*Australia.*

## ABSTRACT

Due to manufacturing variability, a fraction of items produced fail to conform to the design specification. The performance of such items is inferior compared to those which conform. As a result, non-conforming items have a significant impact on the expected warranty service cost when items are sold with warranty. This cost can be reduced through effective quality control. In this chapter we develop a model which uses testing (for weeding out non-conforming items) and lot sizing (to reduce the occurrence of non-conforming items) for improving quality when items are produced in lots. Unfortunately, the reduction in the expected warranty servicing cost is achieved at the expense of increased manufacturing cost. The model examines a quality improvement scheme which achieves a balance between these two costs.

**Key words:** quality control, warranty, lot sizing, life testing

# 1 INTRODUCTION

A warranty is a contractual agreement which requires the manufacturer to rectify all failures occurring within the warranty period. The rectification can include repair, replacement or cash refund. Warranty is important for both consumers and manufacturers. It protects the consumer against non-conforming or poor quality items that function unsatisfactorily and it protects the manufacturer against unreasonable claims from consumers. Consequently, warranty has been used by manufacturers as a promotional tool to market their products.

Offering a warranty results in additional cost to the manufacturer. The warranty servicing cost depends on various factors, including the type of warranty, the warranty period, reliability characteristics of the item and warranty servicing strategies. The design decisions determine the reliability of the product (for example, mean time to failure). The warranty servicing cost can be reduced by better design, but this may result in increased manufacturing cost. Due to variability in the manufacturing process, a fraction of items do not conform to specifications and their performance is inferior to that of conforming items. The expected warranty servicing cost for a non-conforming item is considerably more than that for a conforming item. The warranty servicing cost can be reduced by life testing to weed out non-conforming items. Here items are tested for a specified period and items which fail during testing are scrapped. An alternative approach is to reduce the likelihood that an item produced is non-conforming. When items are produced in lots, lot sizing can be used to achieve this.

Here, we develop a new model which uses both lot sizing and life testing to reduce the expected warranty cost per item sold. Both of these actions increase the manufacturing cost per item. The optimal decision with respect to lot sizing and life testing achieves a balance between these two costs.

The outline of the chapter is as follows. In Section 2, a brief review of the relevant literature is presented. In Section 3, we develop the model formulation. Section 4 deals with a preliminary analysis of the model. This is used in Sections 5–7 to derive optimal lot sizing and testing strategies for both repairable and non-repairable products sold with either Free Replacement warranty (FRW) or Pro-rata warranty (PRW) policies.

# 2 LITERATURE REVIEW

In this section we present a brief review of the relevant literature, so that the contribution of the paper can be put in a proper perspective.

## 2.1 Quality Variations and Lot Size

Porteus (1986), and Rosenblatt and Lee (1986), independently proposed mathematical models linking product quality with lot size. The concept that a smaller lot size leads to better quality can be traced to earlier literature - for example, Schonberger (1982), in his analysis of "Just In Time" processes. In both model formulations, the process is checked to ensure that it is in-control before operation on a new lot is commenced. In Porteus' model, if the process is in-control at the start of an item's production, it can change randomly to being out-of-control or continue to be in-control at the end of the item's production. Once the process changes to out-of-control it stays there until all items in the lot are processed. The model assumes that when the process is in-control, only conforming items are produced and, when the process is out-of-control, all items produced are non-conforming. In Rosenblatt and Lee's model, the process stays in-control for a random duration before switching. Once the switch occurs, as in Porteus' model, the process continues to stay out-of-control until all items in the lot are processed. As a result, the number of non-conforming items in a lot is a random variable with a mean which is a function of the lot size. In both these models, non-conforming items are non-operational and hence can be detected by testing for a very short time period, after which they are reworked to become operational. The performance of an item over time, subsequent to the sale, is of no consequence.

The Porteus model has been extended by many researchers to include additional variables; for example, Keller and Noori (1988), deal with uncertain demand and Chand (1989), incorporates learning effects.

## 2.2 Warranties and Warranty Analysis

Many different types of warranty policies have been formulated and analyzed. Blischke and Murthy (1992), have proposed a taxonomy to categorize them. Of particular interest to this paper are the free replacement warranty (FRW) and the pro-rata warranty (PRW). These are the two most commonly offered warranties. Descriptions of these two policies are given in the next section.

When an item fails under warranty, the manufacturer incurs additional cost in the form of repair/replacement cost (FRW policy) or refund of a fraction of the original sale price (PRW policy). Many different models have been developed to calculate the expected warranty cost per item sold. A review of these models can be found in Blischke (1990), and Murthy and Blischke (1992a, 1992b). For further details of modeling and analysis for warranty costs, see Blischke and Murthy (1994).

## 2.3 Warranty and Quality Improvement

The authors of this paper have developed a variety of models which link warranty with quality improvement. In this sub-section, we give a brief review of these models.

Murthy et al.(1993), deal with the case where the process is in steady state and each item produced is either conforming or non-conforming, so that lot sizing is of no consequence. The quality improvement scheme to reduce warranty cost involves testing each item for a period $T$ (also called burn-in time). Items which fail during testing are scrapped and the others are released for sale. The rationale for this is that, since a non-conforming item has a higher failure rate (compared to a conforming item), it is more likely to fail during testing and so to be subject to being weeded out. The paper derives the optimal value for $T$ which achieves a trade off between the reduction in expected warranty cost and the additional cost incurred due to testing. If the optimal time is zero, it implies that no testing is the optimal strategy.

Djamaludin et al.(1994), use a model similar to that proposed by Porteus for modeling quality variations. In contrast to Porteus, the model assumes that not all items are conforming when the process is in-control and not all items are non-conforming when the process is out-of-control, and that non-conforming items are operational but with inferior characteristics as in Murthy et al.(1993). Quality improvement is achieved through lot sizing and items are released for sale with no testing.

Djamaludin et al.(1995), deal with a model similar to that in Djamaludin et al.(94), except that it also involves testing a fraction of items in some lots. Here, at the end of each lot production, the state of the process is assessed. If it is found to be in-control, then the lot is released with no testing. However, if the state is found to be out-of-control, the last $K$ items in the lot are tested for a period $T$. Those which fail during the testing are scrapped and the others (not tested and those which survive the test) are released for sale. The model involves three variables - the lot size ($L$), number of items tested in the lot ($K$) if the process state is out-of-control at the end of lot production, and the duration of testing ($T$). The optimal values for these are obtained based on minimizing the asymptotic total cost per item where the total cost is the sum of the manufacturing (which includes production, testing and scrapping) cost and the warranty servicing cost.

The model studied here is similar to the above model but differs in two ways. Firstly, it does not assume that the process state is known at the end of a lot production. This is more realistic. Secondly, the scheme for testing items is different as indicated in the next section.

# 3 MODEL FORMULATION

We assume that there is a constant demand for the product. This demand is met by producing items in lots of size $L$, with $1 \leq L \leq L_U$, where $L_U$ is the upper limit to reflect practical limitations on the lot size. We assume that the time horizon is sufficiently large so that it can be approximated as being infinite. Hence we consider the asymptotic case where the number of lots produced tends to infinity.

The production cost associated with a lot is given by

$$C_S + L \times C_m \tag{13.1}$$

where $C_S$ is the setup cost to ensure that the process state is in-control at the start of each lot and $C_m$ is the material and labor cost to produce a single item.

## 3.1 Changes in Process State

The model assumes that items are produced in lots of size $L$. The process is always in-control at the start of production of each lot and can change to out-of-control in an unpredictable manner. During the production of an item, the probability that the process changes from in-control to out-of-control is $(1 - q)$ and stays in-control is $q$. Once the process is out-of-control, it stays there until the completion of the lot.

If the process is in-control at the end of the production of an item, the item produced is conforming with probability $\theta_1$ and non-conforming with probability $(1 - \theta_1)$. Similarly, if the process is out-of-control at the end of the production of an item, the item produced is conforming with probability $\theta_2$ and non-conforming with probability $(1 - \theta_2)$, with $\theta_1 > \theta_2$ implying that an item produced with the process out-of-control is more likely to be non-conforming than one produced with the process in-control.

## 3.2 Characterization of Conforming and Non-Conforming Items

Let $F_1(t)$ $[F_2(t)]$ denote the failure distribution function for conforming [non-conforming] items. Let $f_j(t)$, $\overline{F}_j(t)$ and $r_j(t)$ denote the density function, survivor function and the failure rate associated with $F_j(t), j = 1, 2$. These are related as follows: $f_j(t) = dF_j(t)/dt$, $\overline{F}_j(t) = 1 - F_j(t)$ and $r_j(t) = f_j(t)/\overline{F}_j(t)$.

A non-conforming item has a higher failure rate than a conforming item, that is, $r_1(t) < r_2(t), 0 \leq t < \infty$. This implies

1. $F_2(t) > F_1(t)$ for $t \geq 0$, and
2. $\int_0^\infty \overline{F}_1(t)dt > \int_0^\infty \overline{F}_2(t)dt$, that is, mean time to failure for non-conforming items is smaller than that for conforming items.

## 3.3 Testing to Weed out Non-Conforming Items

Since non-conforming items are operational but have a higher failure rate, the only way to weed them out is through testing (or burn-in) for a period $T$. Items which fail during testing are scrapped. Even if all items are tested, there is no guarantee that all non-conforming items are weeded out. As $T$ increases, a greater fraction of non-conforming items are weeded out. However, this also increases the fraction of conforming items which get scrapped due to failures in the testing period. In addition, testing involves additional cost. As a result, 100% testing is not optimal. We use a scheme which involves a sequential decision rule for testing items. The number of items tested in a lot can vary from 1 to $(L - K + 1)$. The characterisation of the rule requires sequential numbering (1 through $L$) of items in the order in which they are produced in a lot.

After the production of a lot, item $L$ (the last item) is life tested for a period $T$. If it survives, then it and the remaining items in the lot are released with no further testing. On the other hand, if it fails during the test, then it is scrapped and item $K$ ($1 \leq K < L$) is life tested for duration $T$. If item $K$ survives, then it and the remaining items in the batch are released with no further testing. However, if item $K$ fails, then it is scrapped and items $(K+1)$ to $(L-1)$ are life tested for duration $T$. Those which survive the test are released along with the first $(K-1)$ items which are released with no testing.

The rationale for this is as follows. If item $L$ survives the test, it is more likely that the process is still in-control and hence the number of non-conforming items in the lot is small. In this case, testing to weed these out is not worthwhile. If item $L$ fails, it is possible that the process state has changed from in-control to out-of-control. By testing item $K$, we obtain more information. If item $K$ fails the test, then it is more likely that the process change occurred before it was produced. As a result, the number of non-conforming items in those from $(K+1)$ to $(L-1)$ can be high so testing these to weed them out is a good strategy. If item $K$ survives the test, the change in process state would be more likely to have occurred after item $K$ was produced and by not testing any other items, one is hopefully taking only a small risk that a certain number of non-conforming items are released.

Let $\eta_i$ denote the number of items tested in lot $i$. This is a random variable and can take the values 1, 2 and $(L-K+1)$. Let $\varepsilon_i$ denote the number of items which fail during testing and are scrapped. This is also a random variable. The number of items released from lot $i$ is given by $(L-\varepsilon_i)$.

An item released can be one of the four types below, with their failure distributions as indicated:

Type A: Conforming and not tested $[F_1(t)]$

Type B: Non-conforming and not tested $[F_2(t)]$

Type C: Conforming and survived testing $[\hat{F}_1(t)]$

Type D: Non-conforming and survived testing $[\hat{F}_2(t)]$

where

$$\hat{F}_j(t) = \{F_j(t+T) - F_j(T)\}/\overline{F}_j(T) \tag{13.2}$$

for $1 \leq j \leq 2$. Let $N_{Ai}$, $N_{Bi}$, $N_{Ci}$ and $N_{Di}$ denote respectively the number of Type A, B, C and D items in lot $i$. These are random variables.

The testing cost per item is assumed to be of the form $a_1 + a_2T$, with $a_1 \geq 0$ and $a_2 > 0$. This implies that the cost increases linearly with the duration of the test.

The total testing cost for lot $i$ is a random variable since the number of items tested ($\eta_i$) is uncertain. It depends on the lot size, the testing scheme and the testing time, and is given by

$$\eta_i(a_1 + a_2T) + \varepsilon_i C_{sc} \tag{13.3}$$

where $C_{sc}$ is the cost of scrapping a unit.

## 3.4 Warranty Policies and Servicing

We consider the following types of warranties with warranty period $W$:

*Free Replacement warranty* (FRW) where the manufacturer agrees to replace a failed item with a new one, or repair the failed item, should the failure occur within the warranty period $W$. This is done at no cost to the customer.

*Pro-rata warranty* (PRW) where the manufacturer refunds a fraction of the original purchase price to the consumer should the item fail within the warranty period $W$.

The warranty servicing cost depends on whether the items are repairable or not and the type of repair used in the case of repairable items. We consider the following three cases:

Case I: Items repairable and sold with FRW policy,

Case II: Items not repairable and sold with PRW policy,

Case III: Items not repairable and sold with FRW policy.

For Case I, we assume minimal repair. Under minimal repair, the failure rate after repair is the same as that just before failure. See Murthy (1991), for more details. In Case III, failed items are replaced by new ones.

The warranty servicing cost for lot $i$ depends on the numbers ($N_{Ai}$, $N_{Bi}$, $N_{Ci}$ and $N_{Di}$) of the four types of items released and their failure distributions, as well as the warranty. These are influenced by the lot size and the testing scheme. Let $C_{Wi}(L, K, T)$ denote this cost for lot $i$.

## 3.5 Optimal Control Strategy

The total cost for lot $i$ is the sum of production, testing and warranty servicing costs and is given by

$$C_{Ti}(L,K,T) = C_{Mi}(L,K,T) + C_{Wi}(L,K,T) \tag{13.4}$$

where

$$C_{Mi}(L,K,T) = (C_S + LC_m) + [\eta_i(a_1 + a_2T) + \varepsilon_i C_{sc}] \tag{13.5}$$

represents the total manufacturing cost (the sum of production cost and testing cost) - see Equations (13.1) and (13.3)) for lot $i$.

The asymptotic cost per item released is given by

$$C_A(L,K,T) = \lim_{n\to\infty} \frac{\sum_1^n C_{Ti}(L,K,T)}{\sum_1^n (L - \varepsilon_i)}. \tag{13.6}$$

Since the lots are statistically similar and $(C_{Ti}(L,K,T), \varepsilon_i)$ are statistically independent over $i$ with $L - E[\varepsilon_i] > 0$, it follows from the weak law of large numbers (Heathcote (19071)), that

$$C_A(L,K,T) = \frac{E[C_{Ti}(L,K,T)]}{L - E[\varepsilon_i]}. \tag{13.7}$$

The optimal quality improvement strategy is given by $L^*$, $K^*$ and $T^*$ which minimize $C_A(L,K,T)$ subject to the constraints $1 < L^* \leq L_U$, $1 \leq K^* < L^*$ and $T^* \geq 0$.

## 3.6 Additional Assumptions

1. All failures under warranty result in claims and the claims are made as soon as the items fail.
2. All claims are valid and failures are rectified as per warranty terms.
3. The time to rectify (replace or repair) is relatively small compared with the mean time between failures. Hence, these times are treated as being zero.
4. The cost associated with each claim is modeled by a single variable representing all the costs associated with servicing (that is, handling, labor, material and so on).

# 4 PRELIMINARY ANALYSIS

In this section, we carry out the analysis to obtain expressions for the expected values of $\eta_i$, $\varepsilon_i$ and $C_{Ti}(L, K, T)$.

Let $N_i$ denote the number of items in lot $i$ produced with the process in-control. If $N_i = L$, then no item is produced with the process out-of-control and if $N_i < L$, then the last $(L - N_i)$ items are produced with the process out-of-control. It is easily shown that $N_i$ has a truncated geometric distribution.

The probability that item $L$ in lot $i$ fails during testing depends on $N_i$. If $N_i = L$ $[N_i < L]$, then the probability is given by $\rho_1(T)$ $[\rho_2(T)]$, where $\rho_j(T)$ $(j = 1, 2)$ is given by

$$\rho_j = \rho_j(T) = \theta_j F_1(T) + (1 - \theta_j)F_2(T). \tag{13.8}$$

Similarly, the probability that item $K$ fails during testing is given by $\rho_1$ if $N_i \geq K$ and by $\rho_2$ if $N_i < K$.

Using the distribution of $N_i$ and conditional expectations, the expected values of the number of items tested $(\eta_i)$ and the number of items to fail when tested $(\varepsilon_i)$, can be shown to be

$$E[\eta_i] = 1 + \rho_2 + (L - K - 1)\rho_2^2 + (\rho_1 - \rho_2)[q^L + (L - K - 1)(\rho_1 q^L + \rho_2 q^K)] \tag{13.9}$$

and

$$\begin{aligned} E[\varepsilon_i] &= \rho_2 + \rho_2^2 + (L - K - 1)\rho_2^3 \\ &\quad + (\rho_1 - \rho_2)[\{1 + \rho_1 + (L - K - 1)\rho_1^2\}q^L \\ &\quad + \{\rho_2 + (L - K - 1)\rho_2^2\}q^K + \rho_1\rho_2(q^{K+1} - q^L)/(1 - q)]. \end{aligned} \tag{13.10}$$

(Note: We have omitted the details of the derivation and interested readers can find it in Djamaludin (1993). The expected total manufacturing cost (see Equation (13.5)) is given in terms of Equations (13.9) and (13.10) by

$$E[C_{Mi}(L, K, T)] = C_S + LC_m + E[\eta_i](a_1 + a_2 T) + E[\varepsilon_i]C_{sc} \tag{13.11}$$

To compute the expected values of $C_{Wi}(L, K, T)$, we need the expected values of the numbers of different types of released items. Again based on conditional expectations, we have

$$E[N_{Ai}] = \tau_1\theta_1 + \tau_2\theta_2 \quad (13.12)$$
$$E[N_{Bi}] = \tau_1(1-\theta_1) + \tau_2(1-\theta_2) \quad (13.13)$$
$$E[N_{Ci}] = \tau_3\theta_1 + \tau_4\theta_2 \quad (13.14)$$
$$E[N_{Di}] = \tau_3(1-\theta_1) + \tau_4(1-\theta_2) \quad (13.15)$$

where

$$\begin{aligned}
\tau_1 &= \{q/(1-q)\}(1-q^L) - \tau_3, \\
\tau_2 &= L - \{q/(1-q)\}(1-q^L) - \tau_4, \\
\tau_3 &= \{1+\rho_1+(L-K-1)\rho_1^2\}q^L + \rho_2(q^K - q^L) \\
&\quad +\rho_1\rho_2[\{q/(1-q)\}(q^K - q^L) - (L-K)q^L], \\
\tau_4 &= (1-q^L) + \rho_2(1-q^K) + \rho_1\rho_2[(L-K-1)q^K \\
&\quad -\{q/(1-q)\}(q^K - q^{L-1})] + \rho_2^2(L-K-1)(1-q^K)
\end{aligned}$$

Note: $\tau_1$ [$\tau_2$] is the expected number of items produced when the process is in [out of] control and not tested. $\tau_3$ [$\tau_4$] is the expected number of items produced when the process is in [out of] control and tested.

In the next three sections we give expressions for $E[C_{Wi}(L,K,T)]$ for the three cases (see Section 3.4) and illustrate the effect on $C_A(L,K,T)$ by changing $W$. In addition, comparisons are made with the following two cases:

(i) Suppose that the lot size is not a variable and is set equal to $L_U$ (the upper limit). As a result, the quality can be improved only through $K$ and $T$. Let $K_U^*$ and $T_U^*$ denote the optimal item to be tested second and the optimal duration of the life testing and $C_A(L_U, K_U^*, T_U^*)$ the corresponding asymptotic cost per item released.

(ii) Suppose that no testing is employed. In this case, the only way to improve quality is through lot sizing. Let $L_0^*$ denote the optimal $L$ under this condition and let $C_A(L_0^*, 0, 0)$ denote the corresponding asymptotic cost per item released. Note that this case corresponds to the model formulation of Djamaludin et al.(1994), and is a different scheme from that under consideration here.

Let $R_1$, $R_2$ and $R_3$ denote the following percentage reductions in the costs;

$$\begin{aligned}
R_1 &= 100[C_A(L_U,0,0) - C_A(L^*,K^*,T^*)]/C_A(L_U,0,0) \\
R_2 &= 100[C_A(L_U,0,0) - C_A(L_U,K_U^*,T_U^*)]/C_A(L_U,0,0) \quad (13.16) \\
R_3 &= 100[C_A(L_U,0,0) - C_A(L_0^*,0,0)]/C_A(L_U,0,0)
\end{aligned}$$

where $C_A(L_U, 0, 0)$ is the asymptotic cost per item released when lot sizing and testing are not done and with $L = L_U$.

# 5 ANALYSIS OF MODEL: CASE - I [FRW POLICY - MINIMAL REPAIR]

In this policy, items are minimally repaired at no cost to the customer. Consequently, the expected repair cost per item under warranty is different for each of the four types of items released and are given by (from Murthy (1991),

$$\begin{aligned} \omega_A &= C_R \int_0^W r_1(t)dt && \text{[Type A item]} \\ \omega_B &= C_R \int_0^W r_2(t)dt && \text{[Type B item]} \\ \omega_C &= C_R \int_0^W \hat{r}_1(t)dt && \text{[Type C item]} \\ \omega_D &= C_R \int_0^W \hat{r}_2(t)dt && \text{[Type D item]} \end{aligned} \tag{13.17}$$

As a result, $E[C_{Wi}(L, K, T)]$ is given by

$$E[C_{Wi}(L, K, T)] = E[N_{Ai}]\omega_A + E[N_{Bi}]\omega_B + E[N_{Ci}]\omega_C + E[N_{Di}]\omega_D \tag{13.18}$$

where $C_R$ is the cost of each repair and includes material, labor and handling costs.

The asymptotic cost per item released is then obtained by substituting Equations (13.8) - (13.15) and Equations (13.17) - (13.18) into Equation (13.7), which is obviously a complicated function of $L$, $K$ and $T$. Although it is not possible to give an analytical characterisation of the optimal values ($L^*$, $K^*$, $T^*$), they can be obtained by numerical methods.

**Example 1**

Assume that the failure distributions for both conforming and non-conforming items are exponential with parameters $\lambda_1$ and $\lambda_2$, respectively. Let $L_U = 100$ and the nominal values for the parameters be: $C_S$ = \$500.00, $C_m$ = \$10.00, $a_1$ = \$1.00, $a_2$ = \$1.00/year, $C_{sc}$ = \$0.50, $C_R$ = \$3.00, $q = 0.95$, $\theta_1 = 0.95$,

$\theta_2 = 0.15$, $\lambda_1 = 0.1$ and $\lambda_2 = 10.0$. From this, it can be seen that the mean time to failure is ten years for a conforming item and 0.1 years for a non-conforming item. We consider five different values for the warranty periods: $W = 0$ (corresponding to the product being sold with no warranty), $1, 2, 3, 4$.

$L^*$, $K^*$, $T^*$ and $C_A(L^*, K^*, T^*)$ are obtained by evaluating $C_A(L, K, T)$ for $L = 2, .., L_U$, $K = 1, \cdots, (L-1)$, and $T$ incremented in steps of 0.001 from 0 to 1 for the nominal values of the parameters and $W = 0, 1, 2, 3, 4$. This was also done for the case where no lot sizing is done (with $L = L_U$), but the testing is carried out and for the case where no testing is carried but lot sizing is done. Table (1) shows the optimal values for the different cases, the asymptotic cost when no lot sizing or testing is carried out (with $L = L_U$) and the percentage reductions.

| $W$ | 0 | 1 | 2 | 3 | 4 |
|---|---|---|---|---|---|
| $L_U$ | 100 | 100 | 100 | 100 | 100 |
| $C_A(L_U, 0, 0)$ | 15.000 | 36.057 | 57.115 | 78.172 | 99.229 |
| $L^*$ | 100 | 100 | 38 | 28 | 24 |
| $K^*$ | - | 99 | 15 | 11 | 10 |
| $T^*$ | - | 0.001 | 0.411 | 0.589 | 0.689 |
| $C_A(L^*, K^*, T^*)$ | 15.000 | 36.068 | 52.937 | 63.046 | 71.725 |
| $R_1$ (%) | 0.00 | 0.00 | 7.32 | 19.35 | 27.72 |
| $K_U^*$ | - | 99 | 31 | 29 | 28 |
| $T_U^*$ | - | 0.001 | 0.285 | 0.545 | 0.638 |
| $C_A(L_U, K_U^*, T_U^*)$ | 15.000 | 36.068 | 56.923 | 73.309 | 89.432 |
| $R_2$ (%) | 0.00 | 0.00 | 0.34 | 6.22 | 9.87 |
| $L_0^*$ | 100 | 100 | 36 | 25 | 20 |
| $C_A(L_0^*)$ | 15.000 | 36.057 | 53.856 | 67.489 | 79.259 |
| $R_3$ (%) | 0.00 | 0.00 | 5.71 | 13.65 | 20.13 |

**Table 1** $L^*$, $K^*$, $T^*$, $L_0^*$, $K_U^*$, $T_U^*$ and asymptotic cost/item for Example 1

For low values of $W$, the asymptotic total cost is dominated by the manufacturing cost component and, as $W$ increases, the warranty cost component starts to dominate. Consequently, for $W = 0$ (no warranty cost) and $W = 1$, the optimal lot size is $L_U = 100$ and it is better to test as few items as possible (this includes not testing the $L$ or $K$ items). For larger warranty periods ($W = 2, 3, 4$), the increase in warranty cost is slowed by releasing fewer non-conforming items. This is achieved by reducing the lot size ($L^*$) and increasing the duration ($T^*$) of the testing. As well, an earlier item is tested second (that is, $K^*$ is reduced) to increase the chance of eliminating non-conforming items. For the given setup cost ($C_S$), it is better to restart than test large numbers of items, so the lot size ($L^*$) is reduced more quickly than $K^*$. This helps to control the increase in the manufacturing cost brought about by reducing $L^*$ and testing. Obviously, as the warranty period increases, so does the asymptotic cost per released item.

The optimal testing duration $T^*$, for $W = 4$, is 0.689 year. One would not carry out the testing for this length of time. One way of reducing this time is to use accelerated testing, so that testing under accelerated conditions for a period $T_A$ is equivalent to testing for a period $T$ normally. The scaling factor is given by the ratio $(T_A/T)$. By selecting this ratio to equal $10^{-3}$, the optimal testing time (under accelerated condition) becomes roughly 2.15 days.

The percentage reductions, $R_1$, $R_2$ and $R_3$ (see Equation (13.16)), show respectively the savings obtained by employing lot sizing and life testing, just life testing, or just lot sizing, over employing neither. For the cases where there is a saving ($W = 2, 3, 4$), the saving is greater when employing both lot sizing and life testing rather than when employing just one. This will always be true when comparing employing both with employing just life testing since the latter is a sub-model of the former. It may not be true when comparing employing both with employing just lot sizing as the latter is outside the domain of the optimization for the former. For example, if $W = 1$, then the best strategy is to do no life testing at all. The choice of model parameters (including $L_U$) will determine what the situation will be. Finally, it should be observed that, as the warranty period increases (so that the warranty cost dominates), the percentage reductions increase. Thus, lot sizing and life testing become more effective.

It is also worth considering the effect on the optimization caused by changing key model parameters. We focus on the probability for the process staying in-control ($q$), the probability of producing a conforming item while the process is in control ($\theta_1$) and the setup cost ($C_S$). The key factors in this analysis are (a) the homogeneity of the process over time and (b) the dominating cost. If the process is likely to go out of control early or late (near the $L_U$ item), then neither lot sizing nor life testing are worthwhile, since the process is almost homogeneous. This can be seen from Figure (1) where plots of the optimal values ($L^*$, $K^*$, $T^*$, $C_A(L^*, K^*, T^*)$) against $q$ are given. The effectiveness of the schemes depends heavily on the dominant cost, as can be seen by the plots for different values of $W$ in Figure (1). (Note: For $W = 1$ and $q = 0.99$, we have $K^* = 40$. As q decreases, no testing is very quickly the best option, so $K^*$ has no value. Consequently, no plot of $K^*$ versus $q$ is given for $W = 1$. The same situation occurs in Figures (2) and (3). For $W = 1$, in the former, no testing is optimal and, in the latter, $K^* = 9$ for $C_S = 100$ but no testing is optimal very quickly as $C_S$ increases.)

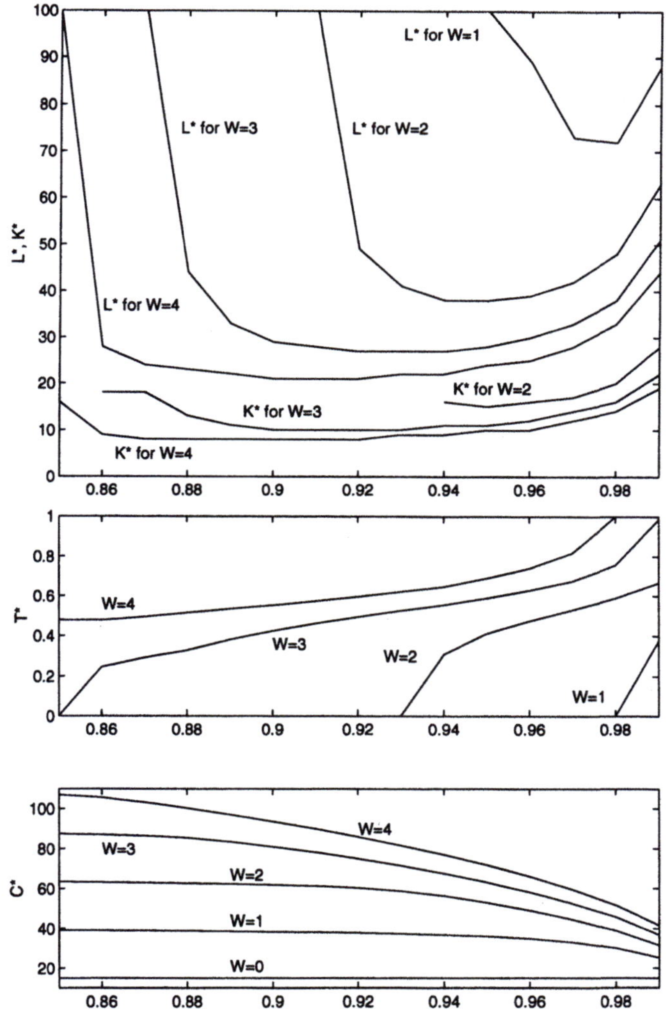


**Figure 1** $L^*$, $K^*$, $T^*$ and $C_A(L^*, K^*, T^*)$ versus $q$

A similar effect occurs when $\theta_1$ and $\theta_2$ are similar - again lot sizing and life testing are not worthwhile. As the difference between them increases, these methods become more effective (see Figure (2)).

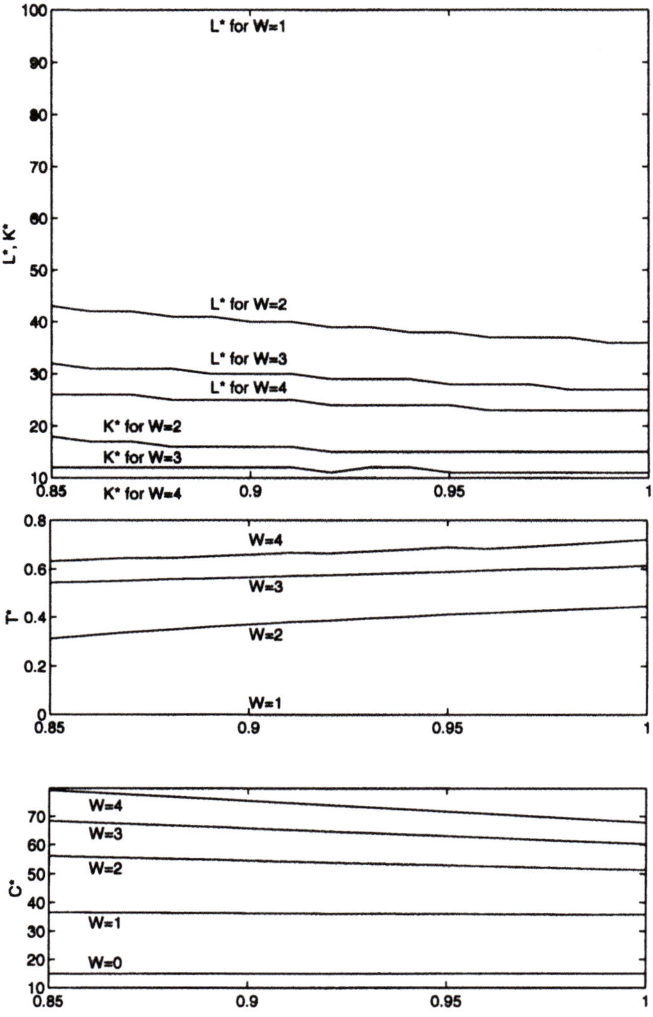


**Figure 2** $L^*$, $K^*$, $T^*$ and $C_A(L^*, K^*, T^*)$ versus $\theta_1$

This can also be seen in Figure (3). As the setup cost increases, the warranty cost becomes less critical so both lot sizing and life testing become less effective.

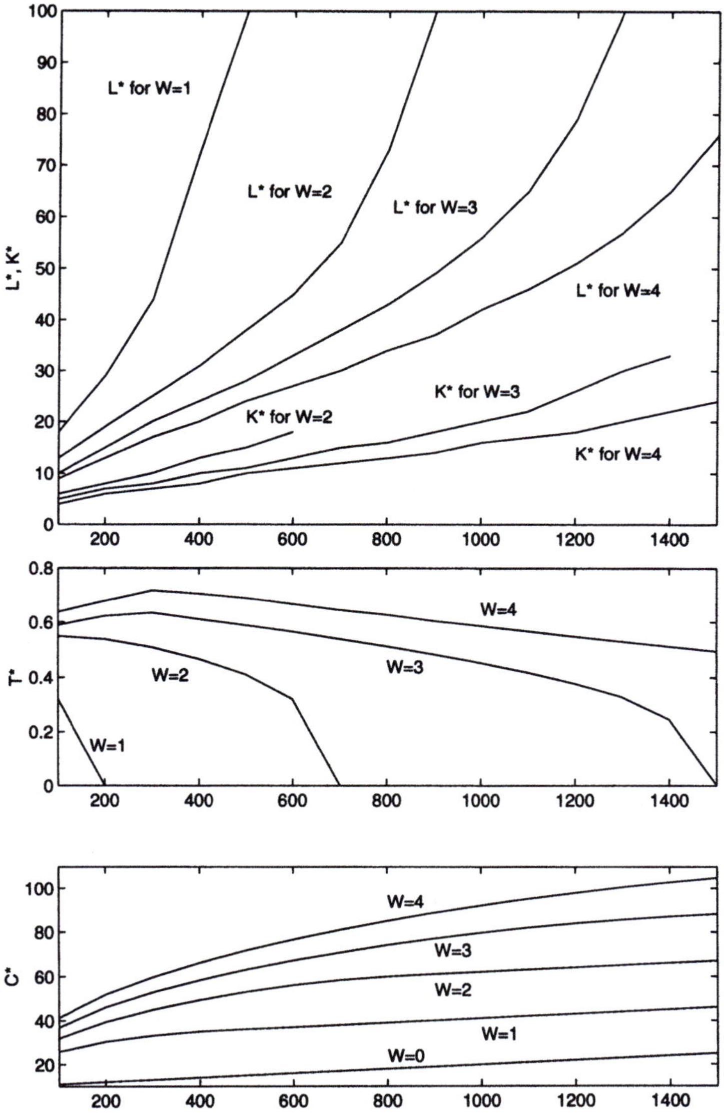


**Figure 3** $L^*$, $K^*$, $T^*$ and $C_A(L^*, K^*, T^*)$ versus $C_S$

# 6 ANALYSIS OF MODEL: CASE - II [PRW POLICY - LINEAR REBATE]

In this policy, a pro-rata refund is paid if the item fails under warranty. If $S$ denotes the sale price, then under linear rebate the refund is $kS[1 - \beta t/W]$, with $0 < k \leq 1$ and $0 < \beta \leq 1$.

The expected refund per item sold is different for each of the four types of items released. They are given by

$$\begin{aligned} \omega_A &= kS\int_0^W (1-\beta t/W)f_1(t)dt + C_h F_1(W) && \text{[Type A item]} \\ \omega_B &= kS\int_0^W (1-\beta t/W)f_2(t)dt + C_h F_2(W) && \text{[Type B item]} \\ \omega_C &= kS\int_0^W (1-\beta t/W)\hat{f}_1(t)dt + C_h \hat{F}_1(W) && \text{[Type C item]} \\ \omega_D &= kS\int_0^W (1-\beta t/W)\hat{f}_2(t)dt + C_h \hat{F}_2(W) && \text{[Type D item]} \end{aligned}$$

where $C_h$ is the handling cost per warranty claim.

As a result, the expected warranty cost for lot $i$ is given by

$$E[C_{Wi}(L,K,T)] = E[N_{Ai}]\omega_A + E[N_{Bi}]\omega_B + E[N_{Ci}]\omega_C + E[N_{Di}]\omega_D. \quad (13.19)$$

and substitution into Equation (13.7) yields $C_A(L,K,T)$. As in Case I, numerical methods need to be used to obtain $L^*$, $K^*$ and $T^*$.

**Example 2**

Assume that the failure distributions of both conforming and non-conforming items are the same as those in Example 1. The nominal values for the parameters $a_1$, $a_2$, $\theta_1$, $\theta_2$, $\lambda_1$, and $\lambda_2$ are as for Example 1, and the nominal values for the other parameters are $C_S = \$250.00$, $C_m = \$5.00$, $C_{sc} = \$0.10$, $S = \$60.00$, $C_h = \$0.50$, $q = 0.96$, $k = 1.0$ and $\beta = 1.0$.

Again, we consider five different warranty periods: $W = 0, 1, 2, 3, 4$. Table (2) shows the optimal values under the different schemes, percentage reductions in costs ($R_1$, $R_2$ and $R_3$) and the corresponding asymptotic costs per item released.

Under the refund linear-rebate policy, failed items incur a single cost (the rebate) and are removed from operation. For released non-conforming items, the rebate increases slowly as $W$ increases for $W > 1$, since these items are likely to fail early, just after sale. However, for released conforming items, the rebate increases more quickly as $W$ increases.

| $W$ | 0 | 1 | 2 | 3 | 4 |
|---|---|---|---|---|---|
| $L_U$ | 100 | 100 | 100 | 100 | 100 |
| $C_A(L_U, 0, 0)$ | 7.500 | 44.537 | 47.456 | 48.992 | 50.142 |
| $L^*$ | 100 | 27 | 27 | 28 | 28 |
| $K^*$ | - | 11 | 11 | 12 | 12 |
| $T^*$ | - | 0.593 | 0.595 | 0.586 | 0.574 |
| $C_A(L^*, K^*, T^*)$ | 7.500 | 33.583 | 36.404 | 38.583 | 40.504 |
| $R_1$ (%) | 0.00 | 24.60 | 23.29 | 21.25 | 19.22 |
| $K_U^*$ | - | 31 | 31 | 31 | 31 |
| $T_U^*$ | - | 0.559 | 0.561 | 0.552 | 0.541 |
| $C_A(L_U, K_U^*, T_U^*)$ | 7.500 | 40.878 | 43.758 | 45.549 | 47.007 |
| $R_2$ (%) | 0.00 | 8.22 | 7.79 | 7.03 | 6.25 |
| $L_0^*$ | 100 | 24 | 24 | 24 | 25 |
| $C_A(L_0^*)$ | 7.500 | 36.426 | 39.270 | 41.299 | 43.038 |
| $R_3$ (%) | 0.00 | 18.21 | 17.25 | 15.70 | 14.17 |

**Table 2** $L^*$, $K^*$, $T^*$, $L_0^*$, $K_U^*$, $T_U^*$ and asymptotic cost/item for Example 2

Under this model, lot size is more effective in reducing the warranty servicing cost than the model in Djamaludin et al.(1994). Thus, as for Policy II in Djamaludin et al.(1994), $T^*$ increases from $W = 1$ to $W = 2$, since the difference in the rebates is large at $W = 2$ (for the given nominal values). As $W$ increases further ($W = 2$ to $W = 3$ and to $W = 4$), the difference in the rebates is smaller. Therefore, $L^*$ increases and $T^*$ decreases.

For each $W$ ($W = 1, 2, 3, 4$), the value of $C_A(L^*, K^*, T^*)$ is smaller than $C_A(L_U, 0, 0)$. The percentage reduction in cost (denoted by $R_1$) shows the savings of using $L^*$, $K^*$ and $T^*$ instead of $L_U$ and no testing.

The value of $C_A(L_U, K_U^*, T_U^*)$ in this example is smaller than the value of $C_A(L_U, 0, 0)$, with the maximum number of items tested and $T^*$ similar to the case of lot sizing and life testing. The percentage reduction in cost by employing life testing is given by $R_2$.

The values of $C_A(L_0^*, 0, 0)$ are also smaller than $C_A(L_U, 0, 0)$ for $W = 1, 2, 3$ and 4. $R_3$ shows this percentage reduction in cost.

Again the percentage reductions show the savings in employing lot sizing and/or life testing. As can be seen from Table (2), their values decrease when $W$

increases. Therefore, lot sizing and life testing are less effective as the warranty period increases. As well, for this example, the saving in cost from employing both life testing and lot sizing is greater than that from employing just lot sizing or just life testing.

The effect on the optimal values caused by changing $q$ is very similar to the previous policy. However, the effects of increasing the length of the warranty period are less here, since increasing $W$ does not increase the warranty cost by as much.

# 7 MODEL ANALYSIS: CASE - III [FRW POLICY - REPLACEMENT BY NEW]

Under this policy, the manufacturer needs to replace all items which fail under warranty with new ones. We assume that all four types of items are pooled together so that an item used for sale (or in replacement) comes from this mixture. As a result, the distribution function for the failure time of an item chosen randomly is given by

$$F(t) = \nu_A F_1(t) + \nu_B F_2(t) + \nu_C \hat{F}_1(t) + \nu_D \hat{F}_2(t) \tag{13.20}$$

where $\nu_A = E[N_{Ai}]/(L - E[\varepsilon_i])$, $\nu_B = E[N_{Bi}]/(L - E[\varepsilon_i])$, $\nu_C = E[N_{Ci}]/(L - E[\varepsilon_i])$ and $\nu_D = E[N_{Di}]/(L - E[\varepsilon_i])$.

Since failed items are replaced by new ones, the mean number of replacements under warranty per item sold is given by the renewal function $M(t)$ associated with the distribution function $F(t)$. As a result, the expected warranty cost per item sold is given by

$$M(W)\{C_h + E[C_{Mi}(L, K, T)]/[L - E[\varepsilon_i]]\} \tag{13.21}$$

where $C_{Mi}$ is the sum of the asymptotic manufacturing and testing cost given in Equation (13.5) and $M(W)$ is given by

$$M(W) = F(W) + \int_0^W M(W - x) dF(x) \tag{13.22}$$

Again, substitution into Equation (13.7) yields $C_A(L, K, T)$. As in the earlier two cases, one needs to use numerical methods to obtain $L^*$, $K^*$ and $T^*$.

As for Examples 1 and 2, we assume that the failure distributions of both conforming and non-conforming items are exponential with parameters $\lambda_1$ and

| $W$ | 0 | 1 | 2 | 3 | 4 |
|---|---|---|---|---|---|
| $L_U$ | 100 | 100 | 100 | 100 | 100 |
| $C_A(L_U, 0, 0)$ | 14.500 | 70.043 | 85.836 | 99.433 | 112.908 |
| $L^*$ | 100 | 37 | 36 | 36 | 36 |
| $K^*$ | - | 21 | 17 | 17 | 17 |
| $T^*$ | - | 0.208 | 0.240 | 0.247 | 0.251 |
| $C_A(L^*, K^*, T^*)$ | 14.500 | 63.394 | 75.046 | 86.605 | 98.163 |
| $R_1$ (%) | 0.00 | 9.49 | 12.57 | 12.90 | 13.06 |
| $K_U^*$ | - | - | 38 | 35 | 34 |
| $T_U^*$ | - | - | 0.219 | 0.232 | 0.240 |
| $C_A(L_U, K_U^*, T_U^*)$ | 14.500 | 70.043 | 85.497 | 98.735 | 111.946 |
| $R_2$ (%) | 0.00 | 0.00 | 0.39 | 0.70 | 0.85 |
| $L_0^*$ | 100 | 37 | 36 | 36 | 35 |
| $C_A(L_0^*)$ | 14.500 | 63.421 | 75.360 | 87.029 | 98.694 |
| $R_3$ (%) | 0.00 | 9.45 | 12.20 | 12.47 | 12.60 |

**Table 3** $L^*$, $K^*$, $T^*$, $L_0^*$, $K_U^*$, $T_U^*$ and asymptotic cost/ item for Example 3

$\lambda_2$ respectively. Let $L_U = 100$ and let the nominal values of the parameters $\theta_1$ and $\theta_2$ be as in Example 1. The nominal values for the other parameters are $C_S$ = \$450.00, $C_m$ = \$10.00, $C_{sc}$ = \$0.50, $C_h$ = \$5.00, $a_1 = 0.8$, $a_2$ = \$0.8/year, $\lambda_1 = 0.2$ and $\lambda_2 = 10.00$.

From this, it can be seen that the mean time to failure is five years for a conforming item and 0.1 years for a non-conforming item. As before, we consider five different values for the warranty period - $W = 0, 1, 2, 3, 4$.

Table (3) shows the optimal values for each scheme, the percentage reduction in costs ($R_1$, $R_2$ and $R_3$), and the corresponding asymptotic costs per item released. The results here are very similar to Policy I. However, the warranty cost dominates more in this model and lot sizing is much more effective than life testing, as shown by the percentage reductions which again show the savings caused by employing lot sizing and/or life testing.

# 8 CONCLUSION

We have studied a model where lot sizing and life testing are used to control the production of non-conforming items (items which do not meet the design specification), when the process state is unknown at the end of the production of each lot.

As might be anticipated, the expected warranty cost increases with the warranty period. In the case of FRW policy with minimal repair (Case-I), the increase in warranty cost is slowed by releasing fewer non-conforming items through reducing the lot size and increasing the duration of the testing. In the case of PRW policy with linear rebate (Case-II), since failed items incur a single cost (the rebate), the optimal lot size and the duration of testing does not change significantly with the warranty period. Finally, for the case of FRW policy with replacement by new (Case-III), the optimal lot size decreases and the testing time increases as in Case-I but the rate of change is smaller.

The results obtained for the quality improvement scheme discussed above are similar to those in Djamaludin et al.(1994, 1995), where the true state of the process is known. When the state of the process is unknown, the risk of releasing non-conforming items is greater and so the warranty cost is more significant, especially for "large" $W$. Consequently, smaller lot sizes and longer testing periods are required to reduce the warranty servicing cost. In particular, lot sizing is more effective than for the scheme in Djamaludin et al.(1994, 1995). Finally, a larger total cost per item released obviously results from not knowing the state of the process.

## Acknowledgements

The authors thank the editors and the three reviewers for their constructive critical comments on an earlier version of the chapter.

# REFERENCES

[1] Blischke, W.R., "Mathematical Models for Analysis of Warranty Policies", *Mathematical Computational Modelling*, 13, pp 1-16, 1990.

[2] Blischke, W.R., and D.N.P. Murthy, "Product Warranty Management-I: A Taxonomy for Warranty Policies", *Euro. J. Oper. Res.*, 62, pp 127-148, 1992.

[3] Blischke, W.R., and D.N.P. Murthy, *Warranty Cost Analysis*, Marcel Dekker, New York, 1994.

[4] Chand, S., "Lot Sizes and Setup Frequency with Learning in Setups and Process Quality", *European Journal of Operations Research*, 42, pp 190-202, 1989.

[5] Djamaludin, I., *Quality Control For Items Sold With Warranty*, Unpublished Doctoral thesis, the University of Queensland, Australia, 1993.

[6] Djamaludin, I., R.J. Wilson, and D.N.P. Murthy, "Quality Control Through Lot Sizing for Items Sold with Warranty", *International Journal of Production Economics*, 33, pp 97-107, 1994.

[7] Djamaludin, I., R.J. Wilson, and D.N.P. Murthy, "Lot Sizing and Testing for Items with Uncertain Quality", *Mathematical and Computer Modelling*, 22, pp 35-44, 1995.

[8] Heathcote, C.R., *Probability: Elements of the Mathematical Theory*, George Allen and Unwin Ltd., London, 1971.

[9] Keller, G. and H. Noori, "Impact of Investing in Quality Improvement on the Lot Size Model", *Omega Int. J. Mngt. Sci.*, 16, pp 595-601, 1988.

[10] Porteus, E.L., "Optimal Lot Sizing, Process Quality Improvement and Setup Cost Reduction", *Operations Research*, 34, pp 137-144, 1986.

[11] Murthy, D.N.P., "A Note on Minimal Repair", *IEEE Transactions on Reliability*, 40, pp 245-246, 1991.

[12] Murthy, D.N.P., and W.R. Blischke, "Product Warranty Management-II: An Integrated Framework for Study," *European Jr. of Operational Research*, 62, pp 261-281, 1992.

[13] Murthy, D.N.P., and W.R. Blischke, "Product Warranty Management-III: A Review of Mathematical Models," *European Jr. of Operational Research*, 63, pp 1-34, 1992.

[14] Murthy, D.N.P., R.J. Wilson, and I. Djamaludin, "Product Warranty and Quality Control", *Quality and Reliability Engineering International*, 9, pp 431-443, 1993.

[15] Rosenblatt, M.J. and H.L. Lee, "Economic Production Cycles with Imperfect Production Processes," *IIE Transaction*, 18, pp 48-55, 1986.

[16] Schonberger, R.J., *Japanese Manufacturing Techniques: Nine Hidden Lessons in Simplicity*, The Free Press, New York, 1982.

# PART V

## ACCEPTANCE SAMPLING

**Chapter 14:** A Concise Review of Lot-by-Lot Acceptance Sampling by Attributes

# 14

# A CONCISE REVIEW OF LOT-BY-LOT ACCEPTANCE SAMPLING BY ATTRIBUTES


**T.C.E. Cheng[1], M.S.D. Lau[2] and S.O. Duffuaa[3]**

[1] *Office of the Vice President,*
*(Research & Postgraduate Studies),*
*The Hong Kong Polytechnic University,*
*Kowloon,*
*Hong Kong.*

[2] *Department of Actuarial and Management Sciences,*
*University of Manitoba,*
*Winnipeg, Manitoba,*
*Canada R3T 2N2.*

[3] *Department of Systems Engineering,*
*King Fahd University of Petroleum and Minerals,*
*Dhahran,*
*Saudi Arabia.*



## ABSTRACT

This chapter purports to offer some insight into lot-by-lot acceptance sampling by attributes, which is one of the common types of acceptance sampling. When using such a sampling plan, a sample of a predetermined number of items is taken from each submitted lot and inspected by attributes. With the information from the inspected sample, a decision of acceptance or rejection of the submitted lot can be made. The basic ideas of these sampling plans are introduced first and the various versions of each sampling plan are then discussed. Optimization models and methods for determining optimal parameters for each plan are presented. The review is supplemented by reference to the appropriate publications in the literature.



**Key words** : acceptance sampling, military standard, Dodge-Romig tables, stage dependent sampling plans, optimal design of sampling plan

# 1 INTRODUCTION

Lot-by-lot acceptance sampling by attributes is one of the common types of acceptance sampling. With such a sampling plan, a sample of a predetermined number of items is taken from each lot and inspected by attributes. Attributes are quality characteristics which are shown on a "go-not-go" basis and each item in the lot can be classified by attributes such as conforming or nonconforming, good or defective, pass or fail, etc. Based on the information from this sample, the lot may be accepted if it conforms to the required standard or rejected if it does not. Each lot is sampled and either accepted or rejected.

It has been pointed out by Deming (1986) that dependency on inspection to correct quality problems is ineffective and costly and hence inspection should not be used as a long term strategy for quality improvement. It is now well accepted that dependence on inspection or screening is ineffective in the long run and will not build quality into the product, but will only remove defective items. This lead researchers and practitioners to focus more on process control. However, inspection may be an attractive option for removing defective items in a population in the short term due to the advances in automatic inspection equipment and computer control in manufacturing. Inspection will also be utilized at early stages until data and means for process control are in place. Therefore, inspection in general and in particular lot-by-lot acceptance sampling will be a part of any quality control program. This justifies a concise review of lot-by-lot acceptance sampling, which is the purpose of this chapter.

Operating characteristics (OC) curve gives the probability of acceptance of a submitted lot as a function of the fraction defective of the lots. The OC curve is one of the common evaluation techniques and the most important characteristic of sampling plans. It shows the discriminatory power of a sampling plan. There are two types of OC curve: the type A OC curve and the type B OC curve. If the lot is an isolated lot with finite size, a type A OC curve is used. For this situation, the probability of acceptance of the lot should be calculated from the hypergeometric probability distribution. On the other hand, if the lots are taken from a steady flow of items which are produced by a single source, a type B OC curve is used, and the probability of acceptance of the lots should be calculated from the binomial probability distribution. With lot-by-lot acceptance sampling by attributes, a type B OC curve may be used.

The purpose of the paper is to present and discuss the basic ideas of some common lot-by-lot attribute acceptance sampling plans, and to highlight the optimization models and methods used to design and determine optimal parameters for these plans. The rest of the paper is organized as follows: Section 2 presents the review of single sampling, double sampling and multiple sampling plans. Section 3 covers sequential sampling plans, truncated life test plans, chain sampling plans, skip lot sampling plans, dependent stage sampling plans and deferred stage sampling plans. Section 4 outlines the Bayesian approach for the design of sampling plans and Section 5 presents Military Standard 105 and Dodge-Romig Tables. Section 6 concludes this paper.

# 2 SINGLE, DOUBLE, AND MULTIPLE SAMPLING PLAN

This section reviews the single, double and multiple sampling plans. The latter plans can be viewed as a generalization for the single sampling plan.

## 2.1 Single Sampling Plan

In the single sampling plan, some predetermined numbers are decided upon, which are:

$N$ = the lot size,
$n$ = the sample size, and
$c$ = the acceptance number.

When using the single sampling plan by attributes, one sample of size n is taken from the lot of size N and inspected. If there are c or less defective items in the sample, the lot is accepted. If there are more than c defective items in the sample, the lot is rejected. In other words, the acceptance or rejection of the lot depends on the inspection results of a single sample. In the single sampling plan, one curve is required for the OC curve. The shape of the OC curve depends on the values of N, n and c. Different values of these numbers will result in different OC curves, i.e., different producer's and consumer's protection.

Some publications which discuss various versions of the single sampling plan are listed in Table 1. They are divided into four categories. The publications in the first category concern the principles and theory of the single sampling plan. The basic characteristics and properties of the single sampling plan are also discussed in these publications. In the second category, the publications present the designs and determinations of various single sampling plans. The constructions of these single sampling plans are explained. In the third category, the publications consider the effects of inspection error of using the single sampling plan. The type I error, accepting non-conforming items, and type II error, rejecting conforming items, are discussed.

In the last category optimization has been applied in determining the parameters of the single sampling plan which are $n$ and $c$ . The optimization criterion is either statistical or economical. The selection of $n$ and $c$ affect the total cost. Most of the optimization models derive an expected total cost function that consists of the cost of false acceptance, cost of inspection and the cost of false acceptance and $n$ and $c$ are selected in an optimal way to minimize the total expected cost. Bernett et al.(1974) developed necessary conditions for the parameters to be optimal and then they used a simple incremental search procedure over $n$ to determine the optimal $n$ and $c$ for a single sampling plan with and without inspection errors.

## 2.2 Double Sampling Plan

The double sampling plan is more complicated than the single sampling plan because a second sample may be required. Generally, the sample sizes of double sampling plans are smaller and the total number of inspections may be reduced. As a result, the total inspection cost is reduced. The predetermined numbers are

$N$ lot size,
$n_1$ sample size for the first sample,
$c_1$ acceptance number for the first sample,
$r_1$ rejection number for the first sample,
$n_2$ sample size for the second sample,
$c_2$ acceptance number for both samples, and
$r_2$ rejection number for both samples.

When using a double sampling plan by attributes, a first sample of size $n_1$ is taken from a lot of size $N$ and inspected. One of the following three decisions is made after inspection:

1. If $c_1$ or fewer defective items are found in the first sample, accept the lot;
2. If $r_1$ or more defective items are found in the first sample, reject the lot;
3. If more than $c_1$ and fewer than $r_1$ defective items are found in the first sample, a second sample of size $n_2$ is required.

If a second sample is required, $n_2$ items are taken from the same lot which has $N - n_1$ items remaining. One of the following two decisions is made after inspection:

1. If $c_2$ or fewer defective items are found in both samples, accept the lot;
2. If $r_2$ or more defective items are found in both samples, reject the lot.

In other words, the decision of acceptance or rejection of the lot is based on the inspection results from both samples when a second sample is required.

In the double sampling plan, two curves are required for the OC curve. The first one is for the probability of acceptance of a lot after inspecting the first sample if the second sample is not required. In determining this curve, the probability of acceptance of the lot after inspecting the first sample is equal to the probability of having $c_1$ or less defective items in the first sample. If a second sample is needed, the second curve is for the probability of acceptance of that lot after inspecting the second sample. When determining the second curve, the probability of acceptance of the lot after inspecting the second sample is equal to the sum of the probability of acceptance of the lot on the first sample and the probability of acceptance of the lot on the second sample.

Some publications which discuss various versions of the double sampling plan are listed in Table 1. They are divided into four categories. In the first category, the publications present the principles and theory of the double sampling plan; and the basic characteristics and properties of the double sampling plan are also mentioned. In the second category, the publications discuss the designs and determinations of the double sampling plan. The constructions of the double sampling plan are also explained. In the third category, the publications consider the effects of inspection error on using the double sampling plan. The type I and type II errors are discussed. In the last category, Guenther (1971b) calculated the average sample number for the truncated double sampling plan. Baker and Brobst (1978) discussed the conditional double sampling plan.

The design of a double sampling plan is more challenging than the single sampling plan. The objective is to determine the parameter of the plan, $n_1, r_1, c_1, n_2, c_2$ to minimize some statistical or an economical criterion. The method of Lagrange has been used to obtain optimal parameters for the double sampling plan. Stewart et al.(1978) utilized a modified version of the Hooke and Jeeves method to obtain optimal parameters for double sampling plans using an economic criterion. The models developed for single and double sampling plans have few variables, which helps in obtaining the optimal parameters of such plans.

## 2.3 Multiple Sampling Plan

Multiple sampling plans are extensions of the double sampling plan. Instead of requiring two samples in a double sampling plan, a multiple sampling plan may require three or more samples with smaller sample sizes. The technique is similar to that used in the double sampling plan. The following illustration is a multiple sampling plan which requires at most three samples. The predetermined numbers for this plan are

$N$ = lot size,
$n_1$ = sample size for the first sample,
$c_1$ = acceptance number for the first sample,
$r_1$ = rejection number for the first sample,
$n_2$ = sample size for the second sample,
$c_2$ = acceptance number for both first and second samples,
$r_2$ = rejection number for both first and second samples,
$n_3$ = sample size for the third sample,
$c_3$ = acceptance number for all three samples, and
$r_3$ = rejection number for all three samples.

When using multiple sampling plan by attributes, a first sample of size $n_1$ is taken from a lot of size $N$ and inspected. One of the following three decisions is made after inspection:

1. If $c_1$ or fewer defective items are found in the first sample, accept the lot;
2. If $r_1$ or more defective items are found in the first sample, reject the lot; and
3. If more than $c_1$ and fewer than $r_1$ defective items are found in the first sample, a second sample of size $n_2$ is required.

If a second sample is required, $n_2$ items are taken from the same lot which has $N - n_1$ items remaining. One of the following three decisions is made after inspection:

1. If $c_2$ or fewer defective items are found in both first and second samples, accept the lot;
2. If $r_2$ or more defective items are found in both first and second samples, reject the lot; and
3. If more than $c_2$ and fewer than $r_2$ defective items are found in both first and second samples, a third sample of size $n_3$ is required.

If a third sample is required, $n_3$ items are taken from the same lot which has $N - n_1 - n_2$ items remaining. One of the following two decisions is made after inspection:

1. If $c_3$ or fewer defective items are found in all three samples, accept the lot; and
2. If $r_3$ or more defective items are found in all three samples, reject the lot.

In other words, the decision of acceptance or rejection of the lot is based on the inspection results from all three samples when three samples are required.

As a matter of fact, it is possible to make the probability of acceptance of a specific lot under a single sampling plan equal to the probability of acceptance of that lot under an appropriate double or multiple sampling plan. In other words, the selection of a sampling plan does not depend on its effectiveness since all three types of sampling plan can yield the same effectiveness when appropriate predetermined numbers are selected. The effectiveness is the ability to reduce both the producer's and consumer's risks. So, when selecting the type of sampling plan to use, we should consider other factors such as the overall cost of sampling.

Generally, for the same degree of effectiveness, the total number of inspections in a single sampling plan is more than that in a double sampling plan since the decision can sometimes be made in the first sample when using a double sampling plan; therefore, no second sample is needed and the total number of inspections is reduced. Similarly in a multiple sampling plan, the total number of inspections is usually less than that in a double sampling plan since the decision can be made in the first few samples. As the total number of inspections decreases, the inspection cost decreases. So, on average, a multiple sampling plan has a lower inspection cost than the single or double sampling plan. On the other hand, a multiple sampling plan is more complicated than the others so that other costs such as training people and recording results are higher. Since the sampling cost is the sum of all these costs, we should consider the overall sampling cost when selecting the type of sampling plan.

The formation of an OC curve for a multiple sampling plan is an extension of the formation of the OC curve for a double sampling plan. Instead of requiring two curves in a double sampling plan, multiple sampling plans require three or more curves to construct an OC curve.

Furthermore, Bartky (1943) considered the multiple sampling plan with constant probability. U.S. Army Chemical Corps. Eng. Agency (1953) gave the master sampling plan for the multiple sampling plan. Hald (1975) evaluated an approximation to the binomial OC curve by means of the Poisson OC curve for the multiple sampling plan. Schilling, Sheesley and Nelson (1978) explained a general routine for multiple sampling plan evaluation. Baker (1987) gave zero inventories and conditional sampling procedures. Maghsoodloo (1987) considered the inspection error effects on performance measures of a multiple sampling plan. They are listed in Table 1.

The same optimization method used for designing a double sampling plan can be used for a multiple sampling plan.

# 3 STAGE DEPENDENT SAMPLING PLANS

This section contains sampling plans that are designed for destructive or costly inspection. The decision about sampling is either made sequentially, reduced, deferred, truncated or skipped. The focus here is to reduce cost or the sample size. In this section 6 plans are presented. The plans are: Sequential, Truncated, Chain, Skip-lot, Dependent Stage and Deferred Stage sampling plans.

## 3.1 Sequential Sampling Plan

Sequential sampling plans are acceptance sampling plans by attributes for destructive or costly inspections. They were developed by Wald (1947). In this sampling plan, only one item at a time is taken from the lot and inspected. After inspection, we compare the cumulative number of defective items to the acceptance number and rejection number. For this sampling plan, the acceptance number and rejection number are not constant. They are given by the following formulas (Wald, 1973):

$$a_m = \frac{log\left(\frac{\beta}{1-\alpha}\right)}{log\left(\frac{p1}{p0}\right) - log\left(\frac{1-p1}{1-p0}\right)} + m\left[\frac{log\left(\frac{1-p0}{1-p1}\right)}{log\left(\frac{p1}{p0}\right) - log\left(\frac{1-p1}{1-p0}\right)}\right] \tag{14.1}$$

and

$$r_m = \frac{log\left(\frac{1-\beta}{\alpha}\right)}{log\left(\frac{p1}{p0}\right) - log\left(\frac{1-p1}{1-p0}\right)} + m\left[\frac{log\left(\frac{1-p0}{1-p1}\right)}{log\left(\frac{p1}{p0}\right) - log\left(\frac{1-p1}{1-p0}\right)}\right] \tag{14.2}$$

here

$m$ = number of items inspected,
$a_m$= acceptance number when $m$ items are inspected,
$r_m$= rejection number when m items are inspected,
$\alpha$ = producer's risk,

$\beta$ = consumer's risk
$p0$ = fraction defective at the acceptable quality level, AQL, and
$p1$ = fraction defective at the limiting quality level, LQL.

If the cumulative number of defective items is less than the acceptance number, accept the lot. If the number of cumulative defective items is greater than the rejection number, reject the lot. Otherwise, continue the inspection until a decision of accepting or rejecting the lot is made. Theoretically, the sequential sampling plan can continue until all the items in the lot are inspected but, in practice, this sampling plan is truncated when the number of inspected items is equal to three times the sample size of the corresponding single sampling plan. Generally, this sampling plan reduces the number of items inspected, so the inspection cost will decrease for destructive or costly inspections. Detailed information can be found in Wald (1973).

Some other publications which discuss various versions of the sequential sampling plan are listed in Table 1. They are divided into four categories. In the first category, the publications concern the properties and characteristics of the sequential sampling plan. The principles and theory of these sampling plans are also explained. In the second category, two publications discuss the direct method in the sequential sampling plan. In the third category, two publications consider the sequential sampling plan when the failure distributions of tested items are exponential. In the last category, Jackson (1960) gave a bibliography on sequential analysis. Tantaratana (1988) discussed the asymptotic efficiencies of some truncated sequential tests with parallel boundaries.

## 3.2 Truncated Life Test Plan

Epstein (1954) discussed some life test plans which he called truncated life tests. Before these sampling plans start, the sample size, $n$, the rejection number, $r$, and the truncated test time, $T$, beyond which the test will not be run, are determined. Then $n$ items are selected from a submitted lot and simultaneously subjected to life tests. If we let $x_r$ denote a random variable of the time at which the $r$th failure occurs and $T$ the predetermined truncation time, the sampling test will be terminated at $\min(x_r, T)$. If the test is terminated at time $T$, i.e., $T$ less than $x_r$ , the submitted lot is accepted; otherwise, it is rejected. The failed items during the test may or may not be replaced. In the replacement case, less time is required to obtain a given number of failures but more items are needed in the test. In the non-replacement case, more time is required to obtain a given number of failure but fewer items are needed.

Some publications which discuss various versions of the truncated life test plan are listed in Table 1. They are divided into three categories. In the first category, three publications give the principles and theory of the truncated life test plan. In the second category, three publications consider the truncated sequential life tests when the failure distributions of the tested items are exponential. In the last category, three publications calculate the average sample number of using the truncated life test plan. The backward recurrence relation for cost was derived by Champernowne (1953a,b) who also discussed the determination of sequential truncated plans based on the same principal (1969).

## 3.3 Chain Sampling Inspection

Dodge (1955a) developed the chain sampling inspection plan to reduce inspection costs for destructive or costly inspections. When a destructive or costly inspection is encountered, the sample size should be small in order to reduce the inspection cost. If the sample size is small, the acceptance number is usually small and sometimes it is zero. As a matter of fact, the sampling plans have a poor shape of OC curves when the acceptance number is zero because the OC curves will be convex throughout (Montogomery, 1985). Also, the probability of acceptance will decrease rapidly as the fraction defective increases. A better shape for these OC curves can be obtained by using the chain inspection plan. This plan uses the results of several previous inspections, so it is assumed that the lots should have the same quality and that they come from a steady flow of items which are produced by a single source.

Consider an example in which a sample of size $n$ is taken from a submitted lot of size $N$. The lot is accepted if there is no defective item in the sample; and the lot is rejected if there are two or more defective items in the sample. When there is only one defective item in the sample, the lot is accepted only if all $i$ preceding lots of the same size were accepted; otherwise, reject the lot. The values of $i$ and $n$ depend on how effective an OC curve is required. The formula for the OC curves for this plan is

$$P_a = P_0 + P_1(P_a)^i \tag{14.3}$$

with

$P_a$ = probability of acceptance of the lot; and
$P_d$ = probability of having $d$ defective items in the sample, $d = 0, 1, ...,$

Some publications discussing various versions of the chain sampling inspection are listed in Table 1. They are divided into two categories. In the first category, the publications concern the principles and theory of the chain sampling inspection plan. In the last category, the publications explain the constructions and designs of the chain sampling inspection plan.

## 3.4 Skip-Lot Sampling Plan

Dodge (1955b) also developed the skip-lot sampling plan to minimize inspection costs by reducing inspection after the submitted lots have good quality history. When lots are taken from a steady flow of items of the same quality, this plan may be used. The procedure is as follows:

1. Each lot is inspected by a specific sampling plan.
2. When $i$ consecutive lots are accepted, stop inspecting every lot. Then only a sample of a fixed number, $f$, of subsequent lots are selected randomly and inspected using the same sampling plan.
3. Whenever a lot is rejected, go to Step (1).

The values of $i$ and $f$ are related to what AOQL value is required and a table for this sampling plan can be found in Dodge (1955b). Some publications which consider the principles and theory of the skip-lot-sampling plan are listed in Table 1. In addition, Hsu (1980) introduced an economic design of the skip-lot sampling plan and Carr (1982) considered some adjustment for skip-lot plans.

## 3.5 Dependent Stage Attribute Acceptance Sampling Plan

Mogg (1969) developed a type of sampling plan, called the dependent stage attribute acceptance sampling plan, which uses information from prior lots to decide whether to accept or reject the current lot. Later, Wortham and Mogg (1970) also discussed this sampling plan. The advantage of this sampling plan is that it reduces the sample size. The notation used in the dependent stage sampling plan is defined as follows:

$n$ = sample size,
$r$ = the maximum number of allowable defective items from the current sample for unconditional acceptance of the lot,
$b$ = the maximum number of additional defective items for which the decision of acceptance or rejection of the current lot will depend on the acceptance or rejection of prior lots,
$P_{a;k}$ = the probability of accepting lot number $k$, and
$P_{x,n}$ = the probability that there are exactly $x$ defective items in a sample size of $n$ items.

Mogg designated the dependent stage sampling plan by DSSP-r,b with an operating procedure outlined by the following steps;

**Step 1**

At the outset, select a random sample of $n$ items from the first lot submitted and accept the lot if the sample contains $r$ or less defective items.

**Step 2**

For each lot number, record the disposition as to whether it was accepted or rejected.

**Step 3**

Repeat Steps 1 and 2 on subsequent lots for the first $b$ lots.

**Step 4**

For lot $b + 1$, select a random sample of $n$ items and accept the lot if the sample contains $r$ or less defective items. For more than r defective items, the decision to accept or reject the current lot will depend on the historical data, and the following courses of action will dictate the decision;

| | | |
|---|---|---|
| $r + 1$ defective items | = | Accept the current lot if lot number 1 was accepted. |
| $r + 2$ defective items | = | Accept the current lot if lot number 2 was |

| | | |
|---|---|---|
| | | accepted. |
| ... | | |
| $r + b$ defective items | = | Accept the current lot if lot number $b$ was accepted. |
| $r + i$ defective items ($i > b$) | = | Reject the current lot. |

**Step 5**

Repeat Step 4 for each subsequent lot. That is, check the disposition of lot $m-b$ if $r+1$ defective items are observed in the $m^{th}$ lot. Check the disposition of lot $m - b + 1$ if $r + 2$ defective items are observed in the $m^{th}$ lot and so on. Reject the lot if more than $r + b$ defective items are observed, or if the lot checked on the review was rejected. Otherwise, accept the lot.

The properties of dependent stage sampling plans could be described by OC curves. The OC curve for such a sampling plan was developed by evaluating the proportion of lots that will be accepted for a product from a process. Mogg considered some elementary dependent stage sampling plans first and then developed the expression for the general OC curve by induction. He showed that the general expression for the OC curve for the dependent stage sampling plan, DSSP-r,b, is

$$P_{a;k} = \frac{\sum_{i=0}^{r} P_{i,n}}{\left(1 - \sum_{j=r+1}^{r} P_{j,n}\right)} \quad r \geq 0, b \geq 1 \tag{14.4}$$

with

$P_{i,n} = \frac{n!}{i!(n-i)!} p^i (1-p)^{n-i}$,
$P_{a;k}$ = the probability of acceptance of the submitted lot number $k$,
$n$ = the sample size, and
$p$ = the fraction defective of the submitted lots.

Mogg pointed out the limitations of the dependent stage sampling plan as follows:

1. Production is steady so that results on current and preceding lots are indicative of a continuing process.

2. Lots are submitted sequentially in the order of their production.
3. A fixed sample size, $n$, from each lot is assumed.
4. Inspection by attributes is assumed with quality measured by fraction defective $p$ for a binomial distribution.

He also mentioned that sometimes the dependent stage sampling plan may not be desirable since the OC curve changes from lot to lot in the early stages of this plan and does not settle down to a fixed curve until approximately ten lots have been inspected. The reason is that the plan acts as a single sampling plan until $b + 1$ lots are inspected and historical results are considered from that point.

## 3.6 Deferred State Attribute Acceptance Sampling Plan

Baker (1971) developed a type of sampling plan which is called the deferred state attribute acceptance sampling plan. The advantage of this sampling plan is that it reduces the average sample number. This sampling plan uses subsequent lots information for making a decision to accept or reject the current lot. The operating procedure of this type of sampling plan is similar to the dependent stage sampling plan except that the conditional decisions depend on the disposition of future lots instead of past lots. Thus, the formations of OC curves of dependent stage sampling plans and deferred state sampling plans are similar. The deferred state sampling plan provides an indicator for quality degradation. If a large number of defective items are observed in a sample, the probability that the process quality has degraded beyond an acceptable level is high. The indicator concept is based on the assumption that the number of defective items from a sample may truly represent the process quality. The notation used in deferred state sampling plans is as follows:

$n$ = sample size,
$r$ = the maximum number of allowable defective items from the current sample for unconditional acceptance of the lot,
$b$ = the maximum number of additional defective items for which the decision of acceptance or rejection of the current lot will depend on the acceptance or rejection of subsequent lots,
$P_{a;k}$ = the probability of accepting lot number $k$, and

$P_{x,n} =$ the probability that there are exactly $x$ defective items in a sample size of $n$ items.

Baker designated the deferred state sampling plan by DS(r,b) sampling plan with an operating procedure outlined by the following steps;

**Step 1**

For lot number $k$, select a random sample of $n$ items from the submitted lot and determine the number of defective items.

**Step 2**

Accept the lot if the sample contains $r$ or fewer defective items. For more than $r$ defective items, the decision to accept or reject the current lot is dictated by the following courses of action;

| | |
|---|---|
| $r+1$ defective items - | Defer the decision until the disposition of lot number $k+b$ is obtained. If lot number $k+b$ is accepted, then accept lot $k$, otherwise reject lot number $k$. |
| $r+2$ defective items - | Defer the decision until the disposition of lot number $k+b-1$ is obtained. If lot number $k+b-1$ is accepted, then accept lot $k$, otherwise reject lot number $k$. |
| ... | |
| ... | |
| $r+b$ defective items - | Defer the decision until the disposition of lot number $k+1$ is obtained. If lot number $k+1$ is accepted, then accept lot $k$, otherwise reject lot number $k$. |
| $r+i$ defective items $(i > b)$- | Reject lot number $k$. |

**Step 3**

Increment $k$ by 1 and return to Step 1.

The properties of the deferred state sampling plan could be described by the OC curves. The OC curve for such a sampling plan was developed by evaluating the proportion of lots that will be accepted for a product from a process. Baker considered some elementary deferred state sampling plans first and then developed the expression for the general OC curve by induction. He showed that the general expression for the OC curve for the deferred state sampling plan, DS(r,b), is

$$P_{a;k} = \frac{\sum_{i=0}^{r} P_{i,n}}{\left(1 - \sum_{j=1}^{b} P_{r+j,n}\right)}, \quad r \geq 0, b \geq 1 \tag{14.5}$$

where

$P_{i,n}$ = $\frac{n!}{i!(n-i)!}p^i(1-p)^{n-i}$, and
$p$ = the fraction defective of the submitted lots.

Baker also discussed the limitations of the deferred state sampling plan. One of the limitations is that a waiting line may be formed when the lots are in a deferred state, so the cost of storing deferred lots should be considered before a deferred state sampling plan is selected instead of any other sampling plan. In developing the distribution of waiting times Baker used the following notation:

$P_{x,n}$ = the probability that there are exactly $x$ defective items in a sample of $n$ items,
$W$ = the parameter which denotes the number of lots that a deferred lot must wait before disposition,
$P(W = i)$ = the probability that a deferred lot waits for $i$ lots before disposition, and
$E(W)$ = the expected wait which is the expected value of $W$

Baker developed the distribution of waiting times for the DS(0,1) sampling plan. He showed that the probability that a deferred lot waits for i lots before disposition is

$$P(W = i) = P_{1,n}^i(1 - P_{1,n}) \tag{14.6}$$

and the expected wait, $E(W)$, is

$$E(W) = \sum_{i=0}^{\infty} iP(W = i) \tag{14.7}$$

Substituting Equation (14.6) into Equation (14.7) gives

$$E(W) = \frac{P_{1,n}}{1 - P_{1,n}} \tag{14.8}$$

The general equation for the expected waiting time can be obtained by induction.

The deferred state sampling plan has a problem similar to that of the dependent stage sampling plan. In the early stages of a dependent stage sampling plan, several lots have to be sampled under a single sampling plan before the dependent stage concept can be used. The OC curve of the dependent stage sampling plan changes from lot to lot in the early stages of the plan and does not settle down until approximately ten lots have been inspected. With a deferred state sampling plan, the problem is how to make the disposition decision of the final lots when the lots are waiting for disposition of future lots which will not be produced. One solution is to use a single sampling plan for the last $b$ lots. That is, if the number of defective items from the sample is less than $r$, accept the entire lot; otherwise, reject the lot. So no submitted lot will wait for future lots which will not be produced. Thus, the OC curve of the deferred state sampling plan changes from lot to lot in the final stages of the plan; it is not a fixed curve in approximately the last ten lots.

Furthermore, Dean (1971) proposed cost models for the deferred state life test plan to see whether the use of the deferred state life test plan would reduce the overall test cost. Finally, he found that the deferred state life test plan can reduce the overall test cost in some situations and detailed explanations can be found in his publication. Wortham and Baker (1971) also gave the procedures for the deferred state sampling plan. In addition, Wortham and Baker (1976) introduced the multiple deferred state inspection in which decision to accept or reject the current submitted lot depends on some sampling test results of other submitted lots.

# 4 BAYESIAN SAMPLING PLANS

In order to present the Bayesian approach for designing sampling plans, we need the definition of a prior distribution. A prior distribution is the expected distribution of lot quality on which the sampling plan is going to operate. A Bayesian sampling plan is a plan obtained by minimizing expected costs, consisting of inspection, acceptance and rejection costs. The expected cost is obtained with respect to the prior distribution. The Bayesian approach will generate an economically based sampling plan and has drawn considerable attention (e.g., Wetherill and Chin (1975), Guenther (1971), Hald (1960), Hald (1981)). The approach can be applied to any sampling plan to design an optimal inspection plan. Therefore, we can have a single, a double and a multiple sampling plan based on the Bayesian criterion.

Economic multiattribute sampling plans require explicit assessment of the economic consequences associated with each attribute when a decision is made to accept or reject a lot. Ailor et al.(1975) and Schmidt and Bennett (1972) have proposed multiattribute models that include cost of inspection, cost of rejection and cost of acceptance for each attribute. Moskwitz et al.(1984) developed a discrete search algorithm to determine the optimal parameters for the inspection plan. The algorithm is based on an extended pattern search. Case and Chen presented some recent findings in Bayesian attribute single sampling plans (1985). Tang et al.(1986) extended the work of Moskwitz by examining interaction among attributes and the effect of these interactions on an optimal inspection plan. They developed a heuristic solution procedure for the multiattribute economic model. These proposed algorithm was shown to be effective for a large number of attributes.

# 5 MILITARY STANDARD AND DODGE-ROMIG TABLES

In this section we present the Military Standard 105D and Dodge-Romig Tables. The Military Standard can be viewed as an application of single, double and multiple sampling plans and has found wide acceptability.

## 5.1 Military Standard 105 (MIL-STD-105)

In 1949, the Statistical Research Group of Columbia University proposed an acceptance sampling plan for lot-by-lot inspection by attributes called JAN-STD-105 (1949). After revisions, MIL-STD-105A (1950), MIL-STD-105B (1958), MIL-STD-105C (1961), and MIL-STD-105D (1963) were published. In 1989, the latest version was published and called MIL-STD-105E (1989). Generally, this standard is used when the lots are taken from a steady flow of items which are produced by a source, but after some adjustments it can also be used for isolated lots. It is the most common type of lot-by-lot acceptance sampling plan for attribute inspection and it is extensively used in industry for acceptance sampling. This standard is applicable to inspection of incoming materials, products in process, end products, and so on. The aim of this standard is to maintain a satisfactory level of average outgoing quality.

Three types of sampling plans are included in this standard. They are the single, double, and multiple sampling plans. For each type of sampling plan, it provides three types of inspection: normal, tightened, and reduced. Normal inspection is used to inspect the lots at the beginning of inspection. After a certain number of inspections, if the quality is not satisfactory, the tightened inspection is used. On the other hand, if the quality is good, the reduced inspection is used. When a lot is submitted for inspection, the probability of acceptance of the lot under reduced inspection is the highest; and the probability of acceptance of the lot under tightened inspection is the lowest among the three types of inspection. In other words, the risk of accepting a defective lot will be the highest when using reduced inspection. Besides, reduced inspection has the smallest sample size and the tightened inspection has the largest sample size so that the inspection cost will be reduced when using reduced inspection. The use of the different types of inspection may be switched from one to another following the criteria stated in the standard. Also, the procedures for using this sampling plan are given in the standard.

Some publications discussing various versions of the military standard are listed in Table 1. They are divided into four categories. In the first category, the publications give the principles and characteristics of the military standard 105. The properties and theory are also included. In the second category, the publications show the procedures of using the military standard 105. In the third category, three publications present the evolution of the military standard 105. In the last category, Brown and Rutemiller (1973) introduced a cost analysis of sampling inspection under military standard 105. Liebesman (1981a) showed how to select military standard 105D plans based on costs. Chakraborty and Bapaye (1989) explained the effect of inspection error on military standard 105D sampling plans.

## 5.2 Dodge-Romig Tables

Dodge and Romig (1959) developed a set of sampling inspection tables in order to minimise the average total number of inspections. There are two types of sampling inspection tables. The first type is based on Lot Tolerance Percent Defective (LTPD), and the second type is based on average outgoing quality limit (AOQL). For each type of table, single and double sampling plans are available. The tables based on LTPD are used when the submitted lots are homogeneous or when the objective of sampling is to assure an average outgoing quality level. The tables based on AOQL are used when the submitted lots are nonhomogeneous or when the objective of sampling is to assure quality no worse than a given target. Whenever the value of LTPD or AOQL is decided and the fraction defective of incoming lots of size $N$ is known, the sample size $n$ may be read directly from the tables of a single or double sampling plan. In addition, two more publications discussing the principles and theory of the Dodge-Romig Table are listed in Table 1.

# 6 CONCLUSION

In this chapter, we reviewed some common attribute acceptance sampling plans. The review presented and discussed each plan and highlighted the optimization models and methods used to obtain the optimal parameters for each plan.

The review is kept simple; however supplementary references are given for the interested reader. The book by Hald (1981) presents a comprehensive approach to sampling inspection by attribute. However, the book is written for individuals with a high level of maturity in mathematical statistics and does not focus on the issue of optimization.

Optimization models and methods have been employed to design optimal inspection plans. The models seek to minimize the expected total cost resulting from inspection, false rejection and false acceptance. A host of algorithms have been used to obtain optimal parameters for the plans. An example of these algorithms is that of Hooke and Jeeves. Systematic search and in some cases heuristic search procedures have been adopted.

It should be emphasized that process control and process improvement are the backbone for improving produced quality and that inspection does not built-in quality. Inspection should be considered only as a short term method to remove nonconforming items from a population. In this short term period optimal inspection plans should be adopted.

Table 1.
A list of publications of lot-by-lot acceptance sampling plans by attributes.

**Single Sampling Plan**

*Principles and Theory*
Peach and Littauer (1946)
U.S. Army Chemical Corps. Eng. Agency (1953)
Wise (1955)
Hamaker (1958)
Prairie, Zimmer, and Brookhouse (1962)
Hald (1967b)
Dodge (1969a)
Schilling, Sheesley, and Nelson (1978)
Stephens (1978)
Nachlas and Kim (1989)
Soundararajan and Arumainayagam (1989)
Nelson(1991)
*Designs and Determinations*
Grubbs (1949)
Cameron (1952)
Golub (1953)
Horsnell (1957)
Guthrie and Johns (1959)
Hald (1967a)
Guenther (1971a)
Hald (1977)
Guenther (1984)
Ohta and Ichihashi (1988)
Ohta and Kanagawa (1988)
Brooks (1989)
Govindaraju (1990)
Soundararajan and Vijayaraghavan (1990)
*Effects of Inspection Error*
Ayoub, Lambert, and Walvekar (1970)
Minton (1972)
Collins, Case, and Bennett (1973)
Bennett, Case, and Schmidt (1974)
Beaing (1981)
Jaraiedi and Herrin (1985)
*Others*

Hald (1965)
Case and Chen (1985)
Baker (1988)

## Double Sampling Plan

*Principles and Theory*
U.S. Army Chemical Corps. Eng. Agency (1953)
Hamaker and Van Strik (1955)
Schilling, Sheesley, and Nelson (1978)
Case and Chen (1985)
Srivenkataramana and Harishchandra (1985)
*Designs and Determinations*
Horsnell (1957)
Chow, Dickinson, and Hughes (1972)
Hald (1977)
Stewart, Montgomery and Hicks (1978)
Chen (1981)
Hald (1981)
Olorunniwo and Salas (1982)
Govindaraju (1990)
*Effects of the Inspection Error*
Beaing and Case (1981)
Maghsoodloo and Bush (1985)
*Others*
Guenther (1971b)
Baker and Brobst (1978)
Multiple Sampling Plan
Bartky (1943)
U.S. Army Chemical Corps. Eng. Agency (1953)
Hald (1975)
Schilling, Sheesley, and Nelson (1978)
Baker (1987)
Maghsoodloo (1987)

## Sequential Sampling Plan

*Principles and Theory*
Wald (1945)
Anscombe (1946)
Barnard (1946)
Wald (1947)

Hamaker (1953)
Hoel (1955)
Kiefer and Weiss (1957)
Anderson (1960)
Eagle (1964)
Chernoff and Ray (1965)
Tallis and Vagholkar (1965)
Pfanzagle and Schuler (1970)
Schafer and Takenaga (1972)
Wald (1973)
Bryant and Schmee (1979)
Garrison and Hichey (1984)
Kremers (1987)
*Direct Method*
Aroian (1968)
Aroian (1976)
*Exponential Case*
Epstein and Sobel (1955)
Aroian and Robison (1966)
*Others*
Jackson (1960)
Tantaratana (1988)

**Truncated Life Test Plan**

*Principles and Theory*
Champernowne (1953a,b)
Epstein (1954)
Champernowne (1969)
miller (1985)
Mason (1986)
*Truncated Sequential Life Tests in the Exponential Case*
Woodal and Kurkjian (1962)
Aroian (1963)
Aroian (1964)
*Aerage Sample Number*
Burr (1957)
Craig (1968)
Guenther (1971)

**Chain Sampling Inspection**

*Principles and Theory*
Dodge (1955a)
Frishman (1960)
Dodge and Stephens (1964)
Stephens and Dodge (1965)
Dodge and Stephens (1966)
Stephens and Dodge (1967)
Soundararajan and Doraiswamy (1984)
Soundararajan and Arumainayagam (1989)
*Constructions and Designs*
Soundararajan (1978a)
Soundararajan (1978b)
Soundararajan and Govindaraju (1983)
Raju (1990)
Soundararajan and Vijayaraghavan (1990)

**Skip-Lot-Sampling Plan**

*Principles and Theory*
Dodge (1955b)
Perry (1970)
Perry (1973a)
Perry (1973b)
Liebesman and Saperstein (1983)
Jaraiedi and Bem (1989)
Kowalewski and Tye (1990)
Perry (1990)
*Others*
Hsu (1980)
Carr (1982)

**Dependent State Sampling Plan**

Mogg (1969)
Wortham and MOgg (1970)

**Deferred State Sampling Plan**

Baker (1971)
Dean (1971)
Wortham and Baker (1971)

Wortham and Baker (1976)

**Military Standard 105**

*Principles and Theory*
Pabst (1963a)
Pabst (1963b)
Cocca (1964)
Stephens and Larson (1967)
Koyama (1969)
Ohmae and Suga (1969)
Koyama, Ohmae, Suga, and Yamamoto (1970)
Brown and Rutemiller (1975)
Sheesley (1977)
Schilling and Sheesley (1978a)
Schilling and Sheesley (1978b)
Liebesman (1979)
Duncan et al.(1980)
Brush, Cautin, and Lewin (1981)
Liebesman (1981b)
Schilling (1982)
Schilling (1983)
Enell (1984)
Keats and Case (1984)
Liebesman and Hawley (1984)
Bee, Teck, and Keng (1985)
Buswell and Hoadley (1985)
Cocca (1985)
Randhawa (1985)
Nelson, Wall, and Caporal (1986)
Flott (1990)
*Procedures*
JAN-STD-105 (1949)
MIL-STD-105A (1950)
MIL-STD-105B (1958)
MIL-STD-105C (1961)
MIL-STD-105D (1963)
Kaplan and MacDonald (1969)
Yokoh (1969)
Hill (1973)
Hahn and Schilling (1975)
Schilling and Johnson (1980)

Baker (1987)
MIL-STD-105E (1989)
*Evolution*
Keefe (1963)
Dodge (1969b)
Liebesman (1982)
*Others*
Brown and Rutemiller (1973)
Liebesman (1981a)
Chakraborty and Bapaye (1989)

**Dodge-Romig Tables**

Dodge and Romig (1959)
Keats and Case (1984)
Flott (1990)

# REFERENCES

[1] Ailor, R.H., J.W. Schmidt and G.K. Bennet, "The Design of Economic Acceptance Sampling Plans for a Mixture of Variables and Attributes," *AIIE Trans.* 7, pp 370-378, 1979.

[2] Anderson, T. W., "A Modification of the Sequential Probability Ratio Test to Reduce the Sample Size," *Annals of Mathematical Statistics*, 31, pp 165-197, 1960.

[3] Angus, J. E., R. E. Schafer, S. Van Den Berg, and H. C. Rutemiller, "Failure-Free Period Life Tests," *Technometrics*, 27(1), pp 49-56, 1985.

[4] Anscombe, F. J., "Linear Sequential Rectifying Inspection for Controlling Fraction Defective," *Supplement Journal of the Royal Statistical Society*, 8, pp 216-222, 1946.

[5] Aroian, L. A., "Exact Truncated Sequential Tests for the Exponential Density Function," *Proceeding-Ninth National Symposium on Reliability and Quality Control*, pp 470-486, 1963.

[6] Aroian, L. A., "Some Comments on Truncated Sequential Tests for the Exponential Distribution," *Industrial Quality Control*, 21, pp 309-312, 1964.

[7] Aroian, L. A., "Sequential Analysis, Direct Method," *Technometrics*, 10, pp 125-132, 1968.

[8] Aroian, L. A., "Application of the Direct Method in Sequential Analysis," *Technometrics*, 18, pp 301-306, 1976.

[9] Aroian, L. A. and D. E. Robison, "Sequential Life Tests for the Exponential Distribution with Changing Parameter," *Technometrics*, 8, pp 217-227, 1966.

[10] Ayoub, M. M., B. Lambert, and A. G. Walvekar, "Effects of Two Types of Inspection Error on Single Sampling Inspection Plan," *Human Factors Society Conference*, San Francisco, 1970.

[11] Baker, R. C., "Dependent-Deferred State Attribute Acceptance Sampling," Ph.D. Dissertation. Texas A & M University, College Station, Texas, 1971.

[12] Baker, R. C., "Marriage of Zero Inventories and Conditional Sampling Procedures," *Production & Inventory Management*, 28(3), pp 27-30, 1987.

[13] Baker, R. C., "Zero Acceptance Sampling Plans: Expected Cost Increases," *Quality Progress*, 21(1), pp 43-46, 1988.

[14] Baker, R. C. and R. Brobst, "Conditional Double Sampling," *Journal of Quality Technology*, 10(4), pp 150-154, 1978.

[15] Barnard, G. A., "Sequential Tests in Industrial Statistics," *Supplement Journal of Royal Statistical Society*, 8, pp 1-26, 1946.

[16] Bartky, W., "Multiple Sampling with Constant Probability," *Annals of Mathematical Statistics*, 14(4), pp 363-377, 1943.

[17] Beaing, I. and K. E. Case, "A Wide Variety of AOQ and ATI Performance Measures with and without Inspection Error," *Journal of Quality Technology*, 13(1), pp 1-9, 1981.

[18] Bee, G. A., L. Y. Teck, and N. J. Keng, "Military Standard 105D: Its Uses and Misuses," *Quality Assurance*, 11(2), pp 33-38, 1985.

[19] Bennett, G. K., K. E. Case, and J. W. Schmidt, "The Economic Effects of Inspector Error on Attribute Sampling Plan," *Naval Research Logistics Quarterly*, 21(3), pp 431-443, 1974.

[20] Besterfield, D. H., *Quality Control*. Prentice-Hall, Englewood Cliffs, New Jersey, 1986.

[21] Bowker, A. H. and G. J. Lieberman, *Engineering Statistics.* Prentice-Hall, Inc., Englewood Cliffs, New Jersey, 1972.

[22] Brooks, G. H., "Acceptance Sampling Design by Computer - Optimality Concerns," *Computers and Industrial Engineering,* 17(1-4), pp 508-513, 1989.

[23] Brown, G. G. and H. C. Rutemiller, "A Cost Analysis of Sampling Inspection Under Military Standard 105D," *Naval Research Logistics Quarterly,* 20, pp 181-199, 1973.

[24] Brown, G. G. and H. C. Rutemiller, "An Analysis of the Long Range Operating Characteristics of the MIL-STD-105D Sampling Scheme and Some Suggested Modification," *Naval Research Logistics Quarterly,* 22(4), pp 667-679, 1975.

[25] Brush, G. G., H. Cautin, and B. R. Lewin, "Outgoing Quality Distributions for MIL-STD-105D Sampling Plans," *Journal of Quality Technology,* 13(4), pp 254-263, 1981.

[26] Bryant, C. M. and J. Schmee, "Confidence Limits on MTBF for Sequential Test Plans of MIL-STD-781B," *Technometrics,* 21, pp 33-42, 1979.

[27] Burr, I. W., "Average Sample Number Under Curtailed or Truncated Sampling," *Industrial Quality Control,* 13(8), pp 5-7, 1957.

[28] Burr, I. W., *Elementary Statistical Quality Control.* Marcel Dekker, Inc. New York, 1979.

[29] Buswell, G. and B. Hoadley, "QMP/USP: A Modern Alternative to MIL-STD- 105D," *Naval Research Logistics Quarterly,* 32(1), pp 95-111, 1985.

[30] Cameron, J. M., "Tables for Construction and for Computing the Operating Characteristics of Single Sampling Plans," *Industrial Quality Control,* 9(1), pp 37-39, 1952.

[31] Carr, W. E., "Sample Plan Adjustment for Inspection Error and Skip-Lot Plan," *Journal of Quality Technology,* 14, pp 105-116, 1982.

[32] Case, K. E. and S. S. Chen, "Some Recent Findings in Bayesian Attribute Single and Double Sampling," *Naval Research Logistics Quarterly,* 32(1), pp 71-79, 1985.

[33] Chakraborty, S., and M. D. Bapaye, "Effect of Inspection Error on MIL-STD-105D Sampling Plans: Some Observations," *International Journal of Quality & Reliability Management (UK),* 6(2), pp 60-70, 1989.

[34] Champernowne, D.G., "The Economic of Sequential Sampling Procedures for Defectives," *Applied Statistics*, 2, pp 118-129, 1953a.

[35] Champernowne, D.G., *Uncertainty and Estimation in Economics*, Vol. 2, Oliver and Boyd, Edinburgh, 1969.

[36] Chen, S. S., "Modeling and Design of Economically-Based Double Sampling Plans," Unpublished Dissertation. Oklahoma State University, Stillwater, OK, 1981.

[37] Chernoff, H. and S. N. Ray, "A Bayesian Sequential Sampling Inspection Plan," *Annals of Mathematical Statistics*, 36, pp 1387-1407, 1965.

[38] Chow, B., P. E. Dickinson and H. Hughes, "A Computer Program for the Solution of Double Sampling Plans," *Journal of Quality Technology*, 4(4), pp 205-209, 1972.

[39] Cocca, O. A., "MIL-STD-105D, An International Standard for Attribute Sampling," *Industrial Quality Control*, 21(5), pp 249-253, 1964.

[40] Cocca, O. A., "Some Experience Using MIL-STD-105 in the United States Air Force," *Naval Research Logistics Quarterly*, 32(1), pp 11-16, 1985.

[41] Collins, R. D., K. E. Case, and G. K. Bennett, "The Effect of Inspection Error on Single Sample Inspection Plans," *International Journal of Production Research*, 11(3), pp 289-298, 1973.

[42] Craig, C. C., "The Average Sample Number for Truncated Single and Double Attributes Acceptance Sample Plans," *Technometrics*, 10, pp 685-692, 1968.

[43] Dean, A. G., *Cost Models in Reliability Testing and Evaluation*, Ph.D. Dissertation. Texas A & M University, College Station, Texas, 1971.

[44] Deming, E.W., *Out of the Crisis*, MIT Press, Cambridge, MA, U.S.A, 1986.

[45] Dodge, H. F., "Chain Sampling Inspection Plan," *Industrial Quality Control*, 11(4), pp 10-13, 1955a.

[46] Dodge, H. F., "Skip-Lot Inspection Plan," *Industrial Quality Control*, 11(5), pp 3-5, 1955b.

[47] Dodge, H. F., "Notes on the Evolution of Acceptance Sampling Plans, Part II," *Journal of Quality Technology*, 1(3), pp 155-162, 1969a.

[48] Dodge, H. F., "Notes on the Evolution of Acceptance Sampling Plans, Part III," *Journal of Quality Technology*, 1(4), pp 225-232, 1969b.

[49] Dodge, H. F. and K. S. Stephens, *A General Family of Chain Sampling Inspection Plans.* Rutgers-The State University Statistics Centre Technical Report No. N-20, 1964.

[50] Dodge, H. F. and K. S. Stephens, "Some New Chain Sampling Inspection Plans," *Industrial Quality Control*, 23(2), pp 61-67, 1966.

[51] Dodge, H. F. and H. F. Romig, *Sampling Inspection Tables.* John Wiley & Sons, Inc., New York, London, Sydney, 1959.

[52] Duncan, A. J., *Quality Control and Industrial Statistics.* Richard D. Irwin, Inc. Homewood, Illinois, 1974.

[53] Duncan, A. J. et al., "LQL Indexed Plans That Are Compatible with the Structure of MIL-STD-105D," *Journal of Quality Technology*, 12(2), pp 40-46, 1980.

[54] Eagle, E. L., "Reliability Sequential Testing," *Industrial Quality Control*, 20(11), pp 48-52, 1964.

[55] Enell, J. W., "Which Sampling Plan Should I Choose?," *Journal of Quality Technology*, 16(3), pp 168-171, 1984.

[56] Epstein, B., "Truncated Life Tests in the Exponential Case," *Annals of Mathematical Statistics*, 25, pp 555-564, 1954.

[57] Epstein, B. and M. Sobel, "Sequential Life Tests in the Exponential Case," *Annals of Mathematical Statistics*, 26, pp 82-93, 1955.

[58] Flott, L. W., "Sampling and Sample Plans," *Metal Finishing*, 88(10), pp 55-58, 1990.

[59] Freeman, H. A., M. Friedman, F. Mosteller, and W. A. Wallis, *Sampling Inspection.* McGraw-Hill, New York, 1948.

[60] Frishman, F., "An Extended Chain Sampling Plan," *Industrial Quality Control*, 17(1), pp 10-12, 1960.

[61] Garrison, D. and J. J. Hichey, "Wald Sequential Sampling for Attribute Inspection," *Journal of Quality Technology*, 16(3), pp 172-174, 1984.

[62] Golub, A., "Designing Single-Sampling Inspection Plans When the Sample Size Is Fixed," *Journal of American Statistical Association*, 48, pp 278-291, 1953.

[63] Govindaraju, K., "On the Construction of Minimum Average Total Inspection Sampling Plans," *AMSE Review*, 13(2), pp 11-18, 1990.

[64] Grubbs, F. E., "On Designing Single Sampling Inspection Plan," *Annals of Mathematical Statistics*, 20, pp 242-256, 1949.

[65] Guenther, W. C., "On the Determination of Single Sampling Attribute Plans Based Upon a Linear Cost Model and Prior Distribution," *Technometrics*, 13, pp 483-498, 1971a.

[66] Guenther, W. C., "The Average Sample Number for Truncated Double Sample Attribute Plans," *Technometrics*, 13, pp 811-816, 1971b.

[67] Guenther, W. C., "Determination of Rectifying Inspection Plans for Single Sampling by Attributes," *Journal of Quality Technology*, 16(1), pp 56-63, 1984.

[68] Guthrie, D. and M. V. Johns, "Bayesian Acceptance Sampling Procedures for Large Lots," *Annals of Mathematical Statistics*, 30, pp 896-925, 1959.

[69] Hahn, G. R. and E. G. Schilling, "An Introduction to the MIL-STD-105D Acceptance Sampling Scheme," *ASTM Standardization News*, 3(9), pp 20-23, 1975.

[70] Hald, A., "The Compound Hypergeometric Distribution and a System of Single Sampling Inspection Plan based on Prior Distribution and Costs," *Technometrics*, 2, pp 275-340, 1960.

[71] Hald, A., "Bayesian Single Sampling Attribute Plans for Discrete Prior Distributions," *Mat. Fys. Skr. Dan. Vid. Selsk.*, 3(2), Munksgaard, Copenhagen, 1965.

[72] Hald, A., "The Determination of Single Sampling Attribute Plans with Given Producer's and Consumer's Risk," *Technometrics*, 9, pp 401-415, 1967a.

[73] Hald, A., "On the Theory of Single Sampling Inspection by Attributes Based on Two Quality Levels," *Review Int. Statistics Institute*, 35, pp 1-29, 1967b.

[74] Hald, A., "An Approximation to the Binomial OC by Means of the Poisson OC for Multiple Sampling Plans". Preprint No. 16, 1975, Institute of Mathematical Statistics, University of Copenhagen, 1975.

[75] Hald, A., "A Note on the Determination of Attribute Sampling Plans of Given Strength," *Technometrics*, 19, pp 211-212, 1977.

[76] Hald, A., *Statistical Theory of Sampling in Inspection by Attributes*, Academic Press, (London) Ltd, 1981.

[77] Hamaker, H. C., "The Efficiency of Sequential Sampling for Attributes," *Philips Research Reports*, 8, pp 35-36, 1953.

[78] Hamaker, H. C., "Some Basic Principles of Sampling Inspection by Attributes," *Applied Statistics*, 7, pp 149-159, 1958.

[79] Hamaker, H. C. and R. Van Strik, "The Efficiency of Double Sampling for Attributes," *Journal of the American Statistical Association*, 50, pp 830-849, 1955.

[80] Hill, I. D., "The Design of MIL-STD-105D Sampling Tables," *Journal of Quality Technology*, 5(2), pp 80-83, 1973.

[81] Hoel, P. G., "On a Sequential Test for the General Linear Hypothesis," *Annals of Mathematical Statistics*, 26, pp 136-139, 1955.

[82] Horsnell, G., "Economical Acceptance Sampling Schemes," *Journal of Royal Statistical Society*, A, 120, pp 148-201, 1957.

[83] Hsu, J. I. S., "An Economic Design of Skip-Lot Sampling Plan," *Journal of Quality Technology*, 12(3), pp 144-149, 1980.

[84] Jackson, J. E., "Bibliography on Sequential Analysis," *Journal of the American Statistical Association*, 55, pp 561-580, 1960.

[85] JAN-STD-105. "Military Standard-Sampling Procedures and Tables for Inspection by Attributes". Department of Defence, Washington, D. C. 20301, 1949.

[86] Jaraiedi, M. and E. Bem, "A New Procedure for Bulk Material Quality Control," *International Journal of Quality & Reliability Management (UK)*, 6(4), pp 41-49, 1989.

[87] Jaraiedi, M. and G. D. Herrin, "Effect of Human Inspector Error on Sample Plan Design," *Proceedings-Fall Industrial Engineering Conference*, pp 436-439, 1985.

[88] Kaplan, A. and E. MacDonald, "Instantaneous Switching Procedure for MIL- STD-105D," *Journal of Quality Technology*, 1(3), pp 172-174, 1969.

[89] Keats, J. B. and K. E. Case, "Sampling Plans Using Cost Ratios and Lot History," *Annual Quality Congress Transactions*, 38, pp 202-207, 1984.

[90] Keefe, G., "The Evolution of Military Sampling Attributes Standard," *Industrial Quality Control*, 19(10), pp 7-12, 1963.

[91] Kiefer, J. and L. Weiss, "Some Properties of Generalized Sequential Probability Ratio Tests," *Annals of Mathematical Statistic*, 28, pp 57-74, 1957.

[92] Kowalewski, M. J. and J. B. Tye, "Symposium on Statistical Sampling," *ASTM Special Technical Publication*, 1097(92), 1990.

[93] Koyama, T., "Analysis on Dynamic Characteristics of Severity Control in MIL-STD-105D," *International Conference on Quality Control, Yokyo 1969-Yokyo Proceedings*, Union of Japanese Scientists and Engineers, pp 645-648, 1969.

[94] Koyama, T., Y. Ohmae, R. Suga, and T. Yamamoto, "MIL-STD-105D and the Japanese Modified Standard," *Journal of Quality Technology*, 2(2), pp 99-108, 1970.

[95] Kremers, W. K., "Improved Estimator of the Mean for a Sequential Binomial Sampling Plan," *Technometrics*, 29(1), pp 109-112, 1987.

[96] Liebesman, B. S., "The Use of MIL-STD-105D to Control Average Outgoing Quality," Journal of Quality Technology, 11(1), pp 36-43, 1979.

[97] Liebesman, B. S., "Selection of MIL-STD-105D Plans Based on Costs," 1981 *ASQC Quality Congress Transactions*, San Francisco, California, pp 475-484, 1981a.

[98] Liebesman, B. S., "The Characterization of MIL-STD-105D Sampling Plans Using Normalized OC curves," *Frontiers in Statistical Quality Control*, H. J. Lenz, G. B. Wetherill, and P. H. Wilrich, Eds. Physica-Verlag, Weurzbrug-Wien, W. Germany, pp 171-185, 1981b.

[99] Liebesman, B. S., "The Continuing Evolution of the Military Standard 105D Sampling System," *Bell System Technical Journal*, 61, pp 137-157, 1982.

[100] Liebesman, B. S. and F. C. Hawley, "Small Acceptance Number Plans for Use in Military Standard 105D," *Journal of Quality Technology*, 16(4), pp 219-231, 1984.

[101] Liebesman, B. S. and B. Saperstein, "Proposed Attribute Skip-Lot Sampling Program," *Journal of Quality Technology*, 15(3), pp 130-140, 1983.

[102] Maghsoodloo, S., "Inspection Error Effects on Performance Measures of a Multistage Sampling Plan," *IIE Transactions*, 19(3), pp 340-347, 1987.

[103] Maghsoodloo, S. and B. K. Bush, "The Effects of Inspection Error on Double-Sampling by Attributes," *Journal of Quality Technology*, 17(1), pp 32-39, 1985.

[104] Mason, S. A, "Quality Estimates of MTBF," *Annual Quality Congress Transactions*, 1986, pp 307-310, 1986.

[105] MIL-STD-105A, "Military Standard-Sampling Procedures and Tables for Inspection by Attributes". Department of Defence, Washington, D.C. 20301, 1950.

[106] MIL-STD-105B, "Military Standard-Sampling Procedures and Tables for Inspection by Attributes". Department of Defence, Washington, D.C. 20301, 1958.

[107] MIL-STD-105C, "Military Standard-Sampling Procedures and Tables for Inspection by Attributes". Department of Defence, Washington, D.C. 20301, 1961.

[108] MIL-STD-105D, "Military Standard-Sampling Procedures and Tables for Inspection by Attributes". Department of Defence, Washington, D.C. 20301, 1963.

[109] MIL-STD-105E, "Military Standard-Sampling Procedures and Tables for Inspection by Attributes". Department of Defence, Washington, D.C. 20301, 1989.

[110] MIL-STD-781B, "Reliability Tests: Exponential Distribution". Department of Defence, Washington, D.C, 1967.

[111] Minton, G., "Verification Error in Single Sampling Inspection Plan," *Journal of the American Statistical Association*, 67(337), pp 46-54, 1972.

[112] Mogg, J. M., "Dependent Stage Attribute Acceptance Sampling", Ph.D. Dissertation. Texas A & M University, College Station, Texas, 1969.

[113] Montgomery, D. C., *Introduction to Statistical Quality Control.* John Wiley & Sons, New York, 1985.

[114] Moskwitz, H. R., A. Plante, A. Ravindran, and K. Tang, "Multiattribute Bayesian Acceptance Sampling Plan for Screening and Scrapping Rejected Lots," *IIE Trans.*, 1984)

[115] Moskowitz, H.R. and A. Plante, "Effect of Risk Aversion on Single Sample Attribute Inspection Plan", *Management Science*, 10, pp 66-74, 1984.

[116] Nachlas, J. A. and S. I. Kim, "Generalized Attribute Acceptance Sampling Plans," *Journal of Quality Technology*, 21(1), pp 32-40, 1989.

[117] Nelson, L. S., "Power in Comparing Poisson Means: I. One-Sample Test," *Journal of Quality Technology*, 23(1), pp 68-70, 1991.

[118] Nelson, W., G. E. Wall, and P. Caporal, "Inspect Small Samples When Quality Is Poor," *Annual Quality Congress Transactions*, 1986, pp 140-148, 1986.

[119] Ohmae, Y. and R. Suga, "Basic Policy and Scheme of Modified MIL-STD-105D," *International Conference on Quality Control, Yokyo 1969-Yokyo Proceedings*, Union of Japanese Scientists and Engineers, pp 648-653, 1969.

[120] Ohta, H. and H. Ichihashi, "Determination of Single-Sampling-Attribute Plans Based on Membership Functions," *International Journal of Production Research*, 26(9), pp 1477-1485, 1988.

[121] Ohta, H. and A. Kanagawa, "Simplified Design Procedure for Minimax Sampling Plans by Attributes," *International Journal of Production Research*, 26(1), pp 143-156, 1988.

[122] Olorunniwo, F. O. and J. R. Salas, "An Algorithm for Determining Double Attribute Sampling Plans," *Journal of Quality Technology*, 14, pp 166-171, 1982.

[123] Pabst, W. R., "Some Background Consideration of MIL-STD-105D," *17th Annual ASQC Convention and Exhibit*, Chicago, Ill, 1963a.

[124] Pabst, W. R., "MIL-STD-105D," *Industrial Quality Control*, 20, No. 5, 4-9, 1963b.

[125] Pfanza, L.J. and W. Schuler, "The Efficiency of Sequential Sampling Plans based on Prior Distribution", *Technometrics*, 12, pp 299-310, 1970.

[126] Peach, P. and S. B. Littauer, "A Note on Sampling Inspection," *Annals of Mathematical Statistics*, 17, pp 81-85, 1946.

[127] Perry, R. L., "A System of Skip-Lot Sampling Plans for Lot Inspection". Unpublished Ph.D. Dissertation, Rutgers-The State University, New Brunswick, New Jersey, 1970.

[128] Perry, R. L., "Skip-Lot Sampling Plans," *Journal of Quality Technology*, 5(3), pp 123-130, 1973a.

[129] Perry, R. L., "Two-Level Skip-Lot Sampling Plans-Operating Characteristic Properties," *Journal of Quality Technology*, 5(3), pp 160-166, 1973b.

[130] Perry, R. L., "Skip-Lot Sampling. History and Perspective," *ASTM Special Technical Publication*, 1097, pp 14-17, 1990.

[131] Prairie, R. R., W. J. Zimmer, and J. K. Brookhouse, "Some Acceptance Sampling Plans Based on the Theory of Runs," *Technometrics*, 4, pp 177-185, 1962.

[132] Raju, C. "Designing Chain Sampling Plans (CHSP-1) with Fixed Sample Size," *International Journal of Quality & Reliability Management (UK)*, 7(3), pp 59-64, 1990.

[133] Randhawa, S. U., "Simulation Model for Estimating Statistical Parameters for Normal to Reduced Sampling in MIL. STD. 105D," *Proceedings of the Summer Computer Simulation Conference 1985*, pp 637-640, 1985.

[134] Schafer, R. E. and R. Takenaga, "Sequential Probability Ratio Test for Availability," *Technometrics*, 14, pp 123-135, 1972.

[135] Schilling, E. G., "Revised Attributes Acceptance Sampling Standard-ANSI/ASQC Z1.4, 1981)," *Journal of Quality Technology*, 14(4), pp 215-219, 1982.

[136] Schilling, E. G., "New ANSI Versions of MIL-STD-414 and MIL-STD-105D," *Naval Research Logistics Quarterly*, 32(1), pp 5-9, 1983.

[137] Schilling, E. G. and L. I. Johnson, "Tables for the Construction of Matched Single, Double, and Multiple Sampling Plans with Application to MIL-STD-105D," *Journal of Quality Technology*, 12(4), pp 220-229, 1980.

[138] Schilling, E. G. and J. H. Sheesley, "The Performance of MIL-STD-105D Under the Switching Rules, Part I: Evaluation," *Journal of Quality Technology*, 10(2), pp 76-83, 1978a.

[139] Schilling, E. G. and J. H. Sheesley, "The Performance of MIL-STD-105D Under the Switching Rules, Part II: Tables," *Journal of Quality Technology*, 10(3), pp 104-124, 1978b.

[140] Schilling, E. G., J. H. Sheesley, and P. R. Nelson, "GRASP: A General Routine for Attribute Sampling Plan Evaluation," *Journal of Quality Technology*, 10(3), pp 125-130, 1978.

[141] Schmidt, J.W. and G.K. Bennett, "Economic Multiattribute Acceptance Sampling," *AIIE Transactions*, 4, pp 184-199, 1972.

[142] Sheesley, J. H., "A Computer Program to Evaluate Military Standard 105D Sampling Plans with Application of Switching Rules". Technical Report 1300-1444, *General Electric Company*, Lighting Research and Technical Services Operation, Nela Park, Cleveland, Ohio 44112, 1977.

[143] Soundararajan, V., "Procedures and Tables for Construction and Selection of Chain Sampling Plans, (CHSP-1), Part 1," *Journal of Quality Technology*, 10, pp 56-60, 1978a.

[144] Soundararajan, V., "Procedures and Tables for Construction and Selection of Chain Sampling Plans (CHSP-1), Part 2," *Journal of Quality Technology*, 10, pp 99-103, 1978b.

[145] Soundararajan, V. and S. D. Arumainayagam, "An Examination of Some Switching Procedures Used in Sampling Inspection," *International Journal of Quality & Reliability Management (UK)*, 6(5), pp 49-59, 1989.

[146] Soundararajan, V. and P. Doraiswamy, "Chain Sampling Plans (CHSP-1) Indexed by Inflection Point and of Control," *QR Journal*, 11(3), pp 95-97, 1984.

[147] Soundararajan, V. and K. Govindaraju, "Construction and Selection of Chain Sampling Plans CHSP-(0,1)," *Journal of Quality Technology*, 15(4), pp 180-185, 1983.

[148] Soundararajan, V. and R. Vijayaraghavan, "Designing Sampling Inspection Plans When the Sample Size is Fixed," *International Journal of Quality & Reliability Management (UK)*, 7(3), pp 65-75, 1990.

[149] Srivenkataramana, T. and K. Harishchandra, "Some Alternatives to Double Sampling for Product Control," *QR Journal*, 12(2), pp 65-68, 1985.

[150] Stephens, L. J., "A Closed Form Solution for Single Sample Acceptance Sampling Plans," *Journal of Quality Technology*, 10, pp 159-163, 1978.

[151] Stephens, K. S. and H. F. Dodge, "Chain Sampling Inspection Plans-CHSP-0,2 and CHSP-1,2", Rutgers-The State University Statistics Centre Technical Report, N-21, 1965.

[152] Stephens, K. S. and H. F. Dodge, "Evaluation of the Response Characteristics of Chain Sampling Inspection Plans". *Technical Report No. 25, Statistics Center*, Rutgers-The State University, New Brunswick, New Jersey, 1967.

[153] Stephens, K. S. and K. E. Larson, "An Evaluation of the MIL-STD-105D System of Sampling Plans," *Industrial Quality Control*, 23(7), pp 310-319, 1967.

[154] Stewart, R.D., D.C. Montgomery and R. Hrikes, "Choice of Double Sampling Plans Based on Prior Distribution and Costs," *AIIE Trans.*, 10(1), pp 19-30, 1978.

[155] Tallis, G. M. and M. K. Vagholkar, "Formulae to Improve Wald's Approximation for Some Properties of Sequential Tests," *Journal of the Royal Statistical Society*, B(27), pp 74-81, 1965.

[156] Tantaratana, S., "Asymptotic Efficiencies of Some Truncated Sequential Tests with Parallel Boundaries," *IEEE 1988 International Symposium on Information Theory*, 25(13), 1988.

[157] Tang, K. R. Plante and H. Moskwitz, "Multiattribute Bayesian Acceptance Sampling Plans Under Nondestructive Inspection," *Management Science*, 32(6), pp 739-750, 1986.

[158] U. S. Army Chemical Corps. Engineering Agency, "Master Sampling Plans For Single, Duplicate, Double, and Multiple Sampling". Manual No. 2. Army Chemical Center, Maryland, 1953.

[159] Wald, A., "Sequential Tests of Statistical Hypotheses," *Annals of Mathematical Statistics*, 16, pp 117-186, 1945.

[160] Wald, A., *Sequential Analysis. Dover Publications*, Inc., New York, 1947.

[161] Wetherill, G.B., and W.K. Chiu, "A Review of Acceptance Sampling Schemes with Emphasis on the Economic Aspects", *International Statistical Review*, 20, pp 896-925, 1975.

[162] Wald, A., *Sequential Analysis.* John Wiley & Sons, New York, 1973.

[163] Wise, M. E., "Formulae Relating to Single Sample Inspection by Attributes," *Philips Research Reports*, 10, pp 97-112, 1955.

[164] Woodal, R. C. and B. M. Kurkjian, "Exact Operating Characteristic for Truncated Sequential Life Tests in the Exponential Case," *Annals of Mathematical Statistics*, 33, pp 1403-1412, 1962.

[165] Wortham, A. W. and R. C. Baker, "Deferred State Sampling Procedures," *Annals of Assurance Sciences, 1971 Annual Symposium on Reliability*, pp 64-70, 1971.

[166] Wortham, A. W. and R. C. Baker, "Multiple Deferred State Sampling Inspection," *International Journal of Production Research*, 14(6), pp 719-731, 1976.

[167] Wortham, A. W. and J. W. Mogg, "Dependent Stage Sampling Inspection," *International Journal of Production Research*, 8(4), pp 385-395, 1970.

[168] Yokoh, T., "Theoretical Basis of Modified MIL-STD-105D Sampling Scheme," *International Conference on Quality Control, Yokyo 1969-Yokyo Proceedings*, Union of Japanese Scientists and Engineers, pp 653-656, 1969.

# LIST OF REFEREES

F. J. Arcelus, University of New Brunswick, Fredericton, N.B., Canada.

P. K.Benerjee, University of New Brunswick, Fredericton, N.B., Canada.

O. Carlsson, University of Örebero, Örebero, Sweden.

K. E. Case, Oklahoma State University, Stillwater, Oklahoma, U.S.A.

L. Cheng, Oklahoma State University, Stillwater, Oklahoma, U.S.A.

F. Costa, University of Wisconsin, Madison, Wisconsin, U.S.A.

M. Ben Daya, King Fahd U. of Petroleum and Minerals, Saudi Arabia.

S. O. Duffuaa, King Fahd U. of Petroleum and Minerals, Saudi Arabia.

E. A. Elsayed, Rutgers University, Piscataway, New Jersey, U.S.A.

D. Y. Golhar, Western Michigan University, Kalamazoo, Michigan, U.S.A.

W. Makis, University of Toronto, Ontario, Canada.

T. P. McWilliams, Arizona State University West, Phoenix, Arizona, U.S.A.

W. Q. Meeker, Iowa State University, Ames, Iowa, U.S.A.

B. J. Melloy, Clemson University, Clemson, South Carolina, U.S.A.

J. E. Middle, Loughborough University of Technology, Leicestershire, U.K.

D. C. Montgomery, Arizona State University, Tempe, Arizona, U.S.A.

H. Moskowitz, Purdue University, West Lafayette, Indiana, U.S.A.

M. R. Nayebpour, University of St. Thomas, Houston, Texas, U.S.A.

H. Ohta, Osaka University of Prefecture, Sakai, Osaka, Japan.

P. E. Pfeifer, University of Virginia, Charlottesville, Virginia, U.S.A.

E. L. Porteus, Stanford University, Stanford, California, U.S.A.

A. I. Rana, Lahore University of Management Sciences, Lahore, Pakistan.

E. M. Saniga, University of Delaware, Newark, Delaware, U.S.A.

S. Z. Selim, King Fahd U. of Petroleum and Minerals, Saudi Arabia.

G. Tagaras, Aristoteles University of Thessaloniki, Thessaloniki, Greece.

M. Tan, Oklahoma State University, Stillwater, Oklahoma, U.S.A.

K. Tang, Louisiana State University, Baton Rouge, Louisiana, U.S.A.

L. Svoboda, Arizona State University, Tempe, Arizona, U.S.A.

# INDEX

GPSR Compliance

*The European Union's (EU) General Product Safety Regulation (GPSR) is a set of rules that requires consumer products to be safe and our obligations to ensure this.*

*If you have any concerns about our products, you can contact us on ProductSafety@springernature.com*

In case Publisher is established outside the EU, the EU authorized representative is:

Springer Nature Customer Service Center GmbH
Europaplatz 3
69115 Heidelberg, Germany

**Batch number: 10512241**

Printed by Printforce, the Netherlands